PARTICLES AND NUCLEI

PARTICLES AND NUCLEI

Essays in Honor of
the 60th Birthday of
Professor Yoshio Yamaguchi

Edited by **H. Terazawa**

Published by

World Scientific Publishing Co Pte Ltd.
P. O. Box 128, Farrer Road, Singapore 9128
242, Cherry Street, Philadelphia PA 19106-1906, USA

Library of Congress Cataloging-in-Publication Data

Particles and nuclei.

Includes index.

1. Particles (Nuclear physics) 2. Quantum field theory.
3. Nuclear physics. 4. Gravity. 5. Yamaguchi, Yoshio, 1926-
I. Yamaguchi, Yoshio, 1926- II. Terazawa, H. (Hidezumi)
QC793.28.P382 1986 539.7'21 86-24645
ISBN 9971-50-202-X

Printed in Singapore by Fong and Sons Printers Pte. Ltd.

EDITOR'S FOREWORD

January 26, 1986 was the 60th birthday of Professor Yoshio Yamaguchi, who was the Director of Institute for Nuclear Study, University of Tokyo. The publication of this collection of scientific essays on particle and nuclear physics has been planned as one of the projects in honor of his 60th birthday. The editor expresses his sincere gratitude to Professor Victor F. Weisskopf for writing the preface and to all the contributors for their prompt submitting of these essays. He also wishes to thank all the members of the Committee for Celebration of the 60th Birthday of Professor Yoshio Yamaguchi, including Professors J. Ezawa, M. Fukugita, Y. Hirao, S. Homma, A. Hosoya, K. Igi, S. Kato, M. Muraoka, T. Nomura, Y. Ohnuki, N. Sakai, F. Sakata, H. Sugawara and A. Ukawa, especially Professors K. Igi and N. Sakai for their useful help in this publication.

Tokyo, Japan, Summer, 1986

Hidezumi Terazawa

PREFACE

When I was director of CERN from 1961 to 1965, Yoshio and Yoriko Yamaguchi were members of the Division of Theoretical Physics. They contributed to the success of CERN in four ways: first, with their work in theoretical physics; second, through their contacts and discussions with other physicists; third, because of their great and contagious enthusiasm for physics and for CERN; fourth because they were living examples for the international spirit of science, on which CERN was based.

Let me specify these four important contributions which characterize Yamaguchi's beneficial influence on science, not only at CERN but in Japan and all over the world.

His work in physics is anything but specialized. It spreads over all fields of nuclear and particle physics, cosmology, and the theory of gravity. It spans the fields from the interpretation of experiments to highly formal theoretical papers. It contains proposals for future experimentation, but also treatises on new theoretical and mathematical ideas.

Yamaguchi was and is always ready and eager to discuss his own problems and those of his colleagues. He loves and practises the scientific dialogue which is perhaps the most productive activity in our science. I have rarely encountered physicists who seek contacts with colleagues with such eagerness and enthusiasm as he does.

One of his greatest impacts on science is his enthusiasm. His love of knowledge radiates from his personality. Every discussion becomes livelier when he participates. His deep personal involvement in many diverse activities makes our scientific lives more vibrant and more productive.

It was expected that such a personality would play an important role in the organization of science in his own country and all over the world. Indeed his influence helped greatly in the miraculous growth of Japanese physics. But his contribution was not only that of a skilled organizer and administrator, it was his warm and inspiring personality that instilled the kind of spirit which is necessary for a successful scientific community.

His influence went far beyond the Japanese community. He was and is a true citizen of the world. He has perfect knowledge of several foreign languages and a thorough understanding of Western culture in all its aspects. This made him the ideal personality for the promotion of international collaboration. He was and is member of numerous international organizations, such as the International Council of Scientific Unions, and the International Union of Pure and Applied Physics. Whenever physicists of the world gather to discuss some international venture, Yamaguchi would be among them.

He represents in many ways the ideal character suited for a scientist. Thorough knowledge of his field, patience and persistence in his work, eagerness for personal contacts with colleagues, to be at home in any country or culture where science is active, and a warm and human personality open to any positive ideas.

The task of the world scientific community is not only the acquisition of deeper insights into the workings of nature. Because of its many international contacts, it also can and should contribute to a better understanding among nations and thus to a more peaceful world. The scientific community would be better able to do so if there were more Yamaguchi's.

Victor F. Weisskopf

Victor F. Weisskopf

19 August 1985

CONTENTS

I. HADRON PHYSICS

PIONS AND QUARKS IN THE NUCLEON

Hironari Miyazawa

Department of Physics
University of Tokyo

ABSTRACT

The nucleon is composed of three quarks covered by a pion cloud. The effect of the cloud is estimated and removed to examine the quark core. The resulting core magnetic moment is not in accord with the prediction of the SU(6) quark model.

1. INTRODUCTION

Forty years ago nucleons and pions were considered as elementary particles. Certain properties of the nucleon, e. g., magnetic moments, nuclear forces, etc., were regarded as due to the pion cloud around the bare nucleon. This picture was partly successful, at least in a qualitative way. The nucleon magnetic moment, however, was not quite well explained by this description, which gave an unduly large isoscalar anomalous moment.

Then came the quarks. A nucleon consists of three quarks in a color singlet configuration. One of the most remarkable results of this model is the prediction that the ratio of the nucleon magnetic moments should be

$$\mu_p / \mu_n = - 3 / 2 ,$$

which is in excellent agreement with the actual value of 1.54. This fact seems to indicate that properties of the nucleon can be understood in terms of the quark structure only. The one-pion-and-nucleon state is an exotic state consisting of four quarks and one antiquark, so disregarding it may be justified as the first approximation.

However, the pion, interpreted as the Nambu-Goldstone mode, has a very small mass, hence its virtual cloud extends far beyond the quark core. Its existence has been confirmed. In fact, e. g., the peripheral properties of two-nucleon systems are well described by the one-pion-exchange potential[1]. The nucleon consists of an inner core of three quarks surrounded by a pion cloud. We wish to calculate the contribution of the outer pion cloud in order to estimate how much is due to the core. We know that the pion mass is 140 MeV and the Yukawa coupling constant is $f^2 / 4\pi = 0.08$.

2. SIZE OF QUARK CORE

Experimentally, the mean square radius of the proton charge is

$$\langle r^2 \rangle = (0.8 \text{ fm})^2 .$$

Part of this is due to pions, i.e.,[2]

$$\langle r^2 \rangle_\pi = \frac{f^2}{4\pi} \frac{1}{\pi m^2} \int dk \frac{k^4}{\omega^7} \left[2\omega^4 \left(\frac{dv}{dk}\right)^2 + 5 (3\omega^2 - 2k^2) v^2 \right], \quad (1)$$

where $v(k)$ is the form factor of the source function for pion emission, m is the pion mass and

$$\omega = \sqrt{k^2 + m^2}.$$

Let us calculate the pion contribution outside of the source (i. e., where $v(k) = 1$). Suppose $v(k) = 1$ for $k < 2m$. We integrate eq (1) over the range $0 < k < 2m = (0.7 \text{ fm})^{-1}$, thus obtaining

$$\langle r^2 \rangle_\pi = 0.14 \text{ fm}^2 . \quad (k < 2m) \quad (2)$$

Higher order corrections[2] and the S-wave contribution[3] are not significant. The quark core part is therefore

$$\langle r^2 \rangle_q = 0.64 \text{ fm}^2 - 0.14 \text{ fm}^2 = (0.7 \text{ fm})^2 . \quad (3)$$

Thus, in the proton, with rms radius 0.8 fm, the quark core extends to ~ 0.7 fm and the pion cloud account for the rest. The quark core may be regarded as the source of pion emission. It can emit pions of momentum $k < (0.7 \text{ fm})^{-1} = 2\,m$. Thus the cut off $k < 2\,m$ in eq (2) is justified by self-consistency.

3. PION CLOUD AND QUARK CORE

The self energy of the nucleon due to emission and reabsorption of pions is

$$\Delta M_{\pi} = - \frac{f^2}{4\pi} \frac{3}{\pi m^2} \int dk \frac{k^4 v^4}{\omega^2} , \tag{4}$$

This integral is sensitive to the shape of the source. Two cases have been considered, i. e., exponential (i. e., in momentum space, dipole) and gaussian shapes. In each case parameter has been chosen so that eq (3) is satisfied.

$$\Delta M_{\pi} = - 320 \text{ MeV}, \quad (\text{ dipole })$$

$$= - 180 \text{ MeV}. \quad (\text{ gaussian })$$

The core is slightly heavier than the nucleon mass, i. e., 1.1 ~ 1.3 GeV. In the following we consider the gaussian source function which seems reasonable for a wave function under a confining potential.

The magnetic moment due to pions consists of two terms, one due to the pion current and the other, the interaction moment, which arises from the current inside the source. We calculate only the first term, the second being associated with the quark core. In units of nuclear magnetons,

$$\mu_{p\pi} = \frac{f^2}{4\pi} \frac{8}{3\pi} \frac{M}{m} \int dk \frac{k^4 v^2}{\omega^4}$$

$$= 0.67 , \tag{5}$$

which is about one third of the anomalous moment.

As a result of pion emission, the nucleon spin flips:

$$\Delta \sigma_3 = \langle \sigma_3 \rangle_{\uparrow} - 1 = - \frac{f^2}{4\pi} \frac{4}{\pi m^2} \int dk \frac{k^4 v^2}{\omega^3}$$

$$= - 0.46. \tag{6}$$

The above integral is slightly shape dependent. For a dipole type form factor the numerical value is about one third larger in magnitude. Also,

$$\Delta \tau_3 \sigma_3 = \langle \tau_3 \sigma_3 \rangle_{p\uparrow} - 1 = -\frac{f^2}{4\pi}\frac{8}{3\pi m^2}\int dk \frac{k^4 v^2}{\omega^3}$$
$$= -0.31. \tag{7}$$

For nucleon magnetic moments, we have

$$\mu_p = \mu_{qp} + \frac{\mu_{qp} + \mu_{qn}}{2}\Delta\sigma_3 + \frac{\mu_{qp} - \mu_{qn}}{2}\Delta\tau_3\sigma_3 + \mu_\pi ,$$

$$\mu_n = \mu_{qn} + \frac{\mu_{qp} + \mu_{qn}}{2}\Delta\sigma_3 - \frac{\mu_{qp} - \mu_{qn}}{2}\Delta\tau_3\sigma_3 - \mu_\pi .$$

Here μ_{qp} and μ_{qn} are the magnetic moments of the quark cores. From eqs (5), (6) and (7),

$$\mu_{qp} = 4.18 ,$$
$$\mu_{qn} = -0.70. \tag{8}$$

The ratio is $-6 : 1$, which is far from the $-3 : 2$ ratio predicted by the SU(6) quark model. The spin flip given by eq (6) is too large.

However, the renormalization of the axial vector coupling constant is satisfactory. The observable constant is

$$g_{qA} \langle \tau_3 \sigma_3 \rangle = g_A .$$

From the experimental value $g_A = 1.2$ and eq (7), we obtain $g_{qA} = 1.7$ which coincides with the predicted value of $5/3$.

Finally the nuclear potential for distance $r > 1.4$ fm is described by the OPEC (one-pion-exchange-contribution). As was stated before, this is confirmed by experimental results. For $r < 1.4$ fm, the quark cores overlap.

4. CONCLUSION

The nucleon consists of a core of three quarks and a surrounding pion cloud. The properties of the undressed core are revealed by stripping away the pion cloud. The resulting magnetic moment is very different from that predicted by the SU(6) quark model. The estimate presented here is insufficient in the sense that the configuration of the core with spin 3/2 and isospin 3/2 is not taken into account. This effect has been examined by Tanaka[5] and others, but the resulting corrections have not remedied the large discrepancy. Perhaps the picture of a three-quark core and a pion cloud is too naive. More complicated configurations or relativistic effect must be considered. Thus, in the realm of quark theory, nucleon magnetic moment still remains a mystery.

ACKNOWLEDGMENTS

I take great pleasure in expressing my thanks to Professor Yoshio Yamaguchi, whose clear and penetrating physical insight has inspired and guided me for over thirty-eight years. This article is dedicated to Professor Yamaguchi on the occasion of his sixtieth birthday.

Thanks are also due to Professor R. G. Sachs for numerous discussions concerning nucleon moments, and to Mr. E. J. Brody for very helpful comments.

REFERENCES

1) Iwadare, J., Otsuki, S., Tamagaki, R. and Watari, W., Supplement, Prog. Theor. Phys. 3, 32 (1956).

2) Salzman, G., Phys. Rev. 99, 973 (1955).

3) Treiman, S. B. and Sachs, R. G., Phys. Rev. 103, 435 (1956).

4) Ando, F. and Miyazawa, H., Prog. Theor. Phys. 17, 607 (1957).

5) Tanaka, K, Master thesis submitted to Department of Physics, University of Tokyo, March, 1981.

WHAT IS THE SKYRMION LIKE ?*)

Shoichiro Otsuki

Department of Physics, Kyushu University
Fukuoka 812
JAPAN

ABSTRACT

Speculative arguments are given: What is the Skyrmion like and why has no clear evidence for the baryonium been found? ——"THE STRANGE CASE OF DR. SKYRMION AND MR. QUARK"——

I. PROLOGUE

The happy 60th birthday of Professor Yoshio Yamaguchi makes me recall a series of seminars he conducted at Kyoto University in 1952 when I was a new graduate student. Being strongly impressed by his enthusiasm, I wondered why he got so excited. It was because the fresh subject of the seminars, "Polarization" he cleverly chose, polarized himself 100%, and perhaps because a graduate girl student, a senior of mine by one year, was a distinguished leader of the attendance. Some time after the end of the seminars he married our charming leader.

Professor Yamaguchi sometimes bothered us new students by asking extremely difficult questions while hotly urging the answers, such as what at all the ultimate difference between the proton and the neutron is. Although no answer satisfied him I found later this question not so bad. A small number of "new" (strange) particles had been discovered in cosmic rays, suggesting that Heisenberg's classic answer might not always be ultimate. Needless to say, Professor Yamaguchi made a big contribution to establish the concept of flavor symmetry of composite hadrons[1].

I have to confess that the very question has been bothering myself. Recently some people say that the baryon, at least the nucleon, is the Skyrmion, a topological soliton of the nonlinear sigma model, though Skyrme proposed it[2] prior to the quark model in which the baryon is of course the three-quark composite qqq. Apologizing to Professor

*) Sections II ∿ IV are based on collaborations with K.Fujii (Department of Mathematics, Kyushu University) and F.Toyoda (The Second Department, Kinki University)[3∿5].

Yamaguchi for postponing my answer for an indefinite time, I present some speculative arguments: What is the Skyrmion like, and, as a bonus, why has no clear evidence been found for the baryonium, the four quark composite $qq\bar{q}\bar{q}$, contrarily to the naive expectation of the quark model?

II. PHYSICAL AND MATHEMATICAL PRELIMINARIES

Noticing similarities between the quark in the meson and that in the baryon such as the quark counting, and assuming that the quark is confined by a colored oriented string as a specific mode of gluons, Imachi, Toyoda and myself proposed the string-junction model of hadrons almost a decade ago[6]. The three strings in the baryon join one another at a junction. The change of the string orientation at the junction reminds me of the topological kink. Imagine that the quarks and strings (gluons) are hidden within the baryon. Yet there would remain the junction J, which is of topological nature, color singlet and has the baryon number B=1. Hadrons with a variety of topological structure, in particular the baryonium containing a pair of $J\bar{J}$, are expected from the model as Fig.1.

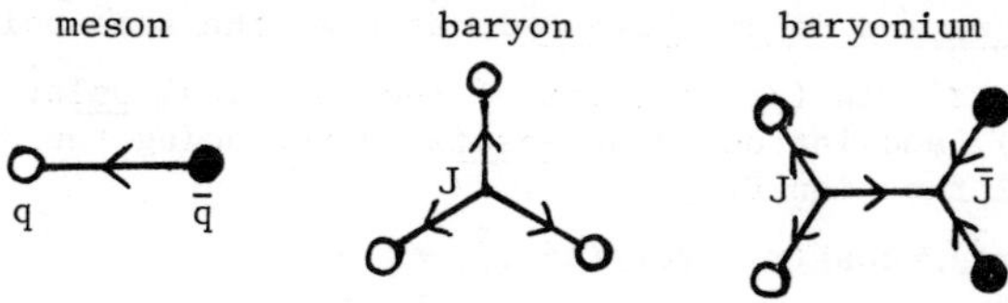

Fig.1. Hadron structure of the string-junction model[6].

The nonlinear sigma model is considered to be the effective theory of QCD where the quark and gluon fields have been integrated away. It originally describes the Nambu-Goldstone boson below the chiral symmetry breaking scale. However, if the model admits the Skyrmion[2], the classical topological soliton with B=1, which the junction is like, then it might not be unnatural that the model could admit another topological configuration corresponding to the baryonium or a pair $J\bar{J}$.

Mathematically, from the homotopy exact sequence

$$\cdots\cdot \to \pi_n(S^1) \to \pi_n(S^3) \to \pi_n(S^2) \to \pi_{n-1}(S^1) \to \cdots\cdot \tag{1}$$

for fiber bundle

$$\begin{array}{l} \qquad\qquad U(1) \approx S^1 \\ \quad \swarrow \\ SU(2) \approx S^3 \\ \downarrow \\ SU(2)/U(1) \approx S^2, \end{array}$$

we have an isomorphism:

$$\pi_n(S^3) \cong \pi_n(S^2) \qquad \text{for } n \geq 3. \tag{2}$$

If the left hand side for n=3 gives the baryon number B of the Skyrmion, it may be interesting to search for a configuration characterized by the right hand side for n=3 giving the Hopf index H.

III. TOPOLOGICAL CONFIGURATIONS

Let $\mathbf{x}=(x^1,x^2,x^3)$ be the Cartesian coordinates, by which

$$z_1 = -\frac{c^2-\mathbf{x}^2}{c^2+\mathbf{x}^2} + i\frac{2cx^3}{c^2+\mathbf{x}^2}, \quad z_2 = i\frac{2c(x^1+ix^2)}{c^2+\mathbf{x}^2} \tag{3}$$

are defined, where c is a parameter with the dimension of length. In terms of z's, let us construct an SU(2) field manifold:

$$U(\mathbf{x}) = \begin{pmatrix} z_1 & -z_2^* \\ z_2 & z_1^* \end{pmatrix} \quad \text{with} \quad z_1 z_1^* + z_2 z_2^* = 1, \tag{4}$$

which forms the well known mapping, $R^3 \to S^3$, and the third homotopy class, $\pi_3(S^3)=Z$ (the Chern-Pontryagin class), gives the baryon number of U under suitable boundary conditions. In term of z's, let us construct an SU(2)/U(1) field manifold:

$$V(\mathbf{x}) = iU(\mathbf{x})\tau_3 U(\mathbf{x})^+ \equiv i\sum_\alpha \tau_\alpha \phi_\alpha \quad \text{with} \quad \phi_\alpha^* = \phi_\alpha, \ \sum_\alpha \phi_\alpha^2 = 1, \tag{5}$$

which forms the Hopf mapping, $R^3 \to S^2$, and $\pi_3(S^2)=Z$ (the Chern-Simons class) gives the Hopf index H of V under suitable boundary conditions. Equation (5) naturally leads to an interpretation that, if U represents a baryon configuration, then the Hopf configuration V does a kind of composite of a baryon and an antibaryon located at the same point[3].

Let us rewrite z's in (3) in terms of the spherical polar coordinates (r,θ,φ), and introduce an ansatz by replacing the dependence on r by a dynamical function F(r):

$$-(c^2-r^2)/(c^2+r^2) \to \cos F(r), \quad 2cr/(c^2+r^2) \to \sin F(r) \tag{6}$$

with boundary conditions in accord with the mapping:

$$F(0) = \pi, \quad F(r) \to 0 \quad \text{as} \quad r \to \infty. \tag{7}$$

Thus, we have the Skyrmion or the hedgehog configuration with B=1:

$$U(\mathbf{x}) = \exp(iF(r)\boldsymbol{\tau}\cdot\hat{\mathbf{x}}). \tag{8}$$

We construct the corresponding polar Hopf configuration with H=1 by (5)[3], but by replacing F(r) by G(r) to avoid confusions:

$$V(\mathbf{x}) = iU_0\tau_3 U_0^+ \quad \text{with} \quad U_0(\mathbf{x}) = \exp(iG(r)\boldsymbol{\tau}\cdot\hat{\mathbf{x}}). \tag{9}$$

We do the same things adopting the toroidal coordinates and obtain the toroidal baryonic configuration $U^{(T)}$[4] while the corresponding toroidal Hopf $V^{(T)}$ was known as "smoke ring"[7]. We could do the same hometasks for other coordinate systems, though the ones apt at compactifying $R^3 \to S^2 \times S^1$ are discussed elsewhere.

Configurations with B, H=K=±1,±2,··· are obtained from $(U)^K$ in the same way[5], with the baryon number

$$B = -\frac{1}{24\pi^2}\int d\mathbf{x}\varepsilon j^{k\ell} \mathrm{Tr}(\textstyle\sum_j^{(K)}\sum_k^{(K)}\sum_\ell^{(K)}) = -\frac{2}{\pi}\int_0^\infty dr(KF')\sin^2(KF) = K, \tag{10}$$

where $\sum_j^{(K)} = (U^+)^K \partial_j (U)^K$, and the last equation is in the polar coordinates (the same last expression of H with $F \to G$, as it should be.)

IV. SYMMETRIES AND THEIR CONSEQUENCES

All the above configurations are proved to satisfy Derrick's theorem[8] for the Lagrangian L of the SU(2) nonlinear sigma model with the Skyrme term[2], so that the dynamical functions are respectively determined to extremize the energy. In particular, the polar Hopf configuration V with H=1 has mass and radius respectively scaled from those of the Skyrmion[3]; $M_V=8\sqrt{2}M_S/3$ and $R_V=\sqrt{2}R_S$.

The Skyrmion U is "spherically symmetric":

$$i\varepsilon_{jk\ell}x^k\partial^\ell U(\mathbf{x}) + [\tau_j/2,\ U(\mathbf{x})] = 0, \tag{11}$$

while the polar Hopf V (together with the toroidal $U^{(T)}$ and $V^{(T)}$) is only axially symmetric:

$$i\varepsilon_{3k\ell}x^k\partial^\ell V(\mathbf{x}) + [\tau_3/2,\ V(\mathbf{x})] = 0. \tag{12}$$

These symmetries imply the following consequences[5].

First, the field equation derived from the Lagrangian L satisfying Derrick's theorem differs from the one derived from the Lagrangian density $\mathcal{L}$ a la Palais' argument[9]. The difference, which is axially symmetric but vanishes after the angular average, gives the decay interaction of the Hopf V. Second, the semi-classical spectrum of the Skyrmion obtained by the collective coordinate method is known to be given by[10]

$$H_S = M_S + (1/8\lambda)(\mathbf{l}_{(3)})^2 \qquad \text{with} \quad \lambda > 0, \tag{13}$$

where $(\mathbf{l}_{(3)})^2$ is the angular momentum squared on S^3 with the eigenvalue $\ell(\ell+2)$ $(\ell=0,1,2,\cdots)$, and the spin and isospin is given as $J=I=\ell/2$. In contrast, the semi-classical spectrum of V, which is given by[11]

$$H_V = M_V + (2/\lambda_1)[\mathbf{J}^2-(\lambda_2/\lambda_1+\lambda_2)(I_3)^2] \tag{14}$$

with $\lambda_1>0>\lambda_2$ and $\lambda_1+\lambda_2>0$, forms no isomultiplet because of (12).

V. CONCLUSIONS

I first remark that the Skyrmion spectrum (13), although isosymmetric, is much restricted compared with that of the quark model and with the observed one. This is because the Skyrmion is "spherically symmetric" only under the simultaneous rotations of space and isospace as (11). A way out of the unwanted restriction may be to regard the Skyrmion U as a relic in the nonlinear sigma model of the junction J. In other words, the Skyrmion, which is classical and stable, serves as a soliton bag <u>for exclusive use of the baryon</u>, while hadronic quantum numbers including the baryon number B are ascribed to the original quarks linked to it. This would enable to avoid a troublesome problem of double counting of the soliton and the hadron composed of quarks.

Our method which treats the hedgehog and the Hopf configurations on an equal footing implies that, if a model admits the former, it does the latter in so far as Derrick's theorem[8] is concerned. Then the polar Hopf configuration V, a relic of the $J\bar{J}$ pair, serves as a soliton bag <u>for exclusive use of the baryonium</u>. Though the lack of isosymmetry (14) is removed by ascribing hadronic quantum numbers to the quarks linked to it, there may be no reason to doubt that the baryonium decays rapidly once its bag gets unstable a la Palais' fashion[9].

These are speculative answers to the questions I asked to myself in Section I. If an experimental revival of narrow peaks of the baryonium will occur as the Skyrmion revived, I would have to revive a selection rule of the string-junction model[6], and search for some dynamical mechanism to suppress the $J\bar{J}$ pair annihilation and creation (approximate conservation of the Hopf index) within a single hadron.

VI. EPILOGUE

In completing my "Strange Case of Dr. Jekyll-Skyrmion and Mr. Hyde-Quark", I add two recent KEK papers to my references. One of them seems to favor and the other disfavor possible baryonium candidates[12]. I hope this would bring joy to Professor Yamaguchi, who taught me to lay stress on experiment in the seminars and made every effort to found KEK.

With cordial gratitude to Professor Yamaguchi and his wife for their encouragement since the seminars, I wish them good health so that he will answer his question in PROLOGUE by himself.

I thank K.Fujii and F.Toyoda for their collaboration.

1) Yamaguchi,Y.,Prog.Theor.Phys.Suppl.11,37(1959).
2) Skyrme,T.H.R.,Proc.Roy.Soc.A260,127(1961).
3) Fujii,K.,S.Otsuki and F.Toyoda,Prog.Theor.Phys.73,524;1287(1985).
4) Fujii,K.,S.Otsuki and F.Toyoda,Contribution Paper to the 1985 Intern. Symp. on Lepton and Photon Int. at High Energies, 19-24 Aug.,1985,Kyoto.
5) Otsuki,S.,Proc. of the 6th INS Winter Seminar on Structure and Forces in Elementary Particle Physics, 16-18 Dec.1985,INS,Tokyo, INS-J-170,128(1985).
6) Imachi,M.,S.Otsuki and F.Toyoda, Prog. Theor. Phys.54,280(1975); 55,551(1976);Prog.Theor.Phys.Suppl.63,49(1978),Chapt.IV.
7) de Vega,H.J.,Phys.Rev.D18,2945(1978). Gipson,J.M. and C.Tze, Nucl.Phys.B183,524(1981). Meissner,J.Phys.Letters 154B,190(1985).
8) Derrick,G.M.,Journ.Math.Phys.5,1253(1964).
9) Palais,R.S.,Commun.Math.Phys.69,19(1979).
10) Adkins,G.S.,C.R.Nappi and E.Witten,Nucl.Phys.B228,552(1983).
11) Bergshoeff,E.A.,R.I.Nepomechie and H.J.Schnitzer,Nucl.Phys.B249, 93(1985). Their supersymmetric Skyrmion is identical with our Hopf V.
12) Tanimori,T.,et al.,Phys.Rev.Letters 55,1835(1985). Chiba,M., et al.,KEK Preprint,May 1986 H.

NOTE ADDED: Section IV shows that recent claim and critical comments by Kundu[13] on our paper[3] are almost nonsense. His compactification of the z axis is incomplete, which gives an infinite mass to what he calls "Hopf" configuration after the z integration, and also yields no difference between the field equations from L and $\mathcal{L}$.

13) Kundu,A., Phys.Letters B171,67(1986).

Multiparticle Production
—— Hydrodynamical Model, Phase Effect of the Secondary Meson Wave, Quark Cascade Model with Recombination Mechanism ——

Chikashi Iso
Department of Physics, Tokyo Institute of Technology

After the establishment of Yukawa's meson theory the phenomenon of the high multiple production of mesons was found in cosmic ray experiment and it takes place with very high probability. To explain this phenomenon several theories were proposed. In 1950 Fermi[1)] proposed to compute multiple production of particle by high energy hadron collision using statistical method. As the interactions of the pion field are strong the cloud surrounding two colliding hadrons is suddenly excited into a furious thermal motion and goes rapidly into a local equilibrium with the volume $(4\pi/3)R^3/E$, where R is the Compton wave length of hadron and E is the incident energy in the CM system.

Landau[2)] assumed that the next period is characterized by the hydrodynamical expansion of the cloud with the equation of state of perfect fluid $p = \varepsilon/3$ (where p and ε are the pressure and the energy density, respectively). This expansion continues till the temperature T cools down to the critical temperature $T_c \simeq \mu c^2$ (equivalent to pion mass) and cloud splids into free particles. It looks that Landau's idea is acceptable theoretically and experimentally. Angular distribution, produced particle ratio, multiplicity etc. can be interpreted successfully by hydrodynamical model.[2,3)] Now the fundamental theory on the strong interaction has changed from meson theory to QCD, but still we can use the similar way of analysis except the phase transition between hadron phase and quark gluon plasma.

Recently many authors[4)] try to modernize the hydrodynamical model. One serve experimental contradiction of the original hydrodynamical model with the data is that the predicted averaged p_T value is larger than the experimental data which can be explained only by the thermal average at the critical temperature $T_c \simeq \mu c^2$. The hydrodynamical pressure makes the average value of p_T larger than the experimental data. So usually people consider only one dimensional solution of the hydrodynamical equation to compare with the data. The average value of p_T may become an important key to check how long hydrodynamical expansion proceeds.

If really the hydrodynamical model is acceptable at all the stage of the expansion of the cloud, all the experimental data on the high energy reactions are covered by the final state interactions at low energy, and it becomes very difficult to see the real high energy behaviors directly from the data. So it is worthwhile to investigate whether the field theory permits the hydrodynamical description for the hadron (quark-gluon) clouds and what deviations from the model predicted by Landau we can expect from the stand point of field theory. The applicability conditions[5)] of the hydrodynamical model consist of local equilibrium and conditions for defining the local system in the cloud, which are prepared by the methods based on quantum statistical mechanics of irreversible processes.[6)]

The temperature dependence of the relaxation time τ_0 and correlation length ξ_0 of the meson fluid are given[5)] in the form,

$$\tau_0 \simeq a/T \quad , \quad \xi_0 \simeq a'/T \quad ,$$

where a and a' depend on the numerical value of the effective coupling constant and we can take to be of order 1. These values may have the same order even if we take QCD instead of the meson theory. The thickness of the initial cloud with volume $(4\pi/3)R^3/E$ is order of $1/T_0{}^2$ and this is too flat and less than the correlation length at high energy. So the interactions in the initial cloud proposed by Fermi and Landau cannot be described by any hydrodynamical equation.[5)] The interactions are, in part, reflected in the initial boundary condition for the subsequent hydrodynamical expansion, but such a boundary condition might be different from that assumed by Landau. Since T decreases to lower value and the system spreads, this guarantees in part the validity of the defining a small cell in the fluid after the size of the system has exceeded the correlation length. In order that the hydrodynamical model is applicable the correlation length ξ_0 and time τ_0 should be much smaller than the characteristic length x_0 and time t_0 for a macroscopic change of the temperature. Using Landau's solutions for deriving x_0 and t_0, we conclude that the hydrodynamical description of the cloud is valid in the central region, but the particles which are emitted in the front region are never in the thermal equilibrium and they may remember the initial interactions in the very small region.

We can also evaluate the heat conductivity κ, shear viscosity $\eta_{(s)}$ and bulk viscosity $\eta_{(v)}$ as follows,[5)]

$$\kappa \simeq \frac{4a}{\pi^2} T^3 \quad , \quad \eta_{(s)} \simeq \frac{2a}{5\pi^2} T^3 \quad , \quad \eta_{(v)} \simeq \frac{4a}{5\pi^2} T^3 \quad .$$

Using these values the produced entropy Σ_{irr} during the hydrodynamical expansion is calculated and is proportional to T_0^2 (or T_i^2, where T_i is the initial value of temperature of the hydrodynamical expansion);

$$\Sigma_{irr} \simeq \frac{5.1aT_0^2}{\pi^2} \left(\frac{T_i}{T_0}\right)^2 [1 + \ell n\left(\frac{T_0}{T_i}\right)^2] \quad .$$

As the total entropy given by Fermi and Landau is proportional to T_0, it is possible that the produced entropy during the hydrodynamical expansion exceeds the original entropy at very high energy hadron reactions.[5)] Recently many authors discuss the entropy production based on the phase transition from quark-gluon plasma phase to hadron phase, and they expect some entropy production. However, as we pointed out, at very high energy the entropy production due to the violation of the assumption of perfect fluid during the hydrodynamical expansion may also exceeds that by phase transition. It is the old but new problem how we can see the real high energy behaviours of the field theory without modification by the final state interaction from the high energy hadron multiple production experiments.

Y. Yamaguchi and L. Van Hove[8] proposed to use unitarity to relate the angular distribution of elastic scattering with the multiparticle production. In order to get the measured sharp angular distribution of the diffraction peak quantitatively, we need not only rapid decrease of the absolute value of the single-particle wave function with the emitted meson angle, but also sharp dependence of the phase[9] of the single-particle wave function with the emitted meson angle. The phase relates directly with the dynamical behaviour of the model; that is, the final state interactions. According to our analysis the hydrodynamical model is acceptable, because owing to long final state interactions the angular dependence of phase becomes reasonably large, and we can get correct angular distribution of the elastic scattering using unitarity relation. On the contrary, fireball model gives flat angular distribution without assuming unreasonably long lifetime of the fireball.

From the above consideration we can divide the produced particles into two parts, one of which is the particles in the front part, that is, in the fragmentation region, and the other of which is the particles in the central region. In the central region the applicability conditions of hydrodynamical model is satisfied after the cloud spreads over a region whose size is the order of the correlation length and the final state interaction can be treated by the statistical mechanics in the form of the hydrodynamical equation. In the fragmentation region we cannot expect thermal equilibrium and hydrodynamical model is not applicable. So we proposed quark cascade model with recombination mechanism.[7] For the forward small p_T scattering, for example, in proton's fragmentation region quark cascade starts from the initial uud quarks with equipartition x (normalized momentum) distribution, and each quark emits mesons successively in forward direction incoherently. Finally three quarks recombine into a baryon and the whole cascade process terminates. The quark cascade in pion's fragmentation region is similar except the initial condition and the different shape of spectrum can be explained from the different number of the constituent quarks. In this way we can predict the particle spectrum in e^+e^-, πN and NN reactions in a unified way, using the same quark cascade function. For the spectrum of the diffraction scattering in recombination mechanism plays an important rule. Field and Feynman[10)] simplified our model neglecting the recombination and this is successfully used to explain e^+e^- and hard scattering reactions. Our model works quite well not only for the single particle spectrum but also for the two particle correlation.[11)] As far as I know two particle correlation cannot[12)] be explained by quark cascade model without recombination or simple recombination model. Besides our model L. Van Hove[13)] proposed another quark cascade recombination model and the relation with our model is discussed in ref. 14). Now the problem is how to found quark cascade recombination model from non-purturbative QCD.

After Lewis, Oppenheimer and Wouthuyson, and Heisenberg started research on the multiparticle production many outstanding physicists worked on this subject. Many interesting subject are pulled out from the high energy reactions, and this will not be done without continuous research on the multiparticle production.

REFERENCES

1) E. Fermi, Prog. Theor. Phys. 5 (1950) 570
2) L. C. Landau, Izv. Akad. Nauk SSSR. Ser. Fiz., 17 (1953) 51
3) S. Amai, H. Fukuda, C. Iso and M. Sato, Prog. Theor. Phys. 17 (1956) 241
4) For example, K. Kajantie, R. Raito and P. V. Ruuskanen, Nucl. Phys. B222 (1983) 152
5) C. Iso, K. Mori and M. Namiki, Prog. Theor. Phys. 22 (1959) 403
6) R. Kubo and K. Tomita, J. Phys. Soc. Japan 9 (1954) 888
7) H. Fukuda and C. Iso, Prog. Theor. Phys. 57 (1977) 483, 1663
8) L. Van Hove, Nuovo Cimento 28 (1963) 798
9) H. Fukuda and C. Iso, Nuovo Cimento 43 (1966) 43
10) R. D. Field and R. P. Feynman, Nucl. Phys. B136 (1978) 1
11) H. Fukuda, Y. Ishihara and C. Iso, Prog. Theor. Phys. 65 (1981) 961; C. Iso, Zeit. f. Phys. C29 (1985) 525
12) ACCMOR Coll., Zeit. f. Phys. C27 (1985) 211
13) L. Van Hove, Zeit. f. Phys. C9 (1981) 145
14) C. Iso and R. Iwai, Zeit. f. Phys. C11 (1981) 103

MULTIHADRON PRODUCTION IN DUAL CHAIN MODELS

Kisei Kinoshita

Physics Dept., Faculty of Education, Kagoshima University
Kagoshima 890, JAPAN

ABSTRACT

A brief overview of cascade and chain models for multihadron production processes in various types of high energy collisions is given.

1. INTRODUCTION

As the quarks and gluons cannot come out isolatedly in our world, the multihadron production turns out to be the overwhelming phenomena not only in high energy hadronic interactions but also in lepton-induced processes.

Since the strong confinement interaction of QCD is soft, in contrast to the hard interaction which takes place only occasionally, the multiparticle final state reveals the jet structure with limited transverse momentum along a jet axis which is the direction of an incident hadron or one of a constituent from hard interaction vertex.

The chain picture for the multiparticle production is proposed at first as the multiperipheral model[1], such as shown in Fig. 1. The extension of the picture to the e+e- annihilation necessitates the quark chain in place of hadronic one, as shown in Fig. 2. Then, it is quite natural to consider,for hadron-hadron collisions, underlying double chain structure in terms of quarks as shown in Fig. 3 [2].

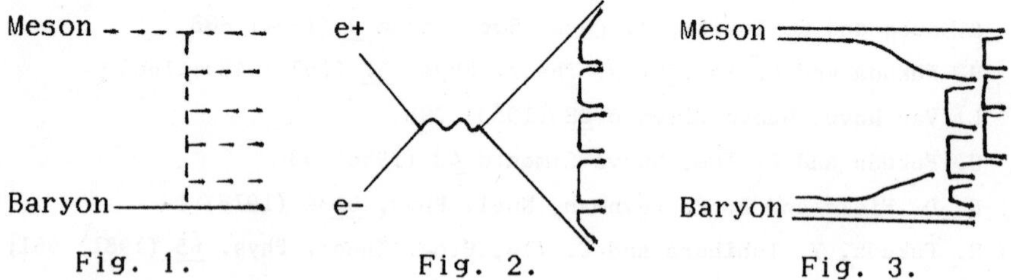

Fig. 1. Fig. 2. Fig. 3.

In this note, we give a short review of the general features of the chain type models and related approaches such as cascade models, and add some remarks.

2. CASCADES IN TERMS OF CONSTITUENTS AND HADRONS

We may incorpolate the cascade relation among the momentum fractions of produced hadrons in each direction in the chain mechanism, as described in Sec. 3. Then, we may adopt two basic alternatives of the cascade relations, namely, in terms of constituents or hadrons. In both cases, we may treat baryon and anti-baryon productions as well as mesons by a slight extension of the quark or meson chains.

On the constituent basis, we may consider the quark-diquark chain model[3] with the cascade relation shown in Fig. 4:

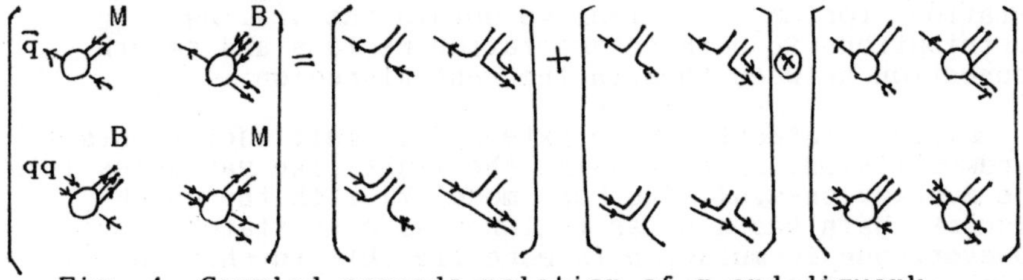

Fig. 4. Coupled cascade relation of q and diquark.

On the other hand, if we limit ourselves to soft processes involving hadrons and nuclei, we may consider hadronic chain model satisfying the cascade relation with the emission kernel as shown below[4]:

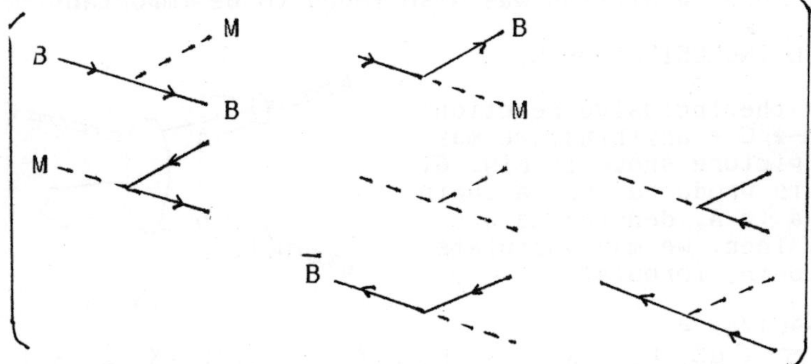

Fig. 5. The cascade kernel in hadronic chain model.

Although the hadronic chain model cannot be applied to quark and gluon jets, in contrast to constituent chain model, it may avoid inessential complications in hadron-nucleus and nucleus-nuclus collisions, and rich enough to recognize various cases of hadron-hadron collisions.

Furthermore, it is possible to treat diffractive component by introducing the Pomeron into the chain.

3. DUAL KINEMATICS

In the hadronization process $a + b \rightarrow$ hadrons, the lightlike fractions of a produced hadron with respect to the directions of a and b in CMS is defined as

$$z_{\pm} = (E \pm p_{\parallel})/\sqrt{s} ,$$

which satisfies the on-shell condition $z_+ z_- = m_T^2/s$. Owing to this condition, we may assume that the hadrons emitted from the a-side satisfy the cascade relation[5] for z_+, while ones from the b-side satisfy the corresponding relation for z_-. Then we obtain the scaling distributions in both fragmentation regions and smooth connection between them in the central region.

This treatment can be extended to multi-jet events in hard collisions by rephrasing the light-like variables into the Sudakov ones[6]. The Lund model[7] with the quark-diquark chain has similar features as ours stated above. However, our formulation is more flexible in choosing appropriate form of the cascade kernel.

Detailed studies of quark-diquark cascade kernel have been done by Noda, Tashiro and the present author, taking into accout the flavor dependence and the production of resonances[8]. The quark-diquark structure of the nucleon with the SU(3) violation was also found to be important.

4. DUAL INCLUSIVE MODEL

For the inclusive reaction $A + B \rightarrow C +$ anything, we may adopt a picture shown in Fig. 6, where C is produced from a chain between a in A, denoted as a/A, and b/B. Then, we may postulate the following formula:

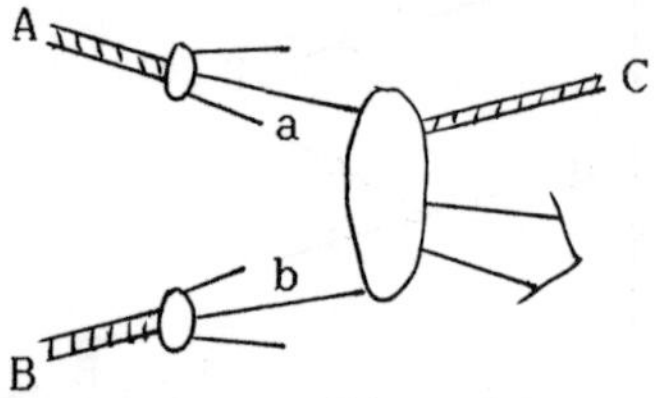

Fig. 6.

$$dN(AB\rightarrow C)/dY = \sum_{a,b} \int dX_+ \, H_{a/A}(X_-) \, dX_- \, H_{b/B}(X_-) \, G_{C/ab}(z_+/X_+, z_-/X_-) ,$$

where H(X) are the distribution functions in each side, and G(Z+,Z-) is the inclusive distribution from the chain. A simple dual parametrization for the latter is

$$G(Z_+,Z_-) = g \, (1-Z_+)^p \, (1-Z_-)^q .$$

This type of inclusive parametrization is utilized in the

multi-chain model of hadron-nucleus and nucleus-nucleus collisions[9], and the gluon jets in heavy quarkonium decay[6] using the Sudakov variables. The interpolation formula and the whole region formula for hadron-hadron collisions[10] can be understood on this basis.

5. EVENT GENERATION

Since the cascade relation with the scaling kernel is the Volterra's integral equation of the second kind, it can be solved analytically and utilized in the inclusive parametrization. However, the explicit solution is available only in some simple cases because of the necessity of the inverse Mellin transformation. Therefore, the Monte-Carlo method was used in the detailed studies of the quark-diquark cascade model[8].

Furthermore, in order to study the full consequences of the dual chain model concerning the multiplicity distribution, energy dependence of the inclusive spectra, correlations, and so on, it is necessary to construct a Monte-Carlo model for the event generation.

A formulation satisfying the exact energy momentum conservation is presented in [4]. The essential point is as follows: starting from a and b denoted as the side 1 and 2 respectively, the hadron emmision proceeds from both sides alternatively, taking the large component of the light-like fraction randomly with the weight given by the cascade kernel. At the same time, the emitted hadron takes away also the small component so as to satisfy the on-shell condition. The emission then follows from the opposite side, up to the stage where the remaining energy-momentum turns out to be less than the two-body threshold. At this stage, the cascade emission is cancelled and replaced by a two-body final state.

5. CASCADE KERNEL

A simple and flexible parametrization of the cascade kernel is the beta distribution for the light-like fraction x of the daughter link with respect to the parent one:

$$P_{ij}(x) = b_{ij}\, x^{A1(i,j)}\, (1-x)^{A(i,j)}/B(A1+1,A+1) \,,$$

where i and j denote the species of the links, b_{ij} is the branching ratio, and B(A1+1,A+1) is the beta function. The emitted hadron takes the light-like fraction of the parent on the average as $\langle 1-x\rangle = (A+1)/(A+A1+2)$.

In the simplist case of single chain and single channel with A = 0, the inclusive distribution is described as

$$EdN/dp_{\parallel} = (A1+1)\ (1-z_{+})^{A1}\ (1+z_{-})^{A1}\ .$$

For the single chain case in general, the parameter A1 determines the $z \sim 1$ behavior of the inclusive distribution of the hadron emitted at the first step of the cascade as $(1-z)^{A1}$. In the multi-chain case, the behavior is determined as a convolution of initial distributions and chains as mentioned in Sec. 4. The height of the central plateau, which determines energy dependence of the average multiplicity, is given by the number of chains times the height given by each chain.

In terms of the constituent chain, there exists the problem whether the properties of the chains are universal or not in soft hadronic reactions and in final states of short distance interactions[2]. There are strong indications that the dynamical properties of the chains are different in soft jets versus hard ones[11]. We need more studies theoretically and phenomenologically on this point. We also leave the studies of the interrelation between the constituent chain versus the hadronic one for future publibcations.

The author would like to express his sincere gratitude to Professor Y. Yamaguchi for his encouragements at various stages of this direction of research.

[1] Amati, D. Fubini, S. and Stanghellini, A., Nuovo Cim. 26, 896 (1962).
[2] Kinoshita, K., Nuovo Cim. 34A, 413 (1976).
[3] Kinoshita, K., Noda, H., Tashiro, T. and Mizouchi, M., Zeit. Phys. C4, 103 (1980).
[4] Kinoshita, K. and Minaka, A., Proc. Workshop on Relativistic Heavy Ion Collisions and New Form of Nuclear Matter, INS, Tokyo, 1985, p.141; Report at the Research meating on Simulation of MCM, INS, 1986.
[5] For example, Field, R. D. and Feynman, R. P., Nucl. Phys. B136, 1 (1978).
[6] Kawabe, T., Ohtani, Y. and Kinoshita, K., Zeit. Phys. C3, 299 (1980).
[7] Andersson, B., Gustafson, Ingelman, G. and Sjostrand, Phys. Report 97, 31 (1983).
[8] Kinoshita, K., Noda, H. and Tashiro, T., Prog. Theor. Phys. 68, 1697 & 2086 (1982); ibid. 74, 1084 (1985). Noda, H. and Tashiro, T., ibid. 73, 158 (1985).
[9] Kinoshita, K., Minaka, A. and Sumiyoshi, H., Prog. Theor. Phys. 61, 165 (1979); ibid. 63, 928 & 1268; Zeit. Phys. C8, 205 (1981), ibid. C11, 347 (1982).
[10] Kinoshita, K. and Noda, H., Proc. 1st INS Symp.,Tokyo, p.403 (1973); Prog. Theor. Phys. 52, 1622 (1974).
[11] Kinoshita, K., Noda, H. and Tashiro, T., Proc. 1981 INS Symp., Tokyo, 1981, p.161.

POMERON PHYSICS

Haruo Fujisaki

Department of Physics, Rikkyo University, Tokyo 171

Our discussion is primarily confined to the asymptopia. From the phenomenological point of view, diffraction at presently available high energies can be reasonably well described in terms of the bare simple-pole pomeron with the intercept at $t = 0$ slightly above unity. From the theoretical point of view, however, the self-consistent explanation of diffraction at the asymptopia inevitably necessitates the clothed physical pomeron with the unit intercept at $t = 0$. The bare pomeron is built up from the normal reggeon through dual topological unitarization. On the other hand, the clothed physical pomeron is generated by multidiffractive unitarization of the bare pomeron. The clothed pomeron is often referred to as the geometrical pomeron (GP). The GP is universal in the sense that the asymptotic behaviour of the clothed pomeron is independent of the fine details of dynamics building up and unitarizing the bare pomeron. All unusual features of the physical pomeron are commonly inherent in universality of the GP, which plays the role of the most typical guiding principle in pomeron physics. If the GP parametrization is continued in t to beyond the lowest threshold, however, t-channel unitarity is seriously violated because of the hard branching nature. It is then of importance to investigate whether or not the GP universality is self-consistently guaranteed not only from the s-channel point of view but also from the t-channel point of view, and how the universal GP dynamically affects normal reggeons through the repeated pomeron exchange. Solutions to these key questions are summarized after geometrodynamical parts of a series of our works on pomeron physics.

The GP partial wave amplitude $f_P(t, J)$ is synthetically written in the scaling form

$$f_P(t, J) = (J - 1)^{\delta-2\nu-1}\zeta(\rho) \; ; \quad \rho = -r_0{}^2 t(J - 1)^{-2\nu} \quad , \qquad (1)$$

where $0 \leq \delta < 2\nu$, $1/2 < \nu \leq 1$ and the scaling function ζ satisfies the asymptotic constraints : $\zeta(\rho \to 0) \sim$ constant ; $\zeta(\rho \to \infty) \sim \rho^{(\delta-2\nu-1)/2\nu}$. Different GP models disagree with each other only over explicit values of δ and ν , and the interpolating form of ζ . It is a matter of course that the exact geometrical scaling is assured for the elastic amplitude if and only if $\delta = 0$. As can easily be seen from Eq.(1) as well as the constraints on ζ , every GP trajectory function $\alpha_P(t)$ is uniquely determined by ν through the moving leading singular surface which reads $\rho \sim$ constant, irrespective of both δ and ζ . Let us postulate $\nu = n/2m$, where n ; $m = 1, 2, 3, \cdots$ and $m < n \leq 2m$. We then obtain

$$\alpha_P^{[j]}(t) = 1 + (r_0{}^2 t)^{1/2\nu} \exp(\pi i j/\nu) \quad , \qquad (2)$$

where $j = 0, 1, 2, \cdots, n-1$. Since $1/2 < \nu \leq 1$, $\alpha_P^{[j]}$ is not regular at $t = 0$, in striking contrast to the bare simple pole pomeron. Let us next turn our attention to the t-dependent motion of $\alpha_P^{[j]}$ in the J-plane.

Except for integral values of $2\nu - \delta$, all branches $\alpha_P^{[j]}$ coalesce into a hard branch point at $J = 1$ as t tends to 0 . If $2\nu - \delta$ is integral, of course, the coalescence at $t = 0$ turns out to be a double pole or a triple pole according as $2\nu - \delta = 1$ or 2 . Suppose $t < 0$. Then all pairs of branches with $\mathrm{Re}\, \alpha_P^{[j]}(t < 0) > 1$ are correctly removed off the physical sheet of the J-plane into unphysical sheets with the aid of the left-hand fixed branch point at $J = 1$ which originates in the factor $(J - 1)^{2\nu}$ with $1/2 < \nu < 1$ of the scaling function ζ . This offers a typical exemplification of the so-called hiding cut mechanism (HCM) of Oehme type. On the other hand, all branches with Re $\alpha_P^{[j]}(t < 0) < 1$ are surely allowed to exist in the complex conjugate form on the physical sheet of the J-plane. If $\nu = 1$, ζ has no HCM but instead uniquely yields the Schwarz type colliding cut pomeron with

$$\alpha_P^{[\pm]}(t) = 1 \pm i r_0 \sqrt{-t} \;; \quad \text{i.e.} \quad \mathrm{Re}\, \alpha_P^{[j=0;1]}(t < 0) = 1 \quad .$$

Suppose $0 < t < t_0$, where t_0 denotes the t-channel lowest threshold. Then the real positive branch $\alpha_P^{[0]}$, which evidently reads the most right left-hand branch point, is guaranteed to exist on the physical sheet of the J-plane so long as $r_0 \leq 1/\sqrt{t_0}$. It is then possible to conclude that the GP universality is automatically consistent with both s-channel unitarity and the real analyticity as the natural consequence of the HCM of Oehme type. If the GP amplitude (1) is continued in t to beyond t_0, then t-channel unitarity is inevitably violated because of the hard branching structure at $J = \alpha_P^{[j]}$. However the pathological incompatibility of the GP with t-channel unitarity can be remedied by the self-consistent introduction of the shielding cut mechanism (SCM) of Oehme type.

Our discussion is confined to the immediate neighbourhood of $t = 0$. We are then legitimately led to a model amplitude $f_{[P]}(t, J)$ with the SCM modification

$$f_{[P]}(t, J) = f_P(t, J) \times \left[\phi_1(t, J) - \frac{\phi_2(t, J)}{\pi} \int_{-\infty}^{\alpha_1(t)} d\ell \, \frac{\lambda(t, \ell) f_P(t, J-\ell+1)}{(\ell-1-i\varepsilon)\phi_2(t, J-\ell+1)} \right]^{-1} , \quad (3)$$

where $\lambda(t, \ell) = ((\alpha_1(t) - \ell)/(\alpha_1(t) + r_1^2 t_0 - \ell))^{1/2}$, $\alpha_1(t) = 1 + r_1^2(t - t_0)$ and $\phi_{1;2}$ are regular, non-vanishing functions. Let us next examine the analytical structure of the model amplitude (3). The $i\varepsilon$ procedure guarantees that $f_{[P]}$ satisfies t-channel elastic unitarity. The endpoint singularity of the ℓ-integral generates the shielding branch point $\alpha_{SC}^{[j]}(t)$, which reads

$$\alpha_{SC}^{[j]}(t) = 1 + r_1{}^2(t - t_0) + (r_0{}^2 t)^{1/2\nu} \exp(\pi i j/\nu) \quad , \tag{4}$$

where $j = 0, 1, 2, \cdots, n-1$. Since $\phi_{1;2}$ are regular non-vanishing, every shielding branch point at $J = \alpha_{SC}^{[j]}$ makes a soft contribution to $f_{[P]}$, irrespective of the fine details of the branching character. All shielding branch points $\alpha_{SC}^{[j]}$ are thus fully consistent with t-channel unitarity. As J tends to $\alpha_P^{[j]}$, then the moving branch point $t_{SC} = \{\alpha_{SC}^{[j]}\}^{-1}(J)$ exactly coincides with t_0. Consequently the cut running from t_{SC} completely shields the t-channel elastic branch cut in the limit of $J \to \alpha_P^{[j]}$. With the aid of the shielding machinery of the branches $\alpha_{SC}^{[j]}$, all pairs of hard branch points $\alpha_P^{[j]}$ are removed from the physical sheet of the J-plane under the continuation of $f_{[P]}$ into the second sheet of the t-plane. The model amplitude (3) thus successfully satisfies the continuity theorem. It is therefore possible to conclude that the shielding branch point $\alpha_{SC}^{[j]}$ correctly satisfies the principal machinery of the SCM of Oehme type and that the GP ansatz is always made compatible with t-channel elastic unitarity and the continuity theorem by the best possible use of the SCM. It is of importance to note that phenomenological consequences of the soft branch point $\alpha_{SC}^{[j]}$ are legitimately negligible at the asymptopia compared with those of the hard GP branch point $\alpha_P^{[j]}$. Accordingly the self-similarity of the GP amplitude (1), which is the most important realization of universality of the clothed physical pomeron, is asymptotically not destroyed by the self-consistent introduction of the SCM of Oehme type.

We are now confronted with an interesting problem: how the GP universality affects normal reggeons in reggeon-pomeron interactions. Both the Mandelstam pinch mechanism and the Gribov reggeon calculus provide us with the standard machinery which yields a typical materialization of the GP universality in reggeon-pomeron dynamics. The leading corrections to any normal reggeon through the repeated pomeron exchange are in fact estimated as the effect of the simultaneous exchange of the GP and the normal reggeon. In order to clarify the principal machinery of the universal GP in the GP-reggeon dynamics, let us remember the discussion on the GP parametrization (1). The t-dependence of the singular surface is uniquely determined through the scaling parameter ρ . The GP is then described as just one moving leading singular surface, irrespective of the detailed branching structure. Therefore the forward scattering amplitude of the GP exchange is asymptotically factorizable in the standard manner as the consequence of the scaling form (1). The input partial wave amplitude $f_R(t, J)$ of the normal reggeon exchange is synthetically expressed by

$$f_R(t, J) = (J - \alpha)^{-1} \zeta_R(\, , \,) \quad ; \quad \rho_R = -\alpha' t (J - \alpha)^{-2\nu_R} \quad , \qquad (5)$$

where $1/2 \leq \nu_R \neq \nu \leq 1$ and ζ_R o ys the asymptotic constraints : $\zeta_R(\rho_R \to 0) \sim$ constant ; $\zeta_R(\rho_R \to \infty) \quad \rho_R^{-1/2\nu_R}$. The reggeon trajectory function $\alpha_R(t)$ then satisfies the moving leading singular surface which reads $\rho_R \sim$ constant . That is, the normal reggeon is controlled by just one moving leading singular surface, irrespective of the fine details of the branching nature. Accordingly factorizability of the input forward scattering amplitude of the normal reggeon exchange is asymptotically guaranteed in the usual sense as the consequence of the scaled form (5). Our purpose is reduced to the examination of the structure of the clothed leading singular surface in the output partial wave amplitude $\tilde{f}_R(t, J)$ which originates from the simultaneous exchange of the GP and the normal reggeon.

The double exchange mechanism of Mandelstam type is postulated between different pairs of quarks and/or antiquarks without the loss of generality. The clothed leading singular surface of $\tilde{f}_R$ is then asymptotically controlled just by the scaling parameter ρ_R or $\bar{\rho}$, according to whether $1/2 \leq \nu_R < \nu \leq 1$ [case (i)] or $1/2 < \nu < \nu_R \leq 1$ [case (ii)], where $\bar{\rho}$ is defined by $\bar{\rho} = -r_0{}^2 t (J - \alpha)^{-2\nu}$. Therefore the clothed trajectory function $\tilde{\alpha}_R(t)$ arising from the GP-reggeon interaction is uniquely determined by ν_R or ν through the moving leading singular surface $\rho_R \sim$ constant or $\bar{\rho} \sim$ constant , respectively, according to the case (i) or (ii). It is almost needless to mention that this result is irrespective of δ , ζ and ζ_R . In any case the output forward scattering amplitude of the clothed reggeon exchange is asymptotically factorizable in the conventional fashion, irrespective of the detailed branching structure. If the branching nature of the trajectory function of the bare reggeon is less [more] singular at $t = 0$ than that of the GP, then the clothed reggeon carries universally the same trajectory function as that of the bare reggeon [the GP with the exception of the intercept at $t = 0$]. The special case of $1/2 < \nu_R = \nu < 1$ is further considered. The standard Mandelstam pinch mechanism uniquely yields the clothed moving leading singular surface $\bar{\rho}_R \sim$ constant , where the scaling parameter $\bar{\rho}_R$ is expressed by

$$\bar{\rho}_R = -\tilde{\alpha}'t(J - \alpha)^{-2\nu} \quad ; \quad \tilde{\alpha}' = \alpha'(1 + (\alpha'/r_0{}^2)^{1/2(1-\nu)})^{2(\nu-1)} \quad .$$

Factorization of the output forward amplitude is ensured in the ordinary manner. Our attention is now directed to the more physical implications of the GP universality.

Let us first examine the case of the bosonic reggeon. We are then naturally led to the input parametrization

$$f_M\,(t,\,J) = \gamma s_0(J - \alpha_M(t))^{-1} \quad ,$$

where $\alpha_M(t) = \alpha + \alpha't$. The criterion (i) is then satisfied and the clothed bosonic reggeon fully remembers the input information on α_M under the GP-boson interaction. The clothed boson trajectory $\tilde{\alpha}_M(t)$ reads universally

$$\tilde{\alpha}_M(t) = \alpha + \alpha't \quad ,$$

which lends active confirmation to the GP universality. Let us next investigate the case of the fermionic reggeon. The input partial wave amplitude $f_F^{[\pm]}(\sqrt{u},\,J)$ of definite parity is then legitimately written in the form

$$f_F^{[\pm]}(\sqrt{u},\,J) = \pm\,\gamma s_0(J - \alpha_F^{[\pm]}(\sqrt{u}))^{-1} \quad ,$$

where $\alpha_F^{[\pm]}(\sqrt{u}) = \alpha_F \pm r_F\sqrt{u} + \alpha_F'u$. Our discussion is restricted to the immediate neighbourhood of $u = 0$ and the use is made of the information on ν in the reggeon field theory. Accordingly we can postulate both $\alpha_F' = 0$ and $1/2 < \nu < 1$ in what follows. The criterion (ii) is then satisfied and the clothed fermionic reggeon commonly forgets the input information on $\alpha_F^{[\pm]}$ and inevitably carries the same trajectory function as that of the GP except for the intercept at $t = 0$ in consequence of the GP-fermion interaction. The clothed fermion trajectories $\tilde{\alpha}_F^{[\pm]}(u)$ of both parities read universally

$$\tilde{\alpha}_F^{[\pm]}(u) = \alpha_F + (r_0{}^2 u)^{1/2\nu} \begin{bmatrix} 1 \\ \exp(-i\pi/\nu) \end{bmatrix} \; ; \quad 1/2 < \nu < 1 \quad ,$$

where we have definitely taken $r_F > 0$ because of transparent parallelism between both parities. The output scaling parameter $\tilde{\rho}_F$, which is defined by $\tilde{\rho}_F = -r_0{}^2 u(J - \alpha_F)^{-2\nu}$, correctly yields the left-hand HCM branch point of Oehme type at $J = \alpha_F$. Our next task is to examine the motion of two opposite parity trajectories $\tilde{\alpha}_F^{[\pm]}$ as a function of u.

Suppose $u < 0$. Then both parity trajectories $\tilde{\alpha}_F^{[\pm]}$ exist in the complex conjugate pair on the physical sheet of the J-plane. Suppose $u > 0$. Then the negative parity MacDowell partner $\tilde{\alpha}_F^{[-]}$ is removed through the HCM branch cut onto an unphysical sheet and there is just the positive parity trajectory $\tilde{\alpha}_F^{[+]}$ on the physical sheet of the J-plane. Therefore only physical fermion states of positive parity are produced in the case of $r_F > 0$. Similarly only negative parity physical states appear in the case of $r_F < 0$. It is of importance to note that the pathological phenomenon of parity doubling associated with the conventional reggeization of fermions is automatically remedied in terms of the HCM of Oehme type as the universal consequence of the GP-fermion interaction. In addition it is worth while to mention that the clothed fermion trajectory is approximately linear in u if and only if the leading singular surface of the GP is not substantially different from that of the bare simple pole pomeron. The limiting case of $\nu = 1$ is further touched upon. The standard Mandelstam pinch mechanism is successfully applicable if and only if $\alpha_F' \neq 0$ but instead yields no HCM of Oehme type. In the special case of $\nu = 1$, in fact, the parity doubling phenomenon either exactly or effectively coincident in mass cannot be avoided in the fermion reggeization through the GP-fermion interaction. The HCM of Oehme type inherent in the GP thus offers a crucial treatment for fermionic schizophrenia of both parities. Consequently our discussion lends considerable confirmation to an interesting argument on fermion-pomeron interaction in the reggeon field theory and affords precision to the so-called Carlitz-Kislinger HCM in the fermion reggeization without parity doubling.

Let us conclude by recapitulating major features of the clothed physical pomeron. First the HCM of Oehme type naturally originates in the scaling form of the GP and automatically guarantees the consistency of the GP universality with s-channel unitarity, the real analyticity and the asymptotic factorizability. Secondly the self-similarity of the GP is made fully compatible with t-channel unitarity by the self-consistent introduction of the SCM of Oehme type. Thirdly the GP universality not only successfully assures linearity of the boson trajectory but also correctly remedies the parity doubling pathology in the fermion reggeization. Both theoretical and phenomenological features of the reggeon-pomeron dynamics are very encouraging at the asymptopia for the self-similar clothed physical pomeron. Fundamental physics underlying the GP universality undoubtedly deserves more than passing consideration.

ASYMPTOTIC CONSTRAINT FOR CHIRAL CONDENSATION OF COULOMB GAUGE QCD

Tomoya Akiba

Department of Physics, Tohoku University
Sendai 980, Japan

ABSTRACT

We study the $^3P_0$ quark-antiquark pair condensation in QCD, representing the gluon-interaction approximately by an instantaneous Coulomb plus Breit potential. Particular attention is paid to non-uniqueness of the solution of the gap equation. The genuine dynamical solution can be characterized by a rapidly convergent behavior at large momentum, as was originally recognized in covariant formulations of Landau gauge. We impose the required behavior and make a numerical study of the condensation.

1. INTRODUCTION

Spontaneous chiral symmetry breaking (χSB) is one of major problems in particle physics. It is believed to be an essential attribute of the QCD vacuum. In the last five years we have seen several papers[1)-4)] that treat χSB in QCD with Coulomb gauge. They owes the underlying general formalism to a pioneering work of Nambu and Jona-Lasinio.[5)] However, instead of four-fermion interaction instantaneous interactions via single-gluon-exchange (SGE) or phenomenological confinement potentials are used.

Prior to Coulomb gauge χSB was extensively studied in covariant formalism with Landau gauge. In particular studies of the ladder-type Schwinger-Dyson (SD) equation[6)-9)] revealed that the gap equation admits multiple of solutions and genuine dynamical solutions among them are characterized by rapidly convergent behaviors at infinite momentum.

The same situation occurs in the formalism of Coulomb gauge.[10] Unless one controls large-momentum effects by a cut-off, it is crucially important to take into account the asymptotic behavior of dynamical solutions. In this article we elucidate a treatment of χSB which respects the constraint. We discuss only the main points, leaving detailed account to a forthcoming paper.[11]

2. RENORMALIZED GAP EQUATION

Let us first construct the presumed ${}^3P_0$ $q\bar{q}$ condensate vacuum $|\Psi>$ by Bogoliubov-Valatin transformation

$$|\Psi> = e^{iX}\ |0> , \tag{1}$$

where $|0>$ denotes the free vacuum. The state $|\Psi>$ is described in terms of an amplitude $\Psi(|p|)$ for $q\bar{q}$ pair of relative momenta p. The gap equation determines this amplitude:

$$\delta v[\Psi] = 0 , \tag{2}$$

where $v[\Psi] = <\Psi|H_{eff}|\Psi>/\delta^3(0)$ is an effective potential. The effective hamiltonian includes the SGE interaction which consists of two parts, i.e., Coulomb and Breit potentials.

The effective potential involves an ultra-violet divergence due to a correction to quark's self-energy. It is possible to subtract the divergence following to a receipt by Adler and Davis.[4] Unfortunately, a finite-renormalization effect remains unsettled. We decided to neglect it together with vertex correction, in accordance with Ward-Takahashi identity. This amounts to starting from the normal-ordered hamiltonian.[1]

3. ASYMPTOTIC BEHAVIOR OF THE DYNAMICAL SOLUTION

In order to see the essential point we simplify the gap equation as follows:

$$F(p) = \frac{G}{2\pi^2}\int \frac{d^3q}{(p-q)^2}\,\frac{F(q)}{[q^2+F(0)^2]^{1/2}} , \tag{3}$$

where $F(p) \simeq 2p\Psi(p)$ and G denotes a coupling constant. To solve Eq.(3) corresponds to the Goldstein problem[12] in covariant formalism. It can be shown that the solution has the asymptotic behaviors $F(P) \sim p^{\alpha_i - 2}$ $(i=1,2)$, where α_i are two roots determined by

$$1 - \frac{G}{1-\alpha}\cot\left(\frac{\pi\alpha}{2}\right) = 0 \tag{4}$$

in the region $0 < \alpha < 2$, i.e., $\alpha_1 \gtrsim 2G/\pi$ and $\alpha_2 \gtrsim 2(1-G/\pi)$.

It turns out that the chiral condensation should be represented by the first-type solution.[7)-9)] The second one can be associated with the following inhomogeneous equation for a vertex of zero incident-momentum:

$$\Gamma(p) = 1 + \frac{G}{2\pi^2}\int_{9<P_c} \frac{d^3q}{(p-q)^2}\,\frac{\Gamma(q)}{[q^2+\Gamma(0)^2]^{1/2}} \,. \tag{5}$$

In fact, after renormalization Eq.(5) reduces to Eq.(3) in the limit of infinite cut-off $P_c \to \infty$.

Since in QCD the running coupling constant decreases at large momentum because of the asymptotic-free nature, we arrive at the result that $\Psi(p) \sim p^{-3}$ except for some logarithmic factor.

4. VARIATIONAL CALCULATION OF Ψ(P)

We now determine $\Psi(p)$ by a variational method. The effective potential $v[\Psi]$ involves the running coupling constant

$$g^2[(p-q)^2] \gtrsim 2/b\,\ln[(p-q)^2/\Lambda^2_{QCD}] \tag{6}$$

where $b = (33-2n_f)/24\pi^2$, n_f being the number of flavors. For the purpose of partly avoiding the infrared divergence and partly simplifying calculation we use the modified formula

$$g^2(p^2_>) = 2/b\,\ln(p^2_>/\Lambda_{QCD} + \xi) \,, \tag{7}$$

where $p^2_>$ is the larger one of either p^2 or q^2. We are interested in possible role of SGE to χSB. Accordingly the parameter ξ must be such as $\xi \gg 1$.

The amplitude $\Psi(p)$ is parametrized as

$$\Psi(p) = f(p)/\{p + [p^2 + f^2(p)]^{1/2}\} \,, \tag{8}$$

$$f(p) \equiv M\Lambda^2_\chi / (p^2 + \Lambda^2_\chi) \,. \tag{9}$$

Note that $\Psi(p)$ satisfies the asymptotic constraint. The parameter M has a physical meaning of the dynamical mass of a quark, while Λ_χ represents a chiral scale. After substitution of Eq.(8) into $v[\Psi]$ the optimum values of M and Λ_χ that make the potential minimum are searched. If M is non-zero at the minima, it signals a realization of Nambu-Goldstone phase.

Calculation is done for two cases: Case 1. $\Lambda_{QCD} = 300$ MeV and $\xi = 12$, Case 2. $\Lambda_{QCD} = 200$ MeV and $\xi = 6$. The parameter ξ is rather arbitrarily adjusted. However, if ξ exceeds a critical value, only the Wigner phase is stable. The critical value is $\xi_c \approx 16$ and 10 for $\Lambda_{QCD} = 300$ and 200 MeV, respectively. The optimum values are listed in Table I.

Table I. Optimum values of parameters and numerical results for pion properties

	M	Λ_χ	f_π	m(in units of MeV)
case 1	320	390	68	3.0
case 2	520	300	71	3.4

5. PION PROPERTIES

Mandula et al.[13),14)] calculated the pion decay constant f_π, the pion mass μ_π, the charge radius of $\pi^\pm$ and so on. They considered only the Coulomb potential, but not the Breit potential. Inclusion of the Breit potential is expected to affect final results significantly. We calculate f_π and μ_π (or quark's current mass).

It turns out that the formula for μ_π^2 given by Mandula et al. is equivalent ot the well-known result of the current algebra[15)]:

$$\mu_\pi^2 = -(2m/f_\pi^2) \langle Vac|\bar{\psi}(0)\psi(0)|Vac\rangle , \tag{10}$$

where m is quark's current mass. ($\bar{\psi}\psi$ on the r.h.s. stands for a single flavor.) An important point is that the asymptotic behavior discussed in §3 yeilds a divergent integral to the vacuum expectation value in (10). It must be renormalized. As a consequence each of m and the expectation value alone is not renormalization-scale invariant. Only their product is. In Table I calculated values of

f_π and m are given. The current mass m is determined so as to reproduce the experimental value of μ_π , in neglect of a logarithmic factor which depends on renormalization-scale.

6. DISCUSSION

The results in Table I are fairly well. (The experimental value of f_π is 93.2 MeV and the estimated value of m is 5 $\sim$ 9 MeV.[15]) On the other hand an analysis of Adler and Davis[4] indicated that a confinement potential is insufficient to reproduce χSB in QCD. These suggest possibly important role of the SGE interaction to χSB. However, it is perphaps premature to draw any conclusion from preliminary analyses.

REFERENCES

1) Finger, J.E. and Mandula, J.E., Nucl. Phys. B199, 168 (1982).
2) Amer, A., Yaouanc, A.Le, Oliver, L., Pene, O. and Raynal, J.-C., Z. Phys. C17, 61 (1983).
3) Yaouanc, A.Le, Oliver, L., Pène, O. and Raynal, J.-C., Phys. Rev. D29, 1233 (1984).
4) Adler, S.L. and Davis, A.C., Nucl. Phys. B244, 469 (1984).
5) Nambu, Y. and Jona-Lasinio, G. Phys. Rev. 122, 345 (1961).
6) Maskawa, T. and Nakajima, H., Prog. Theor. Phys. 52, 1326 (1974).
7) Higashijima, K. and Nishimura, A., Nucl. Phys. B113, 173 (1976).
8) Fukuda, R. and Kugo, T., Nucl. Phys. B117, 250 (1976).
9) Miransky, V.A., Phys. Lett. 165B, 401 (1985).
10) Akiba, T., Prog. Theor. Phys. 74, 641 (1985).
11) Akiba, T. and Kikuchi, H., a preprint TU/86/297.
12) Goldstein, J., Phys. Rev. 91, 1516 (1953).
Kummer, W., Nuovo Cimento 31, 219 (1964); ibid. 34, 1840 (1964).
13) Finger, J., Mandula, J.E. and Weyers, J., Phys. Lett. 96B, 367 (1980).
14) Govaerts, J., Mandula, J.E. and Weyers, J., Phys. Lett. 130B, 427 (1983).
15) Gasser, J. and Leutwyler, H., Phys. Rep. 87, 77 (1982).

HIDDEN LOCAL SYMMETRY VERSUS MASSIVE YANG-MILLS IN THE NONLINEAR CHIRAL LAGRANGIAN

Koichi Yamawaki

Department of Physics, Nagoya University,
Nagoya 464, Japan

ABSTRACT

Based on the hidden local symmetry in the chiral nonlinear sigma model, I present a unified description of the π, ρ and A_1 mesons coupled to the gauge bosons of the Glashow-Salam-Weinberg model, which relates various approaches including the old massive Yang-Mills Lagrangian.

Increasing attention has recently been paid to the nonlinear sigma model in various contexts of modern particle physics. Its origin is certainly traced back to the nonlinear chiral Lagrangian[1)] constructed for the pion transforming nonlinearly[2)] under the chiral $SU(2)_L \times SU(2)_R$. General theory of the nonlinear realization of the chiral groups was extensively studied by many authors[3),4)], including Professor Yamaguchi[5)], and was further generalized to arbitrary groups[6)].

After two decades, the nonlinear chiral Lagrangian is now considered as the low energy effective theory of QCD and has become even more useful and richer in the physical contents, incorporating low lying hadrons in a characteristic way to the nonlinear realization: Preferred roles under dictate of the chiral symmetry have been given to the baryons as topological solitons (Skyrmions)[7] and to the vector mesons as dynamical gauge bosons of the hidden local symmetry.[8)-14)]

In this article, I shall describe an aspect of such an extended version of the low energy effective Lagrangian of QCD coupled to the gauge bosons of the Glashow-Salam-Weinberg model. The key observation is that any nonlinear sigma model based on the coset space G/H is gauge equivalent to a "linear" model possessing a symmetry $G_{global}\times H_{local}$,[10)] H_{local} being the hidden local symmetry, and is further gauge equivalent to another "linear" model having a larger symmetry $G_{global}\times G_{local}$[14)], G_{local} being the "generalized" hidden local symmetry. In the case of QCD where G is $SU(2)_L\times SU(2)_R$ (or $U(3)_L\times U(3)_R$) and H is $SU(2)_V$ (or $U(3)_V$), a crucial assumption was made[8)-11),13),14)] that the gauge bosons of the "generalized" hidden local symmetry, identified with the ρ and A_1 mesons, acquire kinetic term through quantum effects at the composite level as in the 2- and 3-dimensional nonlinear sigma models[12),15)], or through the underlying dynamics QCD itself. This assumption led us to enormously successful phenomenonlogy of the vector[8)-11)] and axialvector mesons[9),13),14)], thus confirming our basic assumption. Thus in this article I simply add the kinetic term of the hidden gauge bosons from the onset.

Let us start with a "linear" model[9)] which is invariant under $G_{global}\times G_{local}$ ($G=SU(2)_L\times SU(2)_R$),[*)] G_{global} being fully gauged with the external gauge fields $\mathcal{L}_\mu(x')\equiv\mathcal{L}^a_\mu(x)\tau^a/2$ and $\mathcal{R}_\mu(x)\equiv\mathcal{R}^a(x)\tau^a/2$ which may contain γ, W and Z bosons.[14)] The dynamical variables in this model are the unitary matrix representations of this group, $\xi_L(x)$, $\xi_R(x)$ and $\tilde{U}(x)$,[**)] which are introduced as

$$U(x) = \xi_L^\dagger(x)\tilde{U}(x)\xi_R^{\cdot}(x) \quad , \tag{1}$$

where $U(x)\equiv\exp(2i\pi(x)/f_\pi)$ with $\pi(x)\equiv\pi^a(x)\tau^a/2$, f_π ($\simeq$93 MeV) being the

*) The following result can be generalized to arbitrary groups.[14)]

**) $\tilde{U}(x)$ was denoted by $\xi_M(x)$ in ref.9).

decay constant of the pion $\pi(x)$. In accord with the well-known transformation property of $U(x)$[4)-6)] such that $U(x) \to U'(x) = g_L(x)U(x)g_R^\dagger(x)$, these variables transform under $G_{global} \times G_{local}$ as $\xi_{L,R}(x) \to \xi'_{L,R}(x) = \tilde{g}_{L,R}(x)\xi_{L,R}(x)g^\dagger_{L,R}(x)$ and $\tilde{U}(x) \to \tilde{U}'(x) = \tilde{g}_L(x)\tilde{U}(x)\tilde{g}_R^\dagger(x)$, where $g_{L,R}(x) \in$ gauged-G_{global} and $\tilde{g}_{L,R}(x) \in G_{local}$. The Lagrangian is given by,[9),14)]

$$\mathcal{L} = a'\mathcal{L}_V + b'\mathcal{L}_A + c'\mathcal{L}_M + d'\mathcal{L}_\pi - \frac{1}{4g^2}\mathrm{Tr}(\tilde{L}_{\mu\nu}^2 + \tilde{R}_{\mu\nu}^2)$$

$$- \frac{1}{4\xi^2}\mathrm{Tr}(\mathcal{L}_{\mu\nu}^2 + \mathcal{R}_{\mu\nu}^2) \ , \qquad (2)$$

where a', b', c' and d' are arbitrary parameters, $\mathcal{L}_{\mu\nu}$ is defined by $\mathcal{L}_{\mu\nu} \equiv \partial_\mu\mathcal{L}_\nu - \partial_\nu\mathcal{L}_\mu - i[\mathcal{L}_\mu, \mathcal{L}_\nu]$ (similarly for $\mathcal{R}_{\mu\nu}$), $\tilde{L}_{\mu\nu} \equiv \partial_\mu\tilde{L}_\nu - \partial_\nu\tilde{L}_\mu - i[\tilde{L}_\mu, \tilde{L}_\nu]$ (similarly for $\tilde{R}_{\mu\nu}$) are the field strength of the gauge bosons, $\tilde{L}_\mu \equiv \tilde{L}_\mu^a\tau^a/2$ ($\tilde{R}_\mu \equiv \tilde{R}_\mu^a\tau^a/2$), of the hidden local symmetry G_{local} and $g(\xi)$ are the gauge coupling constants associated with G_{local} (G_{global}). The four invariants under (gauged-G_{global})$\times G_{local}$ $\times$(parity), $\mathcal{L}_V$, $\mathcal{L}_A$, $\mathcal{L}_M$ and $\mathcal{L}_\pi$ in Eq.(2), are given by [9)]

$$\mathcal{L}_V \equiv f_\pi^2\,\mathrm{Tr}\left(\frac{D_\mu\xi_L\cdot\xi_L^\dagger + \tilde{U}D_\mu\xi_R\cdot\xi_R^\dagger\tilde{U}^\dagger}{2i}\right)^2 , \qquad (3)$$

$$\mathcal{L}_A \equiv f_\pi^2\,\mathrm{Tr}\left(\frac{D_\mu\xi_L\cdot\xi_L^\dagger - \tilde{U}D_\mu\xi_R\cdot\xi_R^\dagger\tilde{U}^\dagger}{2i}\right)^2 , \qquad (4)$$

$$\mathcal{L}_M \equiv f_\pi^2\,\mathrm{Tr}\left(\frac{D_\mu\tilde{U}\cdot\tilde{U}^\dagger}{2i}\right)^2 , \qquad (5)$$

$$\mathcal{L}_\pi \equiv f_\pi^2\,\mathrm{Tr}\left(\frac{D_\mu\xi_L\cdot\xi_L^\dagger - \tilde{U}D_\mu\xi_R\cdot\xi_R^\dagger\tilde{U}^\dagger - D_\mu\tilde{U}\cdot\tilde{U}^\dagger}{2i}\right)^2 , \qquad (6)$$

where $\xi_L \leftrightarrow \xi_R$ and $\tilde{U} \leftrightarrow \tilde{U}^\dagger$ under parity operation and we defined covariant derivatives, $D_\mu\xi_L(x) \equiv \partial_\mu\xi_L(x) - i\tilde{L}(x)\xi_L(x) + i\xi_L(x)\mathcal{L}_\mu(x)$ (similarly for $D_\mu\xi_R(x)$) and $D_\mu\tilde{U}(x) \equiv \partial_\mu\tilde{U}(x) - i\tilde{L}_\mu(x)\tilde{U}(x) + i\tilde{U}(x)\tilde{R}_\mu(x)$.

It is convenient to rewrite Eqs.(3)-(6) in two bases:

$$\mathcal{L}_V = \frac{f_\pi^2}{4}\,\mathrm{Tr}[(\tilde{L}_\mu - \tilde{\mathcal{L}}_\mu) + \tilde{U}(\tilde{R}_\mu - \tilde{\mathcal{R}}_\mu)\tilde{U}^\dagger]^2 = \frac{f_\pi^2}{4}\,\mathrm{Tr}[(L_\mu - \mathcal{L}_\mu) + U(R_\mu - \mathcal{R}_\mu)U^\dagger]^2, \tag{3'}$$

$$\mathcal{L}_A = \frac{f_\pi^2}{4}\,\mathrm{Tr}[(\tilde{L}_\mu - \tilde{\mathcal{L}}_\mu) - \tilde{U}(\tilde{R}_\mu - \tilde{\mathcal{R}}_\mu)\tilde{U}^\dagger]^2 = \frac{f_\pi^2}{4}\,\mathrm{Tr}[(L_\mu - \mathcal{L}_\mu) + U(R_\mu - \mathcal{R}_\mu)U^\dagger]^2, \tag{4'}$$

$$\mathcal{L}_M = \frac{f_\pi^2}{4}\,\mathrm{Tr}(D_\mu\tilde{U}\, D^\mu\tilde{U}^\dagger) = \frac{f_\pi^2}{4}\,\mathrm{Tr}(D_\mu U\, D^\mu U^\dagger)\ , \tag{5'}$$

$$\mathcal{L}_\pi = \frac{f_\pi^2}{4}\,\mathrm{Tr}(\mathcal{D}_\mu\tilde{U}\,\mathcal{D}^\mu\tilde{U}^\dagger) = \frac{f_\pi^2}{4}\,\mathrm{Tr}(\mathcal{D}_\mu U\,\mathcal{D}^\mu U^\dagger)\ , \tag{6'}$$

where we defined

$$\tilde{\mathcal{L}}_\mu(x) \equiv \xi_L(x)\mathcal{L}_\mu(x)\xi_L^\dagger(x) - i\partial_\mu\xi_L(x)\cdot\xi_L^\dagger(x)\ , \tag{7}$$

$$L_\mu(x) \equiv \xi_L^\dagger(x)\tilde{L}_\mu(x)\xi_L(x) + i\xi_L^\dagger(x)\partial_\mu\xi_L(x)\ , \tag{8}$$

and similarly $\mathcal{R}_\mu(x)$ and $R_\mu(x)$, respectively (with the change $\xi_L \to \xi_R$), and covariant derivatives, $\mathcal{D}_\mu\tilde{U}(x) \equiv \partial_\mu\tilde{U}(x) - i\tilde{\mathcal{L}}_\mu(x)\tilde{U}(x) + i\tilde{U}(x)\tilde{\mathcal{R}}_\mu(x)$, $D_\mu U(x) \equiv \partial_\mu U(x) - iL_\mu(x)U(x) + iU(x)R_\mu(x)$ and $\mathcal{D}_\mu U(x) \equiv \partial_\mu U(x) - i\mathcal{L}_\mu(x)U(x) + iU(x)\mathcal{R}_\mu(x)$. We shall call the first basis (with tilde) a "hidden basis" in which everything transforms under G_{local} (e.g., $\tilde{L}_\mu \to \tilde{L}'_\mu = i\tilde{g}_L\partial_\mu\tilde{g}_L^\dagger + \tilde{g}_L\tilde{L}_\mu\tilde{g}_L^\dagger$, $\tilde{\mathcal{L}}_\mu \to \tilde{\mathcal{L}}'_\mu = i\tilde{g}_L\partial_\mu\tilde{g}_L^\dagger + \tilde{g}_L\tilde{\mathcal{L}}_\mu\tilde{g}_L^\dagger$, $\tilde{U} \to \tilde{U}' = \tilde{g}_L\tilde{U}\tilde{g}_R^\dagger$, etc.), while the second basis (without tilde) an "external basis" in which everything transforms under gauged-G_{global} (e.g., $L_\mu \to L'_\mu = ig_L\partial_\mu g_L^\dagger + g_L L_\mu g_L^\dagger$, $\mathcal{L}_\mu \to \mathcal{L}'_\mu = ig_L\partial_\mu g_L^\dagger + g_L\mathcal{L}_\mu g_L^\dagger$, $U \to U' = g_L U g_R^\dagger$, etc.). Note that $\mathcal{L}_\pi$ in the external basis is nothing but the gauged nonlinear chiral Lagrangian[6] which is invariant under G fully gauged with $\mathcal{L}_\mu$ and $\mathcal{R}_\mu$.

Let us first consider the "hidden basis". We fix the gauge as $\tilde{U}(x)=1$,[9] by which G_{local} is reduced to its subgroup $H_{local} = [SU(2)_V]_{local}$ so as to keep the gauge condition $\tilde{U}(x)=1$ unchanged under the transformation $\tilde{g}_L(x) = \tilde{g}_R(x) \equiv \tilde{h}(x) \in H_{local}$. Accordingly, $\xi_{L,R}(x)$ transform as $\xi_{L,R}(x) \to \xi'_{L,R}(x) = \tilde{h}(x)\xi_{L,R}(x)g_{L,R}^\dagger(x)$, while the gauge bosons $\tilde{L}_\mu$, $\tilde{R}_\mu$, $\tilde{\mathcal{L}}_\mu$ and $\tilde{\mathcal{R}}_\mu$ transform in all the same way as $\tilde{L}_\mu(x) \to \tilde{L}'(x) = i\tilde{h}(x)\partial_\mu\tilde{h}^{\dagger\mu}(x) + \tilde{h}(x)\tilde{L}_\mu(x)\tilde{h}^\dagger(x)$, etc., or $\tilde{V}_\mu \equiv (\tilde{R}_\mu + \tilde{L}_\mu)/2 \to \tilde{V}'_\mu$

$=i\tilde{h}\partial_\mu\tilde{h}^\dagger+\tilde{h}\tilde{V}_\mu\tilde{h}^\dagger$ and $\tilde{A}_\mu\equiv(\tilde{R}_\mu-\tilde{L}_\mu)/2\rightarrow\tilde{A}_\mu'=\tilde{h}\tilde{A}_\mu\tilde{h}^\dagger$ (and similarly do $\tilde{\mathcal{V}}_\mu$ $\equiv(\tilde{\mathcal{R}}_\mu+\tilde{\mathcal{L}}_\mu)/2$ and $\tilde{\mathcal{A}}_\mu\equiv(\tilde{\mathcal{R}}_\mu-\tilde{\mathcal{L}}_\mu)/2$, respectively).

Then $b'\mathcal{L}_A+c'\mathcal{L}_M+d'\mathcal{L}_\pi$ in (2) is rewritten as $(b'+c')f_\pi^2\mathrm{Tr}(\tilde{A}_\mu - \frac{b'}{b'+c'}\tilde{\mathcal{A}}_\mu)^2+(\frac{b'c'}{b'+c'}+d')f_\pi^2\mathrm{Tr}(\tilde{\mathcal{A}}_\mu^2)$. Noting that $f_\pi^2\mathrm{Tr}(\tilde{\mathcal{A}}_\mu^2)=f_\pi^2/4\cdot \mathrm{Tr}(\mathcal{D}_\mu U\mathcal{D}^\mu U^\dagger)$ (gauged nonlinear Lagrangian), we normalize the pion kinetic term via simultaneous rescaling, $\sqrt{r}\pi(x)\rightarrow\pi(x)$ and $\sqrt{r}f_\pi\rightarrow f_\pi$, where $r=b'c'/(b'+c')+d'$. Eq.(2) now reads:

$$\mathcal{L}= af_\pi^2\mathrm{Tr}(\tilde{V}_\mu-\tilde{\mathcal{V}}_\mu)^2 + (b+c)f_\pi^2\mathrm{Tr}(\tilde{A}_\mu - \frac{b}{b+c}\tilde{\mathcal{A}}_\mu)^2$$
$$+\frac{f_\pi^2}{4}\mathrm{Tr}(\mathcal{D}_\mu U\mathcal{D}^\mu U^\dagger) - \frac{1}{2g^2}\mathrm{Tr}(\tilde{V}_{\mu\nu}^2+\tilde{A}_{\mu\nu}^2)$$
$$-\frac{1}{2\xi^2}\mathrm{Tr}(\tilde{\mathcal{V}}_{\mu\nu}^2+\tilde{\mathcal{A}}_{\mu\nu}^2)\ , \tag{9}$$

where $\tilde{V}_{\mu\nu}\equiv\partial_\mu\tilde{V}_\nu-\partial_\nu\tilde{V}_\mu-i[\tilde{V}_\mu,\tilde{V}_\nu]-i[\tilde{A}_\mu,\tilde{A}_\nu]$ and $\tilde{A}_{\mu\nu}\equiv\partial_\mu\tilde{A}_\nu-\partial_\nu\tilde{A}_\mu-i[\tilde{V}_\mu,\tilde{A}_\nu]$ $-i[\tilde{A}_\mu,\tilde{V}_\nu]$ (similarly for $\tilde{\mathcal{V}}_{\mu\nu}$ and $\tilde{\mathcal{A}}_{\mu\nu}$), and $a\equiv a'/r$, $b\equiv b'/r$ and $c\equiv c'/r$.

Phenomenologically the only viable choice is $a=b=c=2$,[9),13),14)] which means $a'=b'=c'\equiv x\neq 0$ and $d'=0$. The simplest choice is $x=1$ (not 2 (!)), in which case Eq.(2) takes the simple form,

$$\mathcal{L}=\mathcal{L}_V+\mathcal{L}_A+\mathcal{L}_M-\frac{1}{4g^2}\mathrm{Tr}(\tilde{L}_{\mu\nu}^2+\tilde{R}_{\mu\nu}^2)-\frac{1}{4\xi^2}\mathrm{Tr}(\mathcal{L}_{\mu\nu}^2+\mathcal{R}_{\mu\nu}^2)\ , \tag{10}$$

This is our basic Lagrangian which I propose in this article.

From (10) we can derive a successful $G_{global}\times H_{local}$ Lagrangian[8)], $\mathcal{L}=a\mathcal{L}_V+\mathcal{L}_A-\frac{1}{2g^2}\mathrm{Tr}(\tilde{V}_{\mu\nu}^2)-\frac{1}{4e^2}\mathcal{B}_{\mu\nu}^2$, with a "magic number" $a=2$ (not 1 (!)), where we have restricted $\mathcal{V}_\mu$ to the photon field ($\xi\mathcal{V}_\mu=e\mathcal{B}_\mu\tau^3/2$). For this end we are only interested in the low energy region below the mass of the ρ meson, in which case A_1 meson mass may be regarded as infinite or the kinetic term of the A_1 meson may be ignored, so that, after gauge fixing $\tilde{U}(x)=1$, we can solve away the $\tilde{A}_\mu$ out of (10)[13),14)] via the equation of motion, $\tilde{A}_\mu=\frac{b}{b+c}\tilde{\mathcal{A}}_\mu=\frac{1}{2}\tilde{\mathcal{A}}_\mu$ (see (9)), thus arriving at the Lagrangian $\mathcal{L}=2f_\pi^2\mathrm{Tr}(\tilde{V}_\mu-\tilde{\mathcal{V}}_\mu)^2+f_\pi^2/4\ \mathrm{Tr}(\mathcal{D}_\mu U\mathcal{D}^\mu U^\dagger)$ $-\frac{1}{2g^2}\mathrm{Tr}(\tilde{V}_{\mu\nu}^2)-\frac{1}{4e^2}\mathcal{B}_{\mu\nu}^2$. Hence the "magic number" $a=2$ has been "derived" from the Lagrangian (10) which seems to have more fundamental looking. The presence of $\mathcal{L}_M$ term is crucial in this respect.

Next we consider our Lagrantian (10) in the "external basis"*), in which we have

$$\mathcal{L} = f_\pi^2 \mathrm{Tr}(V_\mu - \mathcal{V}_\mu)^2 + f_\pi^2 \mathrm{Tr}(A_\mu - \mathcal{A}_\mu)^2 + \frac{f_\pi^2}{4} \mathrm{Tr}(D_\mu U D^\mu U^\dagger)$$

$$- \frac{1}{2g^2} \mathrm{Tr}(V_{\mu\nu}^2 + A_{\mu\nu}^2) - \frac{1}{2\xi^2} \mathrm{Tr}(\mathcal{V}_{\mu\nu}^2 + \mathcal{A}_{\mu\nu}^2) \ . \qquad (11)$$

If we identified the ρ and A_1 mesons with V_μ and A_μ*) instead of $\tilde{V}_\mu$ and $\tilde{A}_\mu$, respectively, then the Lagrangian (11) would precisely coincide with the "massive" Yang-Mills Lagrangian[2),16)] except that (11) contains external gauge fields $\mathcal{V}_\mu$ and $\mathcal{A}_\mu$ which may correspond to γ, W and Z bosons. Thus the Lagrangian (11) may be regarded as a generalization of the "massive" Yang-Mills Lagrangian, possessing the typical "mass mixing" between (V_μ, A_μ) and ($\mathcal{V}_\mu$, $\mathcal{A}_\mu$) as well as the "V_μ- and A_μ-dominance" in an obvious manner since the third term contains no external gauge fields $\mathcal{V}_\mu$ and $\mathcal{A}_\mu$. In the absence of $\mathcal{V}_\mu$ and $\mathcal{A}_\mu$, Eq.(11) also coincides with the Lagrangian of Stremnitzer[17)] who, adopting the $_M$ term of ref.9) in a different (perhaps unjustified)**) context from ours in the "hidden basis", arrived at the same

---

*) Since the basis transformations (7) and (8) are not the gauge transformations unless the pions are absorbed into the external gauge bosons (this is the case for the technicolor), there are no a priori reasons to believe that both bases yield the same physics. Also note that G_{local} is completely gauge-fixed and hence no longer exists in the "external basis".

**) It is not legitimate to put $\tilde{A}_\mu=0$[17)]. The same criticism is also applied to a similar attempt[18)] identifying $\tilde{\mathcal{V}}_\mu$ with the ρ meson while setting $\tilde{\mathcal{A}}_\mu=0$ in the Lagrangian with a'=2, b'=1 and c'=d'=0. Their Lagrangian[18)] can be obtained by solving away $\tilde{\mathcal{A}}_\mu$ via equation of motion in the Lagrangian with a'=b'=d'=1 and c'=0, just in the same way as done for $\tilde{A}_\mu$ in our Lagrangian.

Lagrangian as ours in the "external basis".

In any case, only the approach based on the hidden local symmetry has two conceptually independent charge sources, G_{local} and G_{global}, which enables us to introduce γ, W and Z bosons into the π, ρ and A_1 system in a unique manner without any confusion.

Finally, I would comment on the dynamical generation of the kinetic term of $\tilde{V}_\mu$ and $\tilde{A}_\mu$. As we have seen, our Lagrangian can be written in either bases, "hidden" or "external", so that if one proved the dynamical generation of V_μ and A_μ in the "external basis", then it would mean that the kinetic term of hidden gauge bosons, $\tilde{V}_\mu$ and $\tilde{A}_\mu$, has also been generated, or vice versa. In this connection, the Nambu-Jona-Lasinio model does produce the V_μ-pole (up to cut-off)[19], so does the $\tilde{V}_\mu$-pole as well.[12] Recently it has been claimed[20] that the kinetic term of V_μ and A_μ are both generated from the N-JL model (up to cut-off), which would imply the pole generation of $\tilde{V}_\mu$ and $\tilde{A}_\mu$, accordingly.

Details of this article are given in a separate paper.[21]

I would like to thank M. Bando and T. Kugo for very helpful discussions.

References

1) S. Weinberg, Phys. Rev. Lett. 18 (1967) 188.

2) J. Schwinger, Phys. Lett. 24B (1967) 473.

3) S. Weinberg, Phys. Rev. 166 (1968) 1568.

4) For reviews, see S. Gasiorowicz and D. Geffen, Rev. Mod. Phys. 41 (1969) 531; S. Weinberg, Lectures on Elementary Particles and Quantum Field Theory, ed. by S. Deser, M. Grisaru and H. Pendleton (MIT Press, Cambridge, 1970); B.W. Lee, Chiral Dynamics (Gordon and Breach, New York, 1972).

5) Y. Ohnuki and Y. Yamaguchi, Chiral Dynamics, INS Report (in Japanese) INS-TH-YY-67-1 (August 1967);
K. Hiida, Y. Ohnuki and Y. Yamaguchi, Suppl. Prog. Theor. Phys. (1968) 337.

6) S. Coleman, J. Wess and B. Zumino, Phys. Rev. 177 (1968) 2239; C. Callan, S. Coleman, J. Wess and B. Zumino, Phys. Rev. 177 (1968) 2247.

7) T.H.R. Skyrme, Proc. Roy. Soc. A260 (1961) 127; G.S. Addkins, C.R. Nappi and E. Witten, Nucl. Phys. B228 (1983) 552.

8) M. Bando, T. Kugo, S. Uehara, K. Yamawaki and T. Yanagida, Phys. Rev. Lett. 54 (1985) 1215.

9) M. Bando, T. Kugo and K. Yamawaki, Nucl. Phys. B259 (1985) 493.

10) M. Bando, T. Kugo and K. Yamawaki, Prog. Theor. Phys. 73 (1985) 1541.

11) T. Fujiwara, T. Kugo, H. Terao, S. Uehara and K. Yamawaki, Prog. Theor. Phys. 73 (1985) 926.

12) T. Kugo, Soryushiron Kenkyu (in Japanese) 71 (1985) E78; H. Terao, Master Thesis (in Japanese), Kyoto University, 1985; T. Kugo, H. Terao and S. Uehara, Prog. Theor. Phys. Suppl. No.85 (1985) 122.

13) K. Yamawaki, in Proc. 1985 INS Symp. on Composite Models of Quarks and Leptons ed. by H. Terazawa and M. Yasuè.

14) M. Bando, T. Fujiwara and K. Yamawaki, Nagoya Preprint, DPNU-86-23.

15) V. Golo and A.M. Perelomov, Phys. Lett. 79B (1978) 112; A. d'Adda, P. di Vecchia and M. Lüscher, Nucl. Phys. B146 (1978) 63; B152 (1979) 125; I. Ya. Aref'eva and S.I. Azakov, Nucl. Phys. B162 (1980) 298.

16) J. Wess and B. Zumino, Phys. Rev. 163 (1967) 1727.

17) H. Stremnitzer, Maryland Preprint, No.86-20 (1986).

18) Ö. Kaymakcalan and J. Schechter, Phys. Rev. D31 (1985) 1109; J. Schechter, Phys. Rev. D, to be published.

19) Y. Nambu and G. Jona-Lasinio, Phys. Rev. 122 (1961) 345.

20) A. Dhar, R. Shankar and S.R. Wadia, Phys. Rev. D31 (1985) 3256; D. Ebert and H. Reinhardt, Phys. Lett. 173B (1986) 453. See also H. Terazawa, Y. Chikashige and K. Akama, Phys. Rev. D15 (1977) 480.

21) K. Yamawaki, Nagoya Preprint, DPNU-86-24.

Hadron Spectrum in Quenched QCD

Y. Iwasaki

Institute of Physics, University of Tsukuba
Ibaraki 305, JAPAN

In the sixties, the success of the eightfold way for classification of hadrons which was originally proposed by Professor Y. Yamaguchi, but unfortunately was not published led to the proposal of the quark model. Further, the success of the quark model and the discovely of asymptotic freedom built in QCD (Quantum Chromo Dynamics) motivated that QCD is the fundamental theory of hadrons. However, the non-perturbative feature prevented ones to calculate physical quantities such as hadron masses from continuum QCD.

The only known way which makes it possible to calculate physical quantities by non-perturbative method is lattice QCD.[1)] Thus the lattice formulation of QCD provides a fascinating possibility of calculating hadron masses by numerical methods from first principles. However, in order to obtain reliable results by numerical methods one must be very cautious that systematic errors such as due to finite lattice spacing effects and finite lattice size effects are under control. When we consider the capacity of today's computers, we believe it is urgent to derive the correct hadron masses in the quenched (valence) approximation (which neglects dynamical quark loops) in a calculation where such systematic errors are under control. Here we report the results of the calculation of hadron masses which is done with the intention of obtaining the correct hadron masses in the quenched approximation by reducing such systematic errors.

Our primary concern is what the correct masses of the proton and delta Δ are in the quenched approximation, when the masses of the pion and rho are fitted. If one recalls the success of valence quark models in describing the static properties of hadrons as well as the success of the OZI rule in decay and scattering processes of hadrons, one may expect that the quenched approximation will give reasonable values for hadron masses with about 10 % errors. However, the results which have been obtained up to now with the standard one-plaquette action and with the Wilson's quark action are not satisfactory (for the earliest results see refs. 2)-4); for recent results see refs. 5)-7); for a recent result with Kogut-Susskind quark action see ref. 8)): For example, the proton-to-rho mass ratio m_p/m_ρ turned out to be too large. Possible origins for this large discrepancy are finite lattice spacing effects and finite lattice size effects.

In order to reduce finite lattice spacing effects we take a renormalization group (RG) improved lattice SU(3) gauge action

$$S_g = \frac{1}{g^2}\{c_0 \Sigma \,\mathrm{Tr}\,(\text{simple plaquette}) + c_1 \Sigma \,\mathrm{Tr}\,(1\times 2 \text{ rectangular loop})\} \tag{1}$$

with

$$c_1 = -0.331 , \quad c_0 = 1 - 8c_1 , \tag{2}$$

the form of which has been determined by a block spin renormalization group study[9] and an analysis of instántons on the lattice[10]. In the sum over loops, each oriented loop appears once.

In order to reduce finite lattice effects we make our calculations on a $16^3\times48$ lattice at $\beta=2.4$ ($\beta=6/g^2$). The lattice size and the coupling constant are determined from the results of our previous calculations of hadron masses on a $12^3\times24$ lattice[11] and on a $16^3\times32$ lattice[12]. (We have increased the number of configurations up to 30 configurations in the case of the $16^3\times32$ lattice to improve statistics: The results are essentially the same as those in ref. 12)).

We take Wilson fermions[13] for quarks, because it is not clear how to assign quantum numbers for Kogut-Susskind fermions[14]. The action is given by[13]

$$S_q = -\sum_n \bar{\psi}(n)\psi(n) + K \sum_{n,\mu} \{\bar{\psi}(n)(1-\gamma_\mu)U(n,\mu)\psi(n+\hat{\mu})+\bar{\psi}(n+\hat{\mu})(1+\gamma_\mu)U^\dagger(n,\mu)\psi(n)\} \quad (3)$$

with spinor and color indices being suppressed. The K is the hopping parameter which is related to the bare mass by

$$m_0 a = \frac{1}{2}(1/k - 1/k_c) \ . \quad (4)$$

We put $K_u=K_d=K$ as usual.

We use two operators $\pi_1=\bar{u}\gamma_5 d$ and $\pi_2=\bar{u}\gamma_5\gamma_4 d$ for the pion, $\rho_1=\bar{u}\gamma_i d$ and $\rho_2=\bar{u}\gamma_i\gamma_4 d$ for the rho, respectively. Here u and d are u-quark and d-quark fields respectively and i=1,2,3. We use non-relativistic operators for the proton and delta as well as $(u^T\gamma_5 C^{-1}d)u$ for the proton and $(u^T\gamma_\mu C^{-1}u)u$ for the delta, where C is the charge conjugation operator. We use two operators for all of them to investigate finite size effects[5].

Our strategy for extracting masses of hadrons in the quenched approximation is the standard one which has been employed in most previous works[2)-8)]: After generating a set of gauge configurations by Monte Carlo (MC) simulations, we solve quark propagators on given gauge configurations. Then we calculate hadron propagators G(t) for zero momentum states as a function of euclidean time t. (Here $0 \leqq t \leqq 47$.) From behavior of hadron propagators, we determine masses and decay constants of hadrons. Here we calculate hadron masses (especially the masses of the pion, rho, proton and delta) with the RG improved action at β=2.4 on a $16^3\times 48$ lattice for 15 gauge configurations, separated by one hundred sweeps after a thermalization of 1000 sweeps. The gauge configurations are generated by a Cabibbo-Marinari algorithm slightly modified for vector processors. We use periodic boundary conditions for both gauge fields and quark fields. The hopping parameters K we take are 0.14, 0.145, 0.15, 0.1525 and 0.154. The algorithm we use to calculate quark propagators is the minimum residual method with block incomplete LU decomposition, vectorized by a hyperplane method, which is described in ref. 11).

The results for the mesons are as follows: The propagators can be excellently fitted to a single hyperbolic-cosine for a wide range of t (t: coordinate in time direction; $0 \leqq t \leqq 47$ here; the origin of the propagator is t=0). The ground state mass at each K is stable for a change of the t-range ($8\sim 11 \leqq t \leqq 37\sim 40$) for the fit. Further when we make a two-mass fit to the propagators at each K for the range $4\sim 5 \leqq t \leqq 43\sim 44$, the ground state mass is stable. The masses and the errors are estimated by a least mean squares method using SALS system[15)] here and also hereafter for other particles. The masses determined both from π_1 and π_2 (from ρ_1 and ρ_2) agree with each other within one-standard deviation. Furthermore, the propagators on the $12^3\times 24$, $16^3\times 32$ and $16^3\times 48$ at five K's agree with each other within almost one-standard deviation for the t-region which can be compared. Thus we conclude that we are able to determine the ground state mass of the pion and rho at each K with small errors which is free from finite size effects and free also from the contamination of excited states.

Fitting the masses to a quadratic function of 1/K we obtain

$$K_c = 0.1569(2) , \tag{5}$$

$$m_\rho(K=K_c) = 0.426(15) , \tag{6}$$

where K_c is the hopping parameter value where the pion mass vanishes and a is the lattice spacing. We do not distinguish K_{phy} (where the ratio m_ρ/m_π becomes the physical value) from K_c, because K_{phy} is within the statistical error for K_c. Inputting the physical ρ mass we have a^{-1}=1810(60) MeV, which is slightly larger than that obtained previously, because the fitting procedure is different.

Let us now discuss the results for the baryons. The propagators can be fitted to a single exponentially decaying function for a wide range of t. After trying many one-mass fits as well as two-mass fits, we obtain the results for ground state masses of baryons summarized in Table 1 where we also give the results for mesons. The results for the relativistic operators are in excellent agreement with those for the non-relativistic ones at each K. This implies that finite size effects are under control.

Table 1: The ground state mass of the proton and delta as well as the pion and rho at five hopping parameters in lattice unit. The n and r correspond to the non-relativistic and the relativistic operators, respectively.

K		0.1400	0.1450	0.1500	0.1525	0.1540
P	n	1.722 8	1.421 8	1.099 18	0.918 35	0.811 51
	r	1.719 8	1.419 7	1.094 15	0.912 33	0.795 50
Δ	n	1.748 12	1.458 15	1.161 43	1.004 40	0.922 69
	r	1.747 8	1.457 12	1.156 33	1.014 35	0.944 83
π		1.056 4	0.845 4	0.611 4	0.476 5	0.383 6
ρ		1.095 5	0.907 5	0.712 6	0.609 8	0.542 11

Because the K dependence of the baryon masses near K_c is unknown, we make a guess. To do so we use phenomenological mass formulae which reproduce remarkably the physical hadron masses[17)]. The formulae are also used in ref. 18) for comparison with lattice results. They consist of the sum of the quark masses, a constant (different for mesons and baryons) and a term describing the hyperfine spin spliting:

$$M_{baryon} = M_b + \sum_{i=1}^{3} m_i + \xi_b \sum_{i>j} \vec{S}_i \cdot \vec{S}_j / m_i m_j \ , \qquad (7)$$

$$M_{meson} = M_m + \sum_{i=1}^{2} m_i + \xi_m \vec{S}_q \cdot \vec{S}_{\bar{q}} / m_q m_{\bar{q}} , \tag{8}$$

M_b=0.077 GeV, M_m=-0.057 GeV, ξ_b=0.02205 GeV^3 and ξ_m=0.0715 GeV^3. We can obtain the masses of the rho, proton and delta as a function of m_π^2 which are obtained from eqs. (7) and (8) equating all quark masses. We find that our results are in remarkable agreement with the relations obtained. Note that our adjustable constant is only for a^{-1}, which has been already determined. It should be emphasized that any phenomenological model which reproduces hadron masses nicely should give the relations which are essentially identical to those obtained from eqs. (7) and (8). Thus we conclude that our numerical results excellently reproduce the phenomenological results for $0.15 \leqq (m_\pi a)^2 \leqq 1.1$. We think it is natural to assume that the agreement will continue down to $(m_\pi a)^2 \simeq 0.01$. With this assumption we get the physical values for the proton mass (940 MeV) and the delta mass (1230 MeV), inputting the physical values for m_π and m_ρ. The calculation of hadron masses at K closer to K_c would reveal whether this assumption is realized.

Of course, the true values of hadron masses can be obtained after including the effect of dynamical quark loops (See ref.19) for such an attempt). Our results imply that even in the quenched (valence) approximation we can obtain reasonable values for the masses of flavor non-singlet hadrons with at most 10∿15 % errors. This is completely consistent with the success of the valence quark model in describing static properties of hadrons.

We hope that we are able to calculate in the near future flavor non-singlet hadron masses at K_{phy} and further the η meson mass which should be closely related with topological structure of the vacuum.

The work has been done in collaboration with S. Itoh and T. Yoshié. The details of our data and analyses will be published elsewhere. The calculation has been performed with the HITAC S810/10 at KEK. We would like to thank S. Kabe, T. Kaneko, R. Ogasawara and other members of Data Handling Division of KEK for their kind arrangement which made this work possible, and the members of Theory Division, particularly, H. Sugawara and T. Yukawa for their warm hospitality and strong support for this work.

References

1) Wilson K.G., Phys. Rev. D14 2445 (1974).
2) Hamber H. and Parisi G., Phys. Rev. Lett. 47 1972 (1981).
3) Marinari E., Parisi G. and Rebbi C., Phys. Rev. Lett. 47 1795 (1981).
4) Weingarten D.H., Phys. Lett. 109B 57 (1982).
5) Bowler K.C., Chalmers D.L., Kenway A., Kenway R.D., Pawley G.S. and Wallace D.J., Nucl. Phys. B240 FS12 213 (1984).
6) Billoire A., Marinari E. and Petronzio R., Nucl. Phys. B251 FS13 141 (1985).
7) König A., Mütter K.H., Schilling K. and Smit J, Phys. Lett. 157B 421 (1985).
8) Barkai D., Moriarty K.J.M. and Rebbi C., Phys. Lett. 156B 385 (1985).
9) Iwasaki Y., Nucl. Phys. B258 141 (1985); preprint UTHEP-118.
10) Iwasaki Y. and Yoshié T., Phys. Lett. 131B 159 (1983); Itoh S., Iwasaki Y. and Yoshié T., Phys. Lett. 147B 141 (1984).
11) Itoh S., Iwasaki Y., Oyanagi Y. and Yoshié T., Nucl. Phys. B274 33 (1986).
12) Itoh S., Iwasaki Y. and Yoshié T., Phys. Lett. 167B 443 (1986).
13) Wilson K.G., in New phenomena in subnuclear physics (Erice 1975), ed. Zichichi A. (Plenum, New York, 1977).
14) Susskind L., Phys. Rev. D16 3031 (1977).
15) Nakagawa T. and Oyanagi Y., in Recent developments in statistical Inference and Data Analysis, ed. K. Matusita (North Holland Publishing Company, 1980).
16) Kobayashi T., Nakamura N. and Saito K., Prog. Theor. Phys. 48 1618 (1972).
17) Ono, S., Phys. Rev. D17 888 (1978).
18) Bowler K.C. et al., Phys. Lett. 162B 354 (1985).
19) Fukugita M., Oyanagi Y. and Ukawa A., preprint UTHEP-152.
Ficito F., Moriarty K.J.M., Rebbi C. and Solomon S., preprint BNL 37546.

QUANTUM CHROMODYNAMICS AT HIGH TEMPERATURES

Akira Ukawa

Institute of Physics, University of Tsukuba
Ibaraki 305, Japan

1. Introduction

The dynamics of strong interactions at zero temperature and zero baryon number density exhibits two distinctive features, namely the permanent confinement of quarks and the spontaneous breakdown of chiral symmetry. It is now widely believed that both these properties will be lost through a phase transition at sufficiently high temperature and/or high baryon number density. In such unusual environments the matter is expected to form a plasma of liberated quarks and gluons. The existence of such a phase transition, especially if it is strongly first order, might have interesting implications in astrophysical problems such as the possibility of quark stars, supernova explosions, and the evolution of the early universe. The conditions for the formation of such a plasma might be realized during the collision of high energy heavy ions and if so this might lead to an experimental detection of such states in the laboratory.

The possibility that the phase of hadronic matter might undergo a dramatic change as the temperature is raised is already implicit in the result of Hagedorn[1)] who showed that within the bootstrap model of hadrons there exists a limiting temperature beyond which the hadronic matter can not exist. If one takes the quark model point of view, the limiting temperature can be interpreted[2)] naturally as the critical temperature of a phase transition across which the gas of hadrons turns into a plasma of quarks.

A simple argument for the existence of such a deconfining transition runs as follows : At zero temperature one may visualize confinement as due to the formation of a flux tube (string) of color electric field connecting a quark and an antiquark. As the temperature is raised, the string starts to fluctuates thermally. If σ is the string tension (i.e. the energy of the string per unit length) and T the temperature, the free energy of the string of length L will be given by $\sigma \cdot L - T \cdot S(L)$ where S(L) denotes the entropy of the string. Regarding the string as the trajectory of a random walk in steps of ℓ which may be taken to be the typical hadronic length scale $\sim (1\mathrm{GeV})^{-1}$, one easily finds that S(L) asymptotically behaves as $L \cdot \log(2d)/\ell$ with d=3 the space dimension. Thus the thermal motion reduces the string tension to $\sigma_{eff} = \sigma - T \cdot \log(2d)/\ell$. With increasing temperature one may expect σ_{eff} to eventually vanish. If this happens the quarks are no longer bound by a linearly rising potential and so become liberated. This simple argument also gives a rough estimate of the deconfining temperature: $T_c \sim \sigma/\ell \sim$ a few hundred MeV.

One can also give an intuitive argument on what may happen to the chiral symmetry at high temperatures. Assume that there are N_f flavors of massless quarks. Then the QCD action is invariant under $SU(N_f)$ global chiral transformation. If q denotes the quark fields, these transformations give rise to global rotations of the bilinears $S = \bar{q}(1+\gamma_5)q$. If one regards S as an $SU(N_f)$ valued spin variable located at each point of space, the chiral rotations correspond to the global rotations of spins and the spontaneous breakdown of the symmetry simply means that the effective spin system has a spontaneous magnetization at zero temperature. When the temperature is raised, the spins will fluctuate thermally and this will eventually lead to a phase transition above which the magnetization $< S > = 0$. One thus expects that the chiral symmetry will be restored at high temperatures through a phase transition.

Quantum chromodynamics provides a theoretical framework for putting these crude arguments into a more quantitative basis. One difficulty, however, is that the dynamics of QCD at zero temperature which is responsible for confinement and spontaneous breakdown of chiral symmetry is non-perturbative. Hence a proper examination of the finite temperature behavior of QCD could not be started until one had gained an understanding of the dynamics at zero temperature. The lattice formulation of QCD and the application of numerical simulation techniques have yielded a great deal of information on this problem and some of the questions on the statistical mechanics of QCD have been clarified during the past several years. Many problems, especially those related with the chiral symmetry, still remain, however, and are actively studied at the moment.

The purpose of the present article is to review[3)] the analytical and numerical results on the statistical mechanics of QCD as well as to report some recent results of simulations which fully incorporate dynamical quarks. We shall mainly discuss finite temperature behavior;A systematic study of the system at high baryon number density has been started only recently and not much is known at the moment.

2. DECONFINING AND CHIRAL PHASE TRANSITIONS AT FINITE TEMPERATURES

2.1 Deconfining transition in the pure gauge sector

The confinement is a property which arises from the dynamics of the color gauge fields. We therefore start our discussion with the finite temperature behavior of the pure gauge sector.

The thermodynamics of this sector follows from the partition function $Z_G = \mathrm{Tr}(\exp(-H_G/T))$ where T is the temperature and H_G denotes the Hamiltonian of the pure SU(3) gauge system. By a standard trick one can rewrite Z_G as a path integral over the Euclidean space-time of temporal extent 1/T :

$$Z_G = \int [dA_\mu] \, e^{-S_G} , \qquad (1)$$

where A_μ is the gauge field and S_G is the Euclidean gauge action.

A convenient quantity to characterize confinement is the free energy of an infinitely heavy quark F_q. The coupling of such a static quark at the space coordinate $\vec{x}$ to the gauge field is given by the Polyakov line[4)],

$$\Omega(\vec{x}) = \mathrm{tr}\left(P\exp\{i\int_0^{1/T} dt\, A_4(t,\vec{x})\} \right), \tag{2}$$

with P the path ordering and tr the color SU(3) trace. Thus F_q may be computed by

$$\langle\Omega(\vec{x})\rangle = e^{-F_q/T}, \tag{3}$$

where < > represents the Euclidean average of (1). Confinement of quarks means an infinite free energy $F_q = +\infty$ and hence $\langle\Omega\rangle = 0$. Conversely if $\langle\Omega\rangle \neq 0$, F_q is finite and a single isolated quark can exist.

The partition function (1) is obviously invariant under the gauge transformation $V(t,\vec{x})$ periodic in time. It is a non-trivial and important observation[4)] that (1) is also invariant even if $V(t,\vec{x})$ is globally twisted by the center $Z(3)= \{1,\omega 1,\omega^2 1;\omega=\exp 2\pi i/3\}$ of SU(3) : $V(t+1/T,\vec{x})=\zeta V(t,\vec{x})$, $\zeta\varepsilon Z(3)$. This is because the field strength $F_{\mu\nu}$ transforms under the adjoint representation $F_{\mu\nu} \rightarrow VF_{\mu\nu}V^\dagger$ and hence not affected by a global twist by Z(3). However, the Polyakov line Ω does transform according to $\Omega \rightarrow \zeta\Omega$. Thus the pure SU(3) gauge sector has a global Z(3) symmetry and the Polyakov line is an order parameter for this symmetry. If the symmetry is intact, $\langle\Omega\rangle$ necessarily has to vanish and because of (3) the system is in the confining phase. On the other hand if the quarks are not confined ($\langle\Omega\rangle \neq 0$) the Z(3) symmetry is spontaneously broken. In other words, deconfinement in the pure gauge sector can be characterized as a spontaneous breakdown of the center symmetry.

One can gain some insight on the behavior of $\langle\Omega\rangle$ by constructing an effective potential for $\Omega(\vec{x})$. This is most easily done in the lattice formulation of gauge theories.[5)] In this formulation the gauge variable A_μ is replaced by SU(3) valued matrix variable $U_{n\mu}$ which is defined on the link connecting the lattice site n and the neighboring site $n+\hat{\mu}$ in the μ direction and the simplest choice of the action is given by

$$S_G = \frac{1}{g^2} \sum_p tr(1-U_p) + c.c., \tag{4}$$

where U_p denotes the product of four link variables around a plaquette p (i.e. the elementary square of the lattice) and g is the coupling constant. The Polyakov line is given by a product of U_{n4} along a loop which wraps around the lattice in the temporal direction at a fixed spatial site $\vec{n}$: $\Omega = tr(\Pi U_{n4})$. In the strong coupling region $g^2 >> \infty$, one can integrate out U's in the spatial direction. Noting that the temporal extent of the lattice N_t is equal to 1/T in units of the lattice spacing a, one then finds that the partition function is approximately equal to

$$Z_G = \int \prod_{\vec{n}} d\Omega(\vec{n})\, e^{-S_{eff}}, \tag{5}$$

with

$$S_{eff} = -\frac{1}{T_{eff}} \sum \Omega(\vec{n})\Omega(\vec{m})^{\dagger} + c.c. , \tag{6}$$

where the summation in (6) is over the nearest neighbor sites and $T_{eff} = (3g^2)^{1/T}$. This is the partition function of a three dimensional ferromagnet with Z(3) symmetry which generally exhibits a first order phase transition at a non-zero temperature. The first order nature of the transition is a peculiarity of the Z(3) symmetry of the system which allows Ω^3 interactions in the set of relevant operators. Since high temperature T >> 1 of the original gauge system corresponds to low temperature $T_{eff} << 1$ for the spin system and vice versa, this implies that the gauge system will lose the confining property through a first order phase transition[6] at some temperature. We note that the transition will be second order if the gauge group is SU(2) since the effective spin system will then be a Z(2) symmetric one.

Aside from the heuristic argument given above, there exists a rigorous proof that $\langle\Omega\rangle \neq 0$ at sufficiently high temperature.[7] Combined with the results of numerical simulations discussed below, we have a convincing body of evidence that the pure gauge system does not exhibit confinement at high temperature and that this change of phase takes place through a first order phase transition.

The Polyakov line can be used to obtain information on the static potential between a quark and an antiquark. If $\vec{x}$ and $\vec{y}$ are the space coordinates of the quark and antiquark, the potential $F_{q\bar{q}}(\vec{x}-\vec{y})$ is given by

$$< \Omega(\vec{x})\Omega(\vec{y})^{\dagger} > = e^{-F_{q\bar{q}}(\vec{x}-\vec{y})/T} . \tag{7}$$

Consider now the limit $|\vec{x}-\vec{y}|\to\infty$. Due to the clustering property, one generally expects the behavior

$$< \Omega(\vec{x})\Omega(\vec{y})^{\dagger}> \to <\Omega(\vec{x})><\Omega(\vec{y})^{\dagger}> + O(e^{-m|\vec{x}-\vec{y}|}). \tag{8}$$

If Z(3) symmetry is intact, the first term on the lefthand side vanishes and hence due to (7) the potential $F_{q\bar{q}}$ increases linearly with the separation. This of course is the confining potential expected at low temperature. At high temperature, on the other hand, $<\Omega>\neq 0$ which implies an exponential decrease of the potential and hence no confinement.

Let us turn our attention to the question of the nature of the high temperature phase. From the argument given so far, one expects this to be a plasma of interacting gluons. One essential difference from the QED plasma is that the gluon plasma is expected to develop a mass both for the time-time and space-space component of the gauge propagator. The former appears already at order g^2 of the weak coupling expansion, while the latter is expected from the fact that in the limit $T=\infty$ the partition function Z_G reduces to that of a three dimensional SU(3) gauge theory[4)] which is known to have a mass gap. Unfortunately this mass gap is generated non-perturbatively and hence can not be calculated by the weak coupling technique. Indeed absence of the mass generation in the space-space component of the gluon self-energy and the ensuing infrared divergence invalidate [8)] the weak coupling expansion beyond order g^6.

2.2 Coupling dynamical quarks

Let us now include quarks by adding the action

$$S_q = \int d^4x\, \bar{q}(x)(\, \gamma\cdot D - m_q\,)q(x), \quad D_\mu = \partial_\mu - iA_\mu, \tag{9}$$

to the gauge action S_G in (1) and integrating over the quark field q(x).There are two questions raised by this inclusion, the possible effect of quarks on the deconfining transition of the pure gauge sector and the behavior of chiral symmetry at finite temperature.

Let us begin with the first question. An important property of the quark action in this regard is that the term $\bar{q}\gamma_4 D_4 q$ is not invariant under the center Z(3) symmetry.[9] This is due to the fact that the quark fields transform under the fundamental representation of color SU(3), i.e., $q(t,\vec{x})\to V(t,\vec{x})q(t,\vec{x})$, in contrast to the adjoint transformation of the gluon field. The center invariance is therefore explicitly broken by quarks.

One can see the implication of this breaking by computing the quark contribution to the action of the effective spin model discussed above. Again working on a lattice and using the naive discretization of the Dirac operator, one finds that a heavy quark induces an additional term given by

$$\Delta S_{eff} = - 4m_q^{-1/T} \sum_{\vec{n}} (\, \Omega(\vec{n}) + \Omega(\vec{n})^\dagger\,). \tag{10}$$

This shows that the effective Z(3) ferromagnet is perturbed by a symmetry breaking external magnetic field which becomes stronger as the quark mass decreases. In a ferromagnet which exhibits a second order phase transition any perturbation which explicitly breaks the symmetry ,however small, wipes out the singularity. In other words the system varies smoothly as a function of temperature. Our case is different, however, since the unperturbed system undergoes a first order transition. In this case the symmetry breaking interaction has to be sufficiently strong if it were to smooth out the discontinuous jump of the internal energy. In view of the coefficient $m_q^{-1/T}$ in (10), this implies that the first order deconfining transition will remain for sufficiently heavy quarks but will become weaker as m_q decreases. Of course (10) is not applicable for light quarks. Whether the phase transition is wiped out or not in this region can not be answered with such a simple analysis. Recent results from numerical simulations on this question will be discussed in the next section.

Another important question in the light quark mass region is the behavior of chiral symmetry. As mentioned in the introduction one expects this symmetry to be restored at high temperatures and there indeed exists a rigorous proof[10] that this is true. Unfortunately, arguments based on the QCD Hamiltonian which could tell the order of the transition and the effects of finite quark mass are not available (Note that the chiral symmetry is exact only for massless quarks). We mention, however, that a renormalization group analysis[11] based on an effective lagrangian for pseudoscalar mesons indicates a fluctuation induced first order phase transition for three or more flavors of massless quarks. If this turns out to be the case, the existence of a first order transition in reality again depends on the strength of the chiral symmetry breaking due to the actual quark masses.

The arguments described above suggests that there exists a first order phase transition at both ends of the quark mass range $m_q = 0$ and ∞, and that these transitions will extend into the region of finite and non-zero m_q. How far will they extend and how will they be related with each other are the questions which at the moment can be analyzed only by a numerical simulation including dynamical quarks.

Let us make some comments on a finite baryon number density. This may be incorporated by substituting $m_q \to m_q - \mu\gamma_4$ in (9) with μ the chemical potential. The additional term $\mu\bar{q}\gamma_4 q$ again breaks the global Z(3) symmetry. Thus a large chemical potential implies strong breaking of the symmetry and hence leads to a smoothing of the deconfining transition. Non-zero chemical potential does not break chiral symmetry. However, the unbalance of quark and antiquark densities induced by a non-zero μ energetically favors the chirally symmetric vacuum. Thus a phase transition restoring chiral symmetry is also expected.

3. NUMERICAL SIMULATIONS AT FINITE TEMPERATURES

3.1 METHODS

We have already used several times the lattice formulation of QCD in our discussion. In addition to providing a well-defined basis for understanding confinement, it also enables application of simulation techniques for numerically analyzing the strongly interacting dynamics of QCD which so far has not been amenable to analytical treatments. In fact the standard Monte Carlo methods[12)] have been extremely useful for extracting information on the pure gauge sector near the continuum. However, they are not directly applicable to full QCD with quarks because of the anti-commuting property of the quark variables. Among the multitude of proposals[13-18)] made so far to circumvent this difficulty, the Langevin method[16-17)] and its variants[18)] seem most promising at the moment. We now briefly review this method.

In order to simplify the notation, let ϕ_i be the gauge variable with i denoting the link and color indices, and q_a the quark variable with a the site, Dirac and color indices. The quark action $S_q = \bar{q}\cdot D(\phi)\cdot q$ is bilinear in $\bar{q}$ and q and hence can be integrated. If the operator D is positive definite, one can express the resulting quark determinant detD as a Gaussian integral over the pseudo-fermionic variable Y_a with the action $S_Y = Y^\dagger \cdot D^{-1} \cdot Y$. One can thus rewrite the partition function of full QCD as

$$Z_{QCD} = \int \Pi d\phi \Pi dY dY^\dagger \, e^{-S_{tot}} \quad , \quad S_{tot} = S_G + S_Y \, , \tag{11}$$

Both variables ϕ and Y are complex and hence the Langevin equation[19)] which corresponds to the distribution in (11) is given by[16)]

$$\frac{d}{d\tau}\phi_i = -\frac{d}{d\phi_i}S_{tot} + \xi_i \, , \tag{12a}$$

$$\frac{d}{d\tau}Y_a = -\frac{d}{dY_a}S_{tot} + \zeta_a \, , \tag{12b}$$

where τ is the fictitious time and ξ and ζ are the white noise in τ. It is an easy exercise[16] to show that the solution of (12) covers the phase space with the weight $e^{-S_{tot}}$ in the limit $\tau\to\infty$. Thus the expectation values of observables can be computed via the average over the Langevin time τ.

Numerical implementation of the Langevin equation requires a discretization of (12) with respect to τ. Denoting the time step by $\Delta\tau$ and writing $\phi(n\cdot\Delta\tau)\equiv\phi^{(n)}$ and $Y(n\cdot\Delta\tau)\equiv Y^{(n)}$, the naive discretization takes the following form:

$$\phi_i^{(n+1)} = \phi_i^{(n)} - \Delta\tau\cdot\frac{d}{d\phi_i^{(n)}} S_{tot} + \Delta\tau^{1/2}\cdot \xi_i^{(n)}, \tag{13a}$$

$$Y_a^{(n+1)} = Y_a^{(n)} - \Delta\tau\cdot\frac{d}{dY_a^{\dagger(n)}} S_{tot} + \Delta\tau^{1/2}\cdot \zeta_a^{(n)}, \tag{13b}$$

and again one can show easily that this set of discrete equations generate the distribution $e^{-S_{tot}}$ in the limit $n\to\infty$ up to terms of order $\Delta\tau$. In fact one can generalize (13b) in the following manner without spoiling this property:

$$Y_a^{(n+1)} = (\delta_{ab} - \Delta\tau\cdot B_{ab}(\phi^{(n)}))\cdot Y_b^{(n)} + \Delta\tau^{1/2}\cdot C_{ab}(\phi^{(n)})\cdot\zeta_b^{(n)}, \tag{13c}$$

Here the operators B and C are arbitrary as long as they satisfy the condition $2\cdot CC^{\dagger} - BD - DB^{\dagger} = 0$. In particular one can completely remove the variable Y if one takes the choice $B\equiv 1/\Delta\tau$.[17] This choice has the additional advantage that any number of flavors can be simulated without introducing Y's and ζ's for each flavor.

The Langevin method has several notable features. First of all the distribution of the field variables corresponding to the full action of QCD is guaranteed in the limit of $\Delta\tau\to 0$. Secondly the deviation of the distribution due to the finiteness of $\Delta\tau$ in an actual calculation can be theoretically estimated[16-18,20-21] using the Fokker-Planck equation. This is in contrast with the pseudo-fermion[13] and microcanonical[14] methods for which these points are difficult to establish. Another possible advantage is that the entire field configuration is changed at each update (This feature is shared by the microcanonical method).

In a numerical simulation, one is forced to work on a lattice of finite extent and finite lattice spacing a. The information on the continuum physics is obtained in the continuum limit $a \to 0$ and $g \to 0$. The loss of the underlying lattice structure for small enough a and g is signaled by the scaling behavior: For an observable $\mathcal{O}$ of mass dimension ℓ, this is given by

$$\mathcal{O} \sim \text{const} \cdot \Lambda_L^{\ell} , \tag{14}$$

where

$$\Lambda_L = \frac{1}{a} (b_0 g^2)^{-b_0^2/2b_1} \cdot e^{-1/2b_0 g^2} , \tag{15}$$

is the lattice Λ parameter with

$$b_0 = \frac{1}{16\pi^2} (11 - \frac{2}{3} N_f), \quad b_1 = \frac{1}{(16\pi^2)^2} (102 - \frac{38}{3} N_f) \tag{16}$$

for QCD with N_f flavors of quarks.

3.2 Deconfining transition in the pure gauge sector

This problem can be examined with the standard Monte Carlo methods and a large number of simulations has been carried out with increasing lattice size and statistics (See ref.22-23 for a list of references). In fig.1 we show a typical example[22] of the behavior of the Polyakov line $\langle\Omega\rangle$ as a function of $\beta=6/g^2$. The value of $\langle\Omega\rangle$ jumps up from a small to a finite value over a narrow range of β suggesting a first order transition. A more interesting evidence[22] is given in fig.2: The system is apparently going back and forth between the symmetric phase $\langle\Omega\rangle=0$ and the broken phase $\langle\Omega\rangle\neq 0$ with the number of Monte Carlo iterations, indicating that these two states have very similar values of free

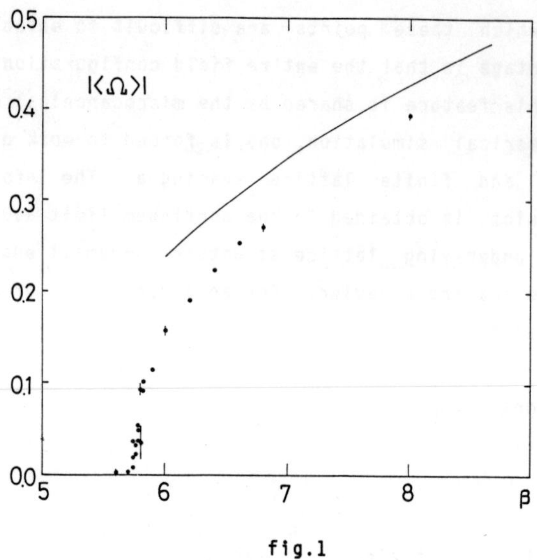
0.5
0.4
0.3
0.2
0.1
0.0
|⟨Ω⟩|
5
6
7
8
β

fig.1

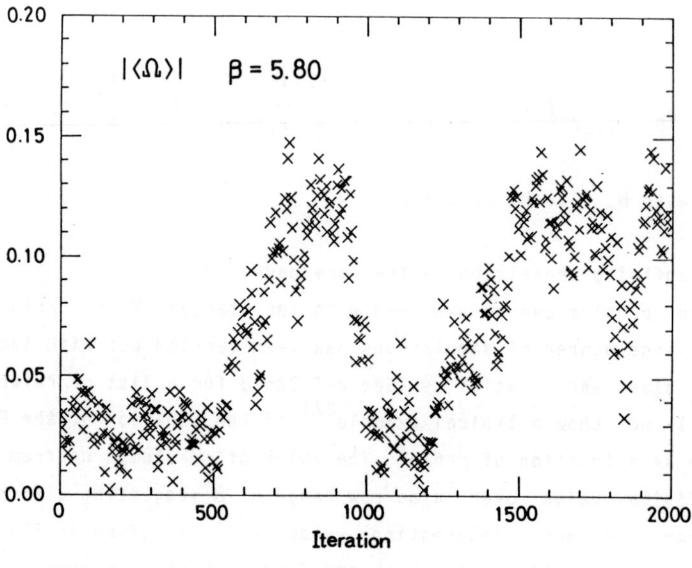
0.20
0.15
0.10
0.05
0.00
|⟨Ω⟩| β = 5.80
0
500
1000
1500
2000
Iteration

fig.2

energy. Such a coexistence of two states is a typical signature of a first order transition.

In order to extract the deconfining temperature T_c, one has to know the dependence of $T_c=1/N_t a$ on the critical coupling β_c and check that it follows the scaling prediction $T_c \sim const \cdot \Lambda_L$. This means the necessity of carrying out simulations with increasingly larger values of N_t. It is still a matter of debate whether the current simulation results have already seen such a scaling behavior. With this cautionary note we quote the best estimate so far[23)] for the deconfining temperature :

$$T_c = (51 \pm 2) \Lambda_L \tag{17}$$

Combining this with an estimate of the zero temperature string tension[24)] $\sqrt{\sigma} \cong 104 \Lambda_L$ gives $T_c \cong 0.49\sqrt{\sigma} \cong 220$MeV with $\sqrt{\sigma}=450$MeV.

One can also extract the static potential at high temperature using (7). The result[22)] shows that the potential has a screened Coulomb form typical of a plasma as expected from the weak coupling analysis but that the approach to the continuum expression is rather slow. The jump of the internal energy density across the transition is another important quantity. Because of the large fluctuations, however, a reliable estimate is not yet available (see ref.25 for some attempts).

3.3 Simulation of full QCD at finite temperature

There now exists a fair number of simulations[26-35)] addressing this problem. However, the nature of the transition is not completely clear in the small quark mass region where a consensus is not yet reached among different groups. We report below the results of our simulation[31,35)] using the Langevin procedure described above and then make some comments on those of other groups.

It should be mentioned at this point that introducing quarks on a lattice suffers from the difficulty of the spectral doubling[36)]. For this reason one is forced to use a quark action either with doubling but with some remnant of chiral symmetry or with explicit breaking of chiral symmetry but without doubling problem. Widely used examples corresponding to these alternatives are the Kogut-Susskind action[37)] and the

Wilson action.[38] One would prefer to use the former for the finite temperature study at small quark masses[26-33] since chiral symmetry is believed to play an important role here. It is nonetheless important to carry out a similar study with the Wilson action[34-35] in order to understand how these actions with quite different chiral properties might lead to the behavior in the continuum which should be universal.

Our Langevin simulation with the Kogut-Susskind action[31] was carried out on a $8^3 \times 4$ lattice with $N_f=4$ flavors. In order to examine the behavior of the system thermal cycle analysis in $\beta=6/g^2$ was carried out for a variety of quark mass values in the range m_q = 0.1-1.0. Our thermal cycle data for the Polyakov line $\langle\Omega\rangle$ and the chiral order

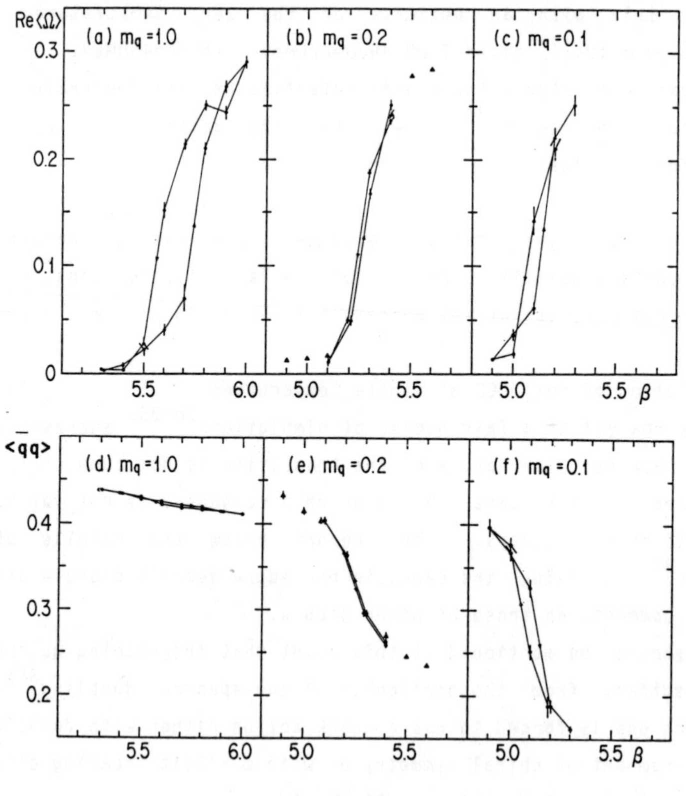

fig.3

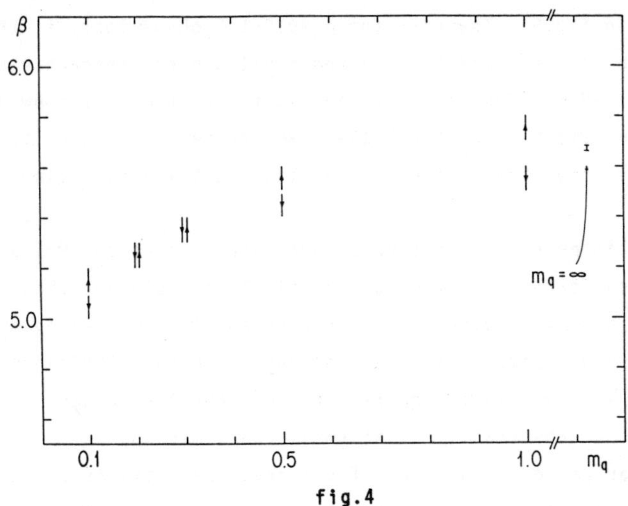

fig.4

parameter $\langle\bar{q}q\rangle$ is shown in fig.3 for several values of m_q. These observables jump or change rapidly over a narrow range of β as the system is heated up or cooled down. Fig.4 summarizes the position of β at which the jump occurred in the heating (upward arrow) and cooling (downward arrow) cycles.

For heavy quarks, e.g. m_q=1.0 (m_q/T_c=4.0), the presence and the location of the hysterisis observed in fig.3 are consistent with the first order deconfining transition of the pure gauge sector ($m_q = \infty$). With decreasing m_q the hysterisis of the thermal cycle weakens and it disappears at $m_q \sim 0.3$. This is consistent with the observation of section 2.2 that heavy quarks induce Z(3) breaking perturbation which should weaken the transition.

As the quark mass is decreased further down to m_q=0.1(m_q/T_c=0.4), however, a hysterisis is again observed. To examine whether this signals the existence of metastable states, the heating and cooling runs at m_q=0.1 and β=5.1 were extended to 40 time units. The value of Re$\langle\Omega\rangle$ at every $\delta\tau$=1.0 for the extended runs is shown in fig.5 in which the open and filled circles are the heating and cooling data, respectively. The relaxation time τ_r for the Polyakov line defined to be that at which the autocorrelation function in Langevin time decreases to 0.1 was τ_r=3~4 for the runs in fig.5. The data for m_q=0.2-1.0 shows that, if there are no metastable states, the heating and cooling runs approach each other already at τ/τ_r~5-6. The Polyakov line in fig 5. does not show any such tendency even after $\tau/\tau_r=40/\tau_r\sim10$. We therefore regard this as an evidence that the transition in the light quark mass region is first order.

Another interesting feature of our data is that the Polyakov line and the chiral order parameter change simultaneously and that the jump of the latter becomes progressively pronounced with decreasing m_q. Thus deconfinement and chiral symmetry restoration occur simultaneously for the quark in the fundamental representation and the latter is the main characteristic of the transition at small quark masses.

To summarize our data we found that there exists a single continuous line of transition throughout the (β, m_q) plane reaching the limit $m_q = 0$ at $\beta \cong 4.9$-5.0. In the middle ($m_q \cong 0.2$-0.3) the transition might not be a genuine phase transition but merely a crossover. For smaller m_q a first order transition is indicated by the persistence of the two phases in fig.5.

Let us now compare our results at the smallest quark mass m_q=0.1 with the ones in the literature[28-30]. Our data for $\langle\Omega\rangle$ and $\langle\bar{q}q\rangle$ agree very well with that in ref.28 (Even at β=5.1 where a rapid change of

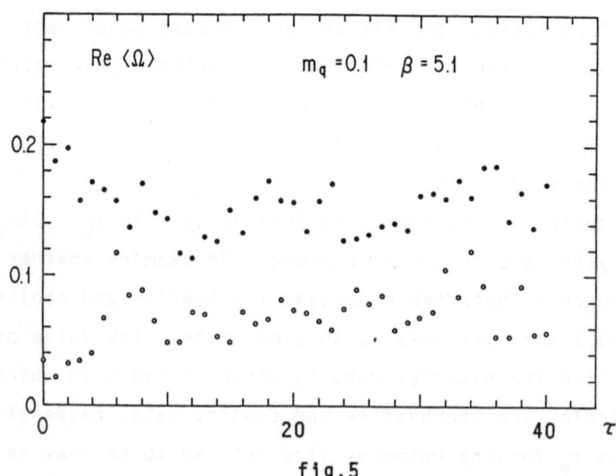

fig.5

physical observables was observed in ref.28). It also agrees well with that in ref.30 except for the transition region,which is shifted upwards by $\Delta\beta \cong 0.06$ in ref.30. These authors do not make any claim on the existence or absence of a first order transition (Some indication of a weak metastability was reported in ref.30. See also ref.29 which reports failure of finding metastable states with the hybrid algorithm[18] at $m_q=0.05$). A merit of our analysis may be the fact that we have carried out a controlled thermal cycle analysis, which is likely to be better than simply comparing runs with a cold and a hot start for detecting metastable states.

Simulations with the number of flavors N_f other than 4 have also been carried out. The order of the transition for the $N_f=3$ case[27] is still not settled but the $N_f=8$ case[30,32] exhibits a clear evidence for a first order transition at small quark mass. Thus the strength of the chiral transition increases with the number of flavors. We also note that for quarks in the adjoint representation the temperature for chiral restoration seems higher than that for deconfinement.[33]

To summarize the data accumulated up to now, there exists a sharp transition for light quarks across which the chiral restoration and the deconfinement takes place simultaneously and the strength of the transition increases with the number of flavors. Whether the transition is a first order one or not for small number of flavors is still not settled.

Let us now discuss the finite temperature behavior with the Wilson quark action which is defined by[36]

$$S_q = \sum_n \bar{q}_n q_n - K \sum_{n,\mu} \{ \bar{q}_n (1-\gamma_\mu) U_{n\mu} q_{n+\hat{\mu}} + \bar{q}_n (1+\gamma_\mu) U^\dagger_{n-\hat{\mu},\mu} q_{n-\hat{\mu}} \}. \quad (18)$$

In the free case $U_{n\mu}\equiv 1$ the hopping parameter K is related to the quark mass m_q by $m_q=\log[1+1/2(1/K-8)]$.

Our Langevin simulation with the Wilson action[35] was carried out on $5^3\times3$ lattice. The number of flavors was chosen to be $N_f=4$ to compare with the results for the Kogut-Susskind action. A large number of points in the (β,K) plane was examined to gain an overall picture of the behavior of the system. A separate spectroscopic analysis on $6^3\times12$

lattice was also made to determine the critical value K_c at which the pion mass vanishes.

Fig.6 shows the Polyakov line as a function of β at fixed values of K. As with the Kogut-Susskind action (see fig.1), a rapid crossover is seen at all the values of K shown, but it gradually becomes less pronounced with increasing K. For $K=10^{-4}$ the rapid change occurs at $\beta \cong 5.5$, which agrees with β_c for the pure gauge system with $N_t=3$. The transition region is shifted towards smaller β with increasing K. These characteristics are in agreement with the perturbative analysis for heavy quarks of section 2.2 and also with the results with the Kogut-Susskind action. A notable difference is that for $K \gtrsim 0.1$ $\langle\Omega\rangle$ takes a finite value at β's even below the transition region and this residual value of $\langle\Omega\rangle$ increases with K. In the Kogut-Susskind case, $\langle\Omega\rangle$ is very small, typically of the order of 10^{-2} or less below the transition region. We have also calculated the gluon internal energy E_g and found that it increases rapidly to a finite value in the same range of β that the Polyakov line does so.

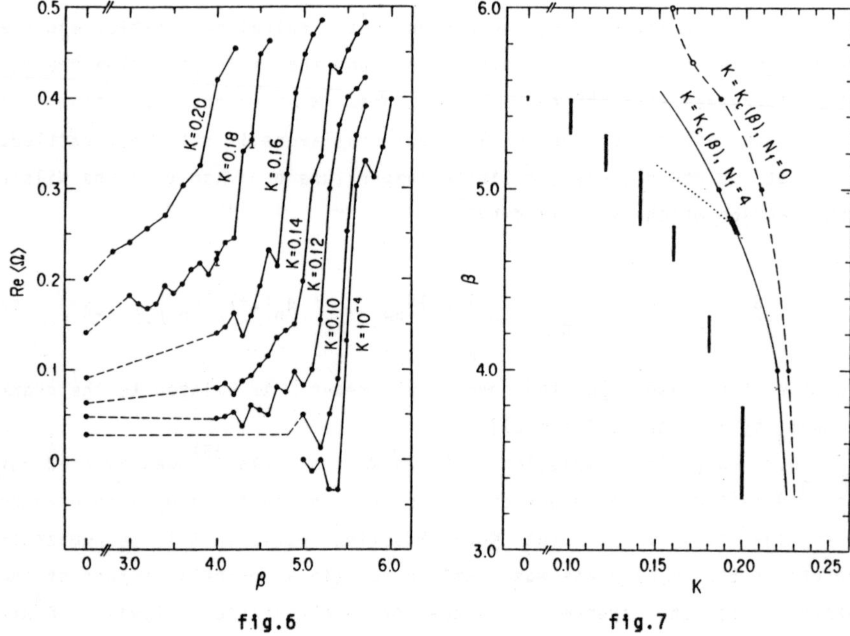

fig.6

fig.7

We show in fig.7 the location of the transition deduced from our analysis. The vertical bars indicate the range of β over which the slope of E_g and $\langle\Omega\rangle$ as a function of β increases from their low value in the small β region to the high value in the transition region. Also shown in the same figure is the value of K_c determined from the pion mass analysis at several β's both for the quenched (N_f= 0) and full (N_f= 4) QCD systems. We observe that the zone of transition does not meet the line of critical hopping parameter $K=K_c(\beta)$ even at $\beta\cong3.5$. This is rather different from what might be expected from the behavior of the system with the Kogut-Susskind action. If one makes a correspondence between the hopping parameter K of the Wilson action and the quark mass m_q of the Kogut-Susskind action by the condition that they give rise to the same value of the pion mass, which in particular means that the critical line $K=K_c(\beta)$ corresponds to m_q=0, then the results for the latter action on $8^3\times4$ lattice[28-31], scaled down to N_t=3 taking account of empirical scaling violation, suggests that the transition region should cross the critical line at $\beta\cong$ 4.75 – 4.85 (see the dotted line and the shaded area in fig.7). Furthermore we have suggested in the Kogut-Susskind case[31] that the transition becomes stronger as the quark mass approaches zero ($K\to K_c$ in the Wilson language), but no symptoms for such a behavior is observed. Of course a precise agreement of values should not be expected since the two actions are expected to produce the same physics only in the continuum limit and we are obviously far from it. However we observe here a signature that the physics is very much different between the two actions. Fig.6 and 7 suggests that the transition line with the Wilson action runs almost parallel with the critical line becoming weaker in strength and reaching the strong coupling limit without crossing it. Our data at β=0 shows that $\langle\Omega\rangle$ increases to a large value toward K_c. This also supports that the phase transition or its remnant reaches the limit β=0 without crossing the line $K=K_c(\beta)$.

This has important implications. If the transition is a real phase transition separating two phases and running down to the strong coupling limit without crossing the critical line, one cannot reach $K=K_c$ on a lattice with finite N_t; the critical line is always in the high temperature phase. Even if the transition turns into a crossover beyond some K, the large values of $\langle\Omega\rangle$ and E_g indicates that the system is practically in the high temperature phase near $K=K_c$

It is clear that the Wilson mass term which breaks chiral symmetry is responsible for the rather striking difference of behavior between the Kogut-Susskind and Wilson actions. Although the effect of this term is expected to become small near the continuum, it seems quite subtle how the Wilson action might produce the sharp chiral transition observed in the Kogut-Susskind case. That this happens is also necessary if the continuum limit with the Wilson action is to be reconciled with the first order chiral transition predicted in the continuum by a renormalization group analysis.[11)]

4. OUTLOOK

In this article we have reviewed the current status of the finite temperature study of QCD. In the pure gauge case the focus now centers on detailed questions such as verification of scaling behavior and extraction of numerical values of physical observables. The full QCD case is in a more preliminary stage. The simulations done so far were carried out on a fairly small lattice and aimed mainly at finding the nature of the finite temperature transition of full QCD but far from complete even in this regard. Simulations on a larger lattice with a weaker coupling is clearly needed to further elucidate this problem and to extract physical predictions as well as for resolve the puzzling difference of behavior between the Kogut-Susskind and Wilson actions.

We have touched upon the question of large baryon number density only briefly in this article. A systematic study has begun only recently and the initial attempts[39-40)] have found some unexpected behavior which is not completely understood. A difficulty of simulations with chemical potential is the complex nature of the quark action. The Monte Carlo methods such as the pseudo-fermion technique, though applicable, is very inefficient for large chemical potential. The

Langevin method can be extended for complex actions in principle[41] and may turn out to be useful.

The study of dynamics of QCD has been relying increasingly heavily on the numerical simulations. This trend will certainly continue in the future. However, current simulations are already severely straining the capability of even the best supercomputers available at present, and a multifold increase of the computer power will be necessary for truly realistic simulations. Of course simulations are basically experiments. More use of theoretical methods such as renormalization group in conjunction with them will be needed for a thorough understanding of the dynamics of QCD.

I would like to dedicate this article to Professor Yoshio Yamaguchi on the occasion of his sixtieth birthday. Ever since I was a graduate student, I have always admired his untiring enthusiasm and curiosity which go far and wide beyond physics. I hope he will be just as active and full of energy in the many more years to come.

REFERENCES

1. Hagedorn, R., Nuovo Cim. Suppl. 3, 147 (1965).
2. Cabbibo, N. and Parisi, G., Phys. Lett. 59B, 67 (1975).
3. See Gross, D., Pisarsky, R. and Yaffe, L. [Rev. Mod. Phys. 53, 43 (1981)] for a review and a list of references up to 1981.
4. Polyakov, A. M., Phys.Lett. 72B, 477 (1978) ;
   Susskind, L., Phys. Rev. D20, 2610 (1979).
5. Wilson, K. G., Phys. Rev. D10, 2445 (1984) ;
   Kogut, J. B. and Susskind, L., Phys. Rev. D11, 395 (1975).
6. Svetitsky, B. and Yaffe,L., Nucl. Phys. B210 [FS6], 423 (1982).
7. Borgs, C. and Seiler, E., Nucl. Phys. B215 [FS7], 125 (1983).
8. Gross, D., Pisarski, R. and Yaffe, L., in ref.3.
9. Banks, T. and Ukawa, A., Nucl Phys. B225 [FS9], 455 (1983) ;
   De Grand, T. and De Tar, C., Nucl. Phys. B225 [FS9], 590 (1983).
10. Tomboulis, E. and Yaffe, L., Commun. Math. Phys. 100, 313 (1985).

11. Pisarski, R. and Wilczek, F., Phys. Rev. D29, 338 (1984).
12. Creutz, M., Phys. Rev. D21, 2308 (1980) ;
Wilson, K. G., in Recent Progress in Gauge Theories, Cargese Lectures 1979, eds. t'Hooft, G. et. al. (Plenum, N. Y. 1980).
13. Fucito, F. et. al., Nucl. Phys. B180 [FS3] 369 (1981).
14. Polonyi, J. and Wyld, H. W., Phys. Rev. Lett. 51, 2257 (1983).
15. Weingarten, D. and Petcher, D., Phys. Lett. 99B, 333 (1982) ;
Scalapino, D. J. and Sugar, R., Phys. Rev. Lett. 46, 519 (1981) ;
Duncan, A. and Furman, M., Nucl. Phys. B190, 767 (1891) ;
Lang, C. and Nicolai, H., Nucl. Phys. B200 [FS4], 135 (1982) ;
Kuti, J., Phys. Rev. Lett. 49, 183 (1982) ;
Zwanziger, D., Phys. Rev. Lett. 50, 1886 (1983).
16. Ukawa, A. and Fukugita, M., Phys. Rev. Lett. 55, 1854 (1985).
17. Batrouni, G. et. al., Phys. Rev. D32, 2736 (1985).
18. Duane, S. Nucl. Phys. B257 [FS14], 652 (1985) ;
Duane, S. and Kogut, J. B.,Illinois preprint Ill-(TH)-86-#15 (1986).
19. Parisi, G. and Wu, Y.-S., Sci. Sin. 14, 483 (1981).
20. Drummond, I. T. et. al. Nucl. Phys. B200, 119 (1983).
21. see also Weingarten, D. [Nucl.Phys. B257 [FS14], 652 (1985)] and Martin, O. C. et. al. [Nucl. Phys. B264, 89 (1986)] for discussion of some related issues.
22. Fukugita, M. ,Kaneko, T. and Ukawa, A., Phys. Lett. 154B, 185 (1985)
23. Kennedy, A. D. et. al., Phys. Rev. Lett. 54, 87 (1985) ;
Gottlieb, S. A. et. al., Phys. Rev. Lett. 55, 1958 (1985) ;
Christ, N. and Terrano, A. E., Phys.Rev. Lett. 56, 111 (1986).
24. Barkai, D., Moriarty, K. J. M. and Rebbi, C., Phys. Rev. D30, 1293 (1984).
25. Kogut, J. et. al., Phys. Rev. Lett. 51, 869 (1983) ;
Celik, T., Engels, J. and Satz, H., Phys. Lett. 129B, 323 (1983) ;
Svetitsky, B. and Fucito, F., Phys. Lett. 131B, 165 (1983).
26. Gavai, R., Lev, M. and Peterson, B., Phys. Lett. 140B, 397 (1984) ;
Fischler, M. and Roskies, R., Phys. Lett. 145B, 99 (1984).
27. Fucito, F., Solomon, S. and Rebbi, C., Phys. Rev. D31, 1460 (1985) ;
Fucito, F. and Solomon, S., Phys. Rev. Lett. 55, 2641 (1985) ;
Gavai, R. V. and Karsch, F., Nucl. Phys. B261, 273 (1985).

28. Polonyi, J. et. al., Phys. Rev. Lett. 53, 644 (1984) ;

29. Duane, S. and Kogut, J., Phys. Rev. Lett. 55, 2774 (1985) ;
Kogut, J. B., Phys. Rev. Lett. 56, 2557 (1986).

30. Gavai, R. V., BNL preprint 37214 (1986) ;

31. Fukugita, M. and Ukawa, A., Kyoto preprint RIFP-642 (1986).

32. Kogut, J. B. et. al., Phys. Rev. Lett. 54, 1475 (1985).

33. Kogut, J. B. et. al., Phys. Rev. Lett. 54, 1980 (1985).

34. Hasenfratz. P., Karsch, F. and Stamatescu, I. O., Phys. Lett. 133B, 221 (1983) ;
Celik, T., Engels, J. and Satz, H., Phys. Lett. 133B, 427 (1983) ;
Nucl. Phys. B256, 670 (1985) ;
Ogilvie, M., Phys. Rev. Lett. 52, 1369 (1984).

35. Fukugita, M., Ohta, S. and Ukawa, A.,KEK preprint KEK-TH 129 (1986).

36. Nielsen, H. B. and Ninomiya, M., Nucl. Phys. B185, 20 (1981).

37. Kogut, J. B. and Susskind, L., in ref. 5 ;
Susskind, L., Phys. Rev. D16, 3031 (1977).

38. Wilson, K. G., in New Phenomenoa in Subnuclear Physics, ed. Zichichi, A., (Plenum, N. Y., 1977).

39. Engels, J. and Satz, H., Phys. Lett. 159B, 151 (1985).

40. Barbour, I. et. al. Illinois preprint Ill-(TH)-86-#23 (1986).

41. Parisi, G., Phys. Lett. 131B, 393 (1983) ;
Klauder, R., Phys. Rev. A29, 2036 (1984).

II. QUARK PHYSICS

TOPONIUM STATES EXPECTED AT TRISTAN

Keiji Igi and Seiji Ono
Department of Physics, University of Tokyo
Tokyo, Japan

1. INTRODUCTION

As is well known, heavy quark systems have a very clean spectroscopy. Therefore, the ψ and Υ families of the $c\bar{c}$ and $b\bar{b}$ bound states are considered as the 'hydrogen atoms' of strong interactions. So far, the search for the corresponding $t\bar{t}$ bound states has not been successful, and the present lower bound of the $t\bar{t}$ ground state has reached 46 GeV. In the very near future, however, the e^+e^- collider such as TRISTAN may have a good chance to find these toponium states. So, it is worthwhile to study various properties of these states, using the flavor-independent interquark potential obtained from $c\bar{c}$ and $b\bar{b}$ systems.

In §2 we explain our $Q\bar{Q}$ potentials which are consistent with QCD expectation and reproduce $c\bar{c}$ and $b\bar{b}$ properties. In §3 we compute spin-dependent forces of $c\bar{c}$ and $b\bar{b}$ states which are very sensitive to the value of $\Lambda_{\overline{MS}}$. Comparison with experiments indicates the prefereed value of $\Lambda_{\overline{MS}}$ to be around 200 MeV. §4 discusses open top states and we show the energies of various thresholds as a function of m_t and compare with the $t\bar{t}$ spectrum. In §5 we examine how the leptonic width of toponium states is influenced by the Z° boson when the $t\bar{t}$ mass is close to the Z° mass.

2. $Q\bar{Q}$ POTENTIAL

Let us start from the $Q\bar{Q}$ potentials which are consistent with QCD and reproduce $c\bar{c}$ and $b\bar{b}$ properties using the Schrödinger equation. As is well known, Buchmüller and Tye[1)] has proposed an elegant potential which accommodates the desired properties for the $Q\bar{Q}$ force. Their potential, however, implies a fairly large $\Lambda_{\overline{MS}}$ of about 500 MeV, which is related to the Regge slope parameter ($\alpha' \sim 1\ \mathrm{GeV}^{-2}$) and so it is not possible to incorporate a reasonably small $\Lambda_{\overline{MS}}$ in this scheme to reproduce correct $c\bar{c}$ and $b\bar{b}$ spectra.

To overcome this problem, we have proposed[2)] to use the following potential in the coordinate space;

$$V(r) = V_{AF}(r) + dre^{-gr} + ar, \tag{2.1}$$

where $$V_{AF}(r) = -\frac{16\pi}{25}\frac{1}{rf(r)}\left(1+\frac{2\gamma_E+\frac{53}{75}}{f(r)} - \frac{462\ln f(r)}{625\ f(r)}\right) \tag{2.2}$$

with $$f(r) = \ln[1/(\Lambda_{\overline{MS}}r)^2 + b] \quad . \tag{2.3}$$

The asymptotic freedom potential $V_{AF}(r)$ above satisfies the two-loop perturbative calculations for $r \to 0$, and becomes zero for $r \to \infty$. We can easily prove that an additional constant b does not change the short-range asymptotic behavior up to the order of $1/\ln(1/\Lambda_{\overline{MS}}^2 r^2)$). The linear-rising potential ar was added to $V_{AF}(r)$. In order to relax the possible range of $\Lambda_{\overline{MS}}$, we also added a term dre^{-gr} which disturbs neither the short- nor long-range part of the potential.

We found that we can change $\Lambda_{\overline{MS}}$ from 100 MeV to 500 MeV keeping the good fit to the $c\bar{c}$ and $b\bar{b}$ spectra. The level splitting e.g., E(2S) - E(1S) increases both with increasing $\Lambda_{\overline{MS}}$ and with increasing m_t. In table 1 we only show level splittings for $\Lambda_{\overline{MS}}$ = 200 MeV. Those for other values of $\Lambda_{\overline{MS}}$ can be found in ref.2. The parameters used here are

$$\begin{aligned} &b = 2, \quad a = 0.1587\ \mathrm{GeV}^2, \quad g = 0.3436\ \mathrm{GeV}, \\ &d = 0.2550\ \mathrm{GeV}^2, \quad m_c = 1.322\ \mathrm{GeV}, \\ &m_b = 4.731\ \mathrm{GeV} \end{aligned} \tag{2.4}$$

(potential J with $\Lambda_{\overline{MS}}$ = 200 MeV in ref.2).

Since the level splittings of $t\bar{t}$ states are sensitive to $\Lambda_{\overline{MS}}$ one might think that it is almost hopeless to predict toponium spectra only from the $c\bar{c}$ and $b\bar{b}$ spectra due to the above ambiguity of $\Lambda_{\overline{MS}}$

Table 1 (a) The excitation E(nS)-E(1S) of $t\bar{t}$ states in MeV.
(b) Hyperfine and fine structure of $t\bar{t}$ states in MeV.

(a)

m_t (GeV)	25	30	35
$E(1S) - m_t$	-648	-734	-813
1S	0	0	0
2S	598	615	634
3S	919	937	958
4S	1142	1160	1182
5S	1313	1331	1353
6S	1453	1471	1493
7S	1574	1590	1610
8S	1683	1696	1714
9S	1783	1792	1809
10S	1876	1882	1896
11S	1965	1967	1978
12S	2050	2048	2056
13S	2131	2125	2130
1P	489	512	536
2P	840	861	884
3P	1080	1101	1123
1D	746	771	796

(b)

m_t (GeV)		25	30	35
$E-E_{COG}$				
1S	$^3S_1$	5.0	5.0	5.0
	$^1S_0$	-14.9	-14.9	-14.9
2S	$^3S_1$	2.1	2.0	1.9
	$^1S_0$	-6.3	-5.9	-5.6
3S	$^3S_1$	1.4	1.3	1.2
	$^1S_0$	-4.2	-3.9	-3.7
4S	$^3S_1$	1.0	1.0	0.9
	$^1S_0$	-3.1	-2.9	-2.7
1P	$^3P_2$	3.6	3.2	3.0
	$^3P_1$	-3.6	-3.3	-3.1
	$^3P_0$	-12.8	-11.6	-10.8
	$^1P_1$	1.9	1.8	1.8
2P	$^3P_2$	2.4	2.2	2.0
	$^3P_1$	-2.4	-2.2	-2.1
	$^3P_0$	-8.7	-7.9	-7.3
	$^1P_1$	1.3	1.2	1.2

value. Let us show, however, that the fine and hyperfine splittings are very sensitive to the value of $\Lambda_{\overline{MS}}$, and hence the allowed range of $\Lambda_{\overline{MS}}$ might be restricted from the data of already known quarkonia, i.e., $c\bar{c}$ and $b\bar{b}$.

3. SPIN DEPENDENT FORCES AND $\Lambda_{\overline{MS}}$

The QCD radiative corrections to the spin-dependent forces has already been calculated[3]. Using this scheme, let us compute the $\Lambda_{\overline{MS}}$ dependence of spin-dependence forces.

The interactions are given by[3)]

$$H_{SS} = \frac{8\pi\alpha_S}{9m^2}\vec{\sigma}_1\cdot\vec{\sigma}_2[\{1 - \frac{\alpha_S}{12\pi}(26 + 9\ln 2)\}\delta(\vec{r}) \; . \; . \; . \; .]$$
$$\equiv C\vec{\sigma}_1\cdot\vec{\sigma}_2$$

$$H_T = \frac{\alpha_S}{3m^2}\frac{3\vec{\sigma}_1\cdot\hat{r}\vec{\sigma}_2\cdot\hat{r} - \vec{\sigma}_1\cdot\vec{\sigma}_2}{r^3}[. \; . \; . \; .]$$
$$\equiv B(3\vec{\sigma}_1\cdot\hat{r}\;\vec{\sigma}_2\cdot\hat{r} - \vec{\sigma}_1\cdot\vec{\sigma}_2) \qquad (3.1)$$

$$H_{LS} = \frac{2\alpha_S}{m^2}\frac{\vec{L}\cdot\vec{S}}{r^3}[. \; . \; . \; .] - \frac{a}{2m^2}\frac{\vec{L}\cdot\vec{S}}{r}$$
$$\equiv A\vec{L}\cdot\vec{S} \; .$$

We choose the renormalization scale so as to minimize the effect of higher-order terms.

Assuming n_f=4 and using non-relativistic wave functions (from our $Q\bar{Q}$ potential), we have computed expectation values of H_{SS}, H_{LS} and H_T (Table 2). These splittings rapidly increase as $\Lambda_{\overline{MS}}$ increases.

From the table 2, we can conclude as follows:

(i) The fine and hyperfine splittings are very sensitive to the short-range potentials and rapidly increase as $\Lambda_{\overline{MS}}$ increases.

(ii) As one may notice from this table, $\Lambda_{\overline{MS}}$ = 400 MeV and 500 MeV

Table 2 The hyperfine and fine splittings in MeV for various $\Lambda_{\overline{MS}}$ predicted by the potentials with b=20. The scheme by Gupta, Radford and Repko[3)] is used.

$c\bar{c}$		S-state $^3S_1 - {}^1S_0$ 1S	2S	1P-state A	B	α_S
$\Lambda_{\overline{MS}}$(MeV)	exp	116±6	95	34.9±0.6	10.0±0.3	
100		117	68	20.8	8.0	0.346
200		140	84	40.5	10.9	0.477
300		160	101	57.9	13.9	0.603
400		195	125	79.6	18.4	0.752
500		245	162	112	25.6	0.939

Table 2 (continued)

$b\bar{b}$	$^3S_1 - {}^1S_0$ 1S	2S	1P-state A	B	2P-state A	B	α_S
exp			14.2±0.3	3.1±0.2			
$\Lambda_{\overline{MS}}$			13±1	2.5±0.3	11±2	1.7±1.0	
100	33.5	18.5	10.3	2.27	6.77	1.52	0.257
200	39.5	21.5	14.3	2.94	9.38	1.98	0.328
300	44.6	23.4	17.6	3.51	12.0	2.46	0.389
400	52.7	26.5	21.1	4.15	14.5	2.94	0.446
500	61.6	30.6	24.5	4.79	17.1	3.46	0.500

are completely ruled out. For $\Lambda_{\overline{MS}} \sim 200$ MeV, we can reproduce fine and hyperfine splittings of $c\bar{c}$ and $b\bar{b}$ states.

Therefore, taking the preferred value $\Lambda_{\overline{MS}} = 200$ MeV, we can predict various properties of $t\bar{t}$ states.

4. OPEN TOP STATES AND $T\bar{T}$ THRESHOLDS

In our previous paper[4)] we derived the following formula of hyperfine splitting for all quarkonium states.

$$[M_{ij}(^3S_1)]^2 - [M_{ij}(^1S_0)]^2 \cong 64\pi\alpha_S|\psi_{ij}(0)|^2/9\mu \,, \tag{4.1}$$

$$\alpha_s(Q^2) = 12\pi/[(33-2n_f)\ln(Q^2/\Lambda^2)] \,,$$

where $\mu = m_i m_j/(m_i+m_j)$ and $\psi_{ij}(0)$ is the wave function at the origin. Since $\alpha_S|\psi_{ij}(0)|^2/\mu$ is approximately constant we are led to the phenomenological relation

$$[M(^3S_1)]^2 - [M(^1S_0)]^2 \cong \text{constant} \,. \tag{4.2}$$

We found that all the known hyperfine splittings can be reproduced by setting $\Lambda = 160$ MeV and $Q = 4\mu$. Here we extend our calculation of ref.4 to the open top states, where the model is identical to the one used in ref.4 (table 1, $\Lambda = 160$ MeV).

As is seen in Table 3, the mass splitting between T and T* is so small that it is comparable to the electromagnetic mass splitting $T^+ - T^0$. The open top systems $T^+ = t\bar{d}$ and $T^0 = t\bar{u}$ are similar to charmed mesons $D^+ = c\bar{d}$ and $D^0 = c\bar{u}$. One can guess that $T^+ - T^0$ will not be very different from $D^+ - D^0$. Both systems have one heavy quark

Table 3 Hyperfine splitting of open top meson

$M(^3S_1) - M(^1S_0)$ (MeV)	T^*-T	$T_s^*-T_s$	$T_c^*-T_c$
m_t = 25 GeV	10.54	11.36	21.4
m_t = 30 GeV	8.79	9.46	17.8
m_t = 35 GeV	7.53	8.13	15.4
m_t = 40 GeV	6.59	7.10	13.2

and one light quark. Charges of the corresponding quarks are the same. The difference $\Delta \equiv T^+ - T^0 - (D^+ - D^0)$ comes from the following effects.

(i) The change induced by $m_d - m_u$ in the color hyper-fine interaction

$$\frac{32\pi\alpha_s}{9} \frac{\vec{S}_Q \cdot \vec{S}_q}{m_Q m_q} \delta^3(\vec{r}) \tag{4.3}$$

(ii) The magnetic interaction

$$-\frac{8\pi}{3}\alpha \cdot \frac{e_Q e_q}{e^2} \cdot \frac{\vec{S}_Q \cdot \vec{S}_q}{m_Q m_q} \delta^3(\vec{r}) \quad . \tag{4.4}$$

The contribution from (i) can be computed by the relations $\Delta(D^+ - D^0) = \frac{3}{4}(1 - m_u/m_d)(D^* - D)$ and $\Delta(T^+ - T^0) = \frac{3}{4}(1 - m_u/m_d)(T^* - T)$. (Here, $\Delta(D^+-D^0)$, $\Delta(T^+-T^0)$ denote only the difference from (i).) Thus, $\Delta(D^+-D^0)$ and $\Delta(T^+-T^0)$ can be of order 1 MeV and 0.1 MeV, respectively. The contribution which comes from (ii) is comparable to or smaller than that from (i).

As for the difference $\Delta^* \equiv T^{*+} - T^{*0} - (D^{*+} - D^{*0})$ one has similar relations $\Delta(D^{*+} - D^{*0}) = (-1/4)(1 - m_u/m_d)(D^* - D)$ and $\Delta(T^{*+} - T^{*0}) = (-1/4)(1 - m_u/m_d)(T^* - T)$. This means that Δ^* is only one third of Δ and has an opposite sign,

$$T^+ - T^0 - D^+ + D^0 = -3(T^{*+} - T^{*0} - D^{*+} + D^{*0}) \ . \tag{4.5}$$

This relation also holds when the effect (ii) is included since we only use $\vec{S}_Q \cdot \vec{S}_q = 1/4$ for $^3S_1$ state and $-3/4$ for $^1S_0$ state. Similar relations can be derived for other system e.g.,

$$B^- - B^0 - K^- + K^0 = -3(B^{*-} - B^{*0} - K^{*-} + K^{*0}) \qquad (4.6)$$

Next we define the center of gravity (COG) of two S-states ($^3S_1$ and $^1S_0$) in order to remove the spin effect as

$$COG(T_s^*,T_s) = (3T_s^* + T_s)/4 \quad , \qquad (4.7)$$
$$COG(T^*,T) = (3T^* + T)/4 \quad .$$

We then define $\Delta_s \equiv COG(T_s^*,T_s) - COG(T^*,T)$. It was pointed out[5] that Δ_s decreases with increasing m_t and approaches a constant value ($\sim$ 60 MeV) for $m_t \to \infty$. Such a decrease is indeed confirmed experimentally.

	exp(MeV)	theory	
$COG(K^*,K) - COG(\rho,\pi)$	177	183	
$COG(F^*,F) - COG(D^*,D)$	102±20	104	(4.8)
$COG(B_s^*,B_s) - COG(B^*,B)$		78	
$COG(T_s^*,T_s) - COG(T^*,T)$ $(m_t = \infty)$		62	

Here a potential model (potential III in ref.5) is used to find the theoretical values. The same potential is used to find the μ dependence of $\psi(0)$ in eq.(4.1). Using the Feynman-Hellman theorem we can show that the decrease of Δ_s as $m_t \to \infty$ can be found in any potential[6].

In Fig.1 we show the open top spectrum assuming $COG(T_s^*,T_s) - COG(T^*,T) = 60$ MeV, $T^{*+} - T^{*0} = D^{*+} - D^{*0} = 2.9$ MeV, $T^+ - T^0 = D^+ - D^0 = 4.7$ MeV.

We now compute the position of the $T\bar{T}$ threshold in $t\bar{t}$ states. For $c\bar{c}$ the $D\bar{D}$ threshold is just above the 2S-state and for $b\bar{b}$ the $B\bar{B}$ threshold is just below the 4S-state. In general the threshold energy becomes higher as the quark mass increases since the binding energies of $Q\bar{Q}$ states get large much more rapidly than those of $Q\bar{q}$ states with increasing m_Q. This is understandable since the reduced mass $\mu = m_Q m_q/(m_Q+m_q)$ approaches a constant value m_q for $m_Q \to \infty$. From the threshold energy of $b\bar{b}$ system one can derive that of $t\bar{t}$ by using the Feynman-Hellman theorem or an explicit potential model. We compute the position of the $T\bar{T}$ threshold in the following way.

(i) We use the experimental data to know the quantity $\delta_b \equiv E(B\bar{B}$ threshold) - $\Upsilon(3S)$.

(ii) Compute the increase of the binding energy ($\Delta_{Q\bar{Q}}$) from $\Upsilon(3S)$ to $t\bar{t}$, 3S and that ($\Delta_{Q\bar{q}}$) from B to T.

(iii) Combining (i) and (ii) one finds $\delta_t \equiv E(T\bar{T}$ threshold) $- E(t\bar{t},3S) = \delta_b - 2\Delta_{Q\bar{q}} + \Delta_{Q\bar{Q}}$.

The reason why we use $\Upsilon(3S)$ is the following. (a) The $c\bar{c}$ states are too far away from $t\bar{t}$, thus large errors might arise in the evaluation (ii). (b) The masses of $\Upsilon(4S)$, $\Upsilon(5S)$, and $\Upsilon(6S)$ cannot be reproduced in a simple potential model. One can guess that these levels are disturbed[7)] by $b\bar{b}$ - $b\bar{b}g$ mixing and/or threshold effects. (c) We do not use (1S) and (2S) either since the effects of hyperfine splitting are larger and cannot be removed reliably, and further $t\bar{t}$, 1S and 2S states move into the central region where the shape of the potential is not very well known.

We now describe the details of our calculation. From experimental values $B^* = 5325$ MeV, $B = 5273$ MeV one finds

$$\delta_b \equiv 2 \times COG(B^*,B) - \Upsilon(3S) = 268.5 \text{ MeV}. \tag{4.9}$$

In order to evaluate how much the binding energy increases from $\Upsilon(3S)$ to $t\bar{t}$, 3S-state ($\equiv\Delta_{Q\bar{Q}}$) we simply use our potential proposed in ref.2 (potential J, $\Lambda_{\overline{MS}}$ = 200 MeV). Then we obtain

$$\begin{aligned} \Delta_{Q\bar{Q}} &= 535 \text{ MeV} \quad \text{for } m_t = 20 \text{ GeV}, \\ &\quad\; 686 \text{ MeV} \quad \text{for } m_t = 30 \text{ GeV}, \\ &\quad\; 795 \text{ MeV} \quad \text{for } m_t = 40 \text{ GeV}. \end{aligned} \tag{4.10}$$

Since the shape of the potential in the region where quarks move inside both $c\bar{c}$ state and $b\bar{b}$ state is well known, the potential dependence of $\Delta_{Q\bar{Q}}$ is expected to be small for a given value of m_t.

One must also take into account the binding energy increase from B to T ($\equiv\Delta_{Q\bar{q}}$). This can be computed e.g., by using a potential model which can reproduce the masses of D, D^*, F, F^*, B, B^* correctly. For m_t = 20 GeV we find $\Delta_{Q\bar{q}}$ = 17 MeV in potential II and 18 MeV in potential III in ref.5. More generally one can use the Feynman-Hellman theorem $\partial E/\partial\mu = -T/\mu$. Here, μ is a reduced mass $\mu = m_q m_Q/(m_q + m_Q)$, and T is the kinetic energy in this energy region[6]. Integrating the Feynman-Hellman theorem from $E(b\bar{u})$ to $E(t\bar{u})$ one finds

$$\begin{aligned}\Delta_{Q\bar{q}} \equiv E(b\bar{u}) - E(t\bar{u}) &= - T_0 \ln(\mu(b\bar{u})/\mu(t\bar{u})) \\ &= - T_0 \ln \frac{m_b(m_t+m_u)}{m_t(m_b+m_u)} \quad .\end{aligned} \tag{4.11}$$

The kinetic energy has an approximately constant value $T_0 \approx 400$ MeV in this energy region. Using $m_b = 5.29$ GeV, $m_u = 336$ MeV one finds

$$\begin{aligned}\Delta_{Q\bar{q}} &= 17 \text{ MeV} \quad \text{for } m_t = 20 \text{ GeV} \\ &= 18 \text{ MeV} \quad \text{for } m_t = 25 \text{ GeV} \\ &= 19 \text{ MeV} \quad \text{for } m_t = 30 \text{ GeV} \ .\end{aligned} \tag{4.12}$$

Thus we put $\Delta_{Q\bar{q}} = 18$ MeV for all m_t. Taking into account all the effects explained above one finds the values of $E(T\bar{T}$ threshold$) - E(t\bar{t},1S)$.

Table 4 The values of $E(T\bar{T}$ threshold$) - E(t\bar{t},1S)$ in MeV.

	$T^0\bar{T}^0$	$T^+\bar{T}^-$	$T^{*0}\bar{T}^{*0}$	$T^{*+}\bar{T}^{*-}$	$T_s\bar{T}_s$	$T_s^*\bar{T}_s^*$
m_t = 20 GeV	1647	1656	1675	1680	1773	1801
30 GeV	1839	1848	1858	1864	1963	1982
40 GeV	2038	2047	2052	2058	2161	2176

In Fig.2 we show the positions of various thresholds as a function of m_t and compare with the $t\bar{t}$ spectrum. The number of bound S-states is 9 for m_t = 30 GeV and 11 for m_t = 40 GeV. Above threshold we expect[8] that the widths of S-states are order of 100 MeV which is much larger than $T^* - T$, and the chance to have a pure T meson factory is very little, while we have a D meson factory at $\psi(3772)$ and a B meson factory at $\Upsilon(4S)$.

5. THE e^+e^- DECAY WIDTH

The leptonic width of toponium states is strongly influenced by Z^0 boson if the mass of $t\bar{t}$ is close to the Z^0 pole. The QCD correction factor $1 - 16\alpha_s/3\pi + \ldots$ for the $t\bar{t}$ state is not so important as those for $c\bar{c}$ and $b\bar{b}$ because of the decrease of α_s. However, it is still as large as $\sim$ 20%.

Including the Z^0 pole effect (A) and the QCD corrections (B) one finds

$$\Gamma_{ee} = \frac{4\alpha^2 e_Q^2 R(0)^2}{M(t\bar{t})^2} A\ B$$

$$B = 1 - 16\ \alpha_S/3\pi \qquad (5.1)$$

$$A = 1 - 3 \cdot \frac{(1-\frac{8}{3}\sin^2\theta_W)^2(-1+4\sin^2\theta_W)(m_{t\bar{t}}^2 - m_Z^2)m_{t\bar{t}}^2}{[(m_{t\bar{t}}^2 - m_Z^2)^2 + m_Z^2\Gamma_Z^2](2\sin 2\theta_W)^4}$$

$$+ \frac{9}{4} \cdot \frac{(1-\frac{8}{3}\sin^2\theta_W)^2[(-1+4\sin^2\theta_W)^2+1]\, m_{t\bar{t}}^4}{[(m_{t\bar{t}}^2 - m_Z^2)^2 + m_Z^2\Gamma_Z^2](2\sin 2\theta_W)^4} .$$

We use the following parameters[9)]

$$m_Z = 94\ \text{GeV}, \qquad \Gamma_Z = 2.5\ \text{GeV}$$

$$\sin^2\theta_W = 0.217 \qquad (5.2)$$

$$\alpha_S(m) = \frac{12\pi}{23\ln\left(\frac{m}{100\ \text{MeV}}\right)^2} , \qquad n_f = 5 .$$

In order to calculate the wave function at the origin we use our model[2)] (potential J, $\Lambda_{\overline{MS}}$ = 200 MeV). In Fig.3, $\Gamma_{ee}(nS)$, $n = 1, \ldots, 13$ are plotted as a function of the mass of toponium ground state M(1S). As is well known Γ_{ee} becomes very large near the Z^0 pole. If the lowest $t\bar{t}$ state is slightly (2-5 GeV) lower than Z^0, Γ_{ee} for higher excited states are relatively enhanced since they are nearer to Z^0. Numerically we find the relation $\Gamma_{ee}(5S) \approx \Gamma_{ee}(6S) \approx \Gamma_{ee}(7S) \approx \cdots$. On the other hand, if the toponium ground state is just above Z^0, Γ_{ee} decreases rapidly as we go to higher excited states.

The leptonic width, Γ_{ee} for $c\bar{c}$ and $b\bar{b}$ states is proportional to the production cross section of $Q\bar{Q}$ states in e^+e^- scattering process. On the other hand, if $t\bar{t}$ and Z^0 happen to have the same mass we only see a dip of $t\bar{t}$ in Z^0 resonance peak because of the interference effect. If $t\bar{t}$ and Z^0 are close each other the interference becomes more complicated[10)]. In these situation Γ_{ee} must be determined in a different way experimentally.

6. SUMMARY

We have studied the properties of $t\bar{t}$ states by using a potential which has a short-range asymptotic behaviour calculated to the fourth order in perturbative QCD. We have also computed the open top spectrum ($t\bar{u}$, $t\bar{d}$, $t\bar{s}$) and studied the number of states below these thresholds.

ACKNOWLEDGEMENT

One of us (K.I.) would like to thank Professor Yoshio Yamaguchi for all the discussions and criticisms on high energy physics for more than fifteen years at the same department.

The computer calculation for this work has been financially supported in part by Institute for Nuclear Study, University of Tokyo.

REFERENCES

1. Buchmüller, W. and Tye, S. -H. H., Phys. Rev. D24, 132 (1982); Buchmüller, W., Grunberg, G. and Tye, S. -H. H., Phys. Rev. Lett., 45 103 (1980).

2. Igi, K. and Ono, S., Phys. Rev. D33, 3349 (1986).

3. Gupta, S. N., Radford, S. F. and Repko, W. W., Phys. Rev. D26, 3305 (1982); Gupta, S. N. and Radford, S. F., Phys. Rev. D25, 3430 (1982).

4. Igi, K. and Ono, S., Phys. Rev. D32, 232 (1985).

5. Ono, S., Phys. Rev. D20, 2975 (1979).

6. Bertlmann, R. A. and Ono, S., Zeit. f. Physik C8, 271 (1980); C10, 37 (1981); Phys. Lett. 96B, 123 (1980).

7. Ono, S., Phys. Rev. D33, 2660 (1980); Zeit. f. Physik C26, 307 (1985).

8. Ono, S., Zeit. f. Physik C8, 7 (1981).

9. Buchmüller, W. et al., Report No. MPI-PTh 85/85, CERN 86-02.

10. Kühn, J. H. and Zerwas, P. M., Phys. Lett. 154B, 448 (1985); Franzini, P. J. and Gilman, F. J., Phys. Rev. D32, 237 (1985); Eichten, E. Fermilab-Conf.-85/29-T.

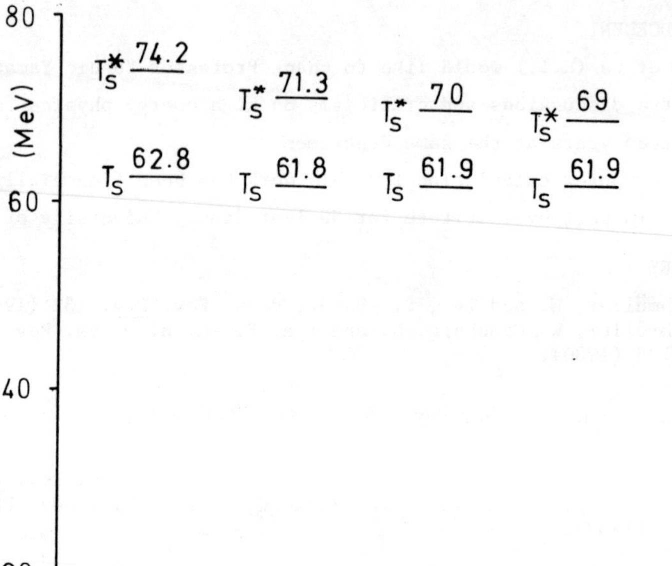

Fig.1 Energy spectrum of open top mesons for various m_t. $T^+ - T^0$ = 4.7 MeV and $T^{*+} - T^{*0}$ = 2.9 MeV.

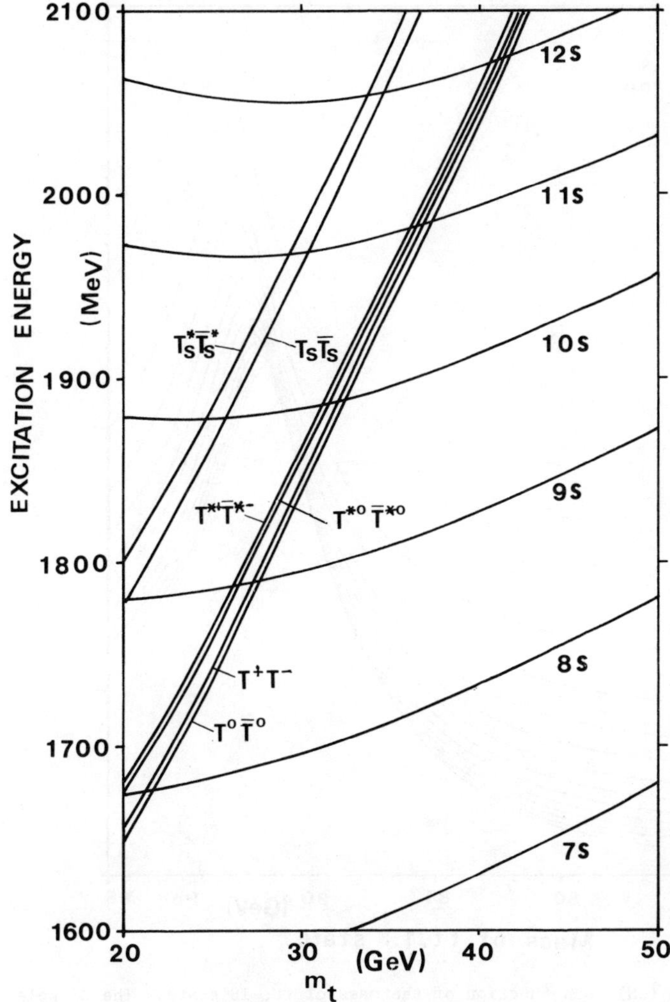

Fig.2 Excitation energy E(nS) - E(1S) for $t\bar{t}$ states vs threshold positions of $T^0\bar{T}^0$, $T^+\bar{T}^-$, $T^{*0}\bar{T}^{*0}$, $T^{*+}\bar{T}^{*-}$, $T_s\bar{T}_s$ and $T_s^*\bar{T}_s^*$ as a function of m_t.

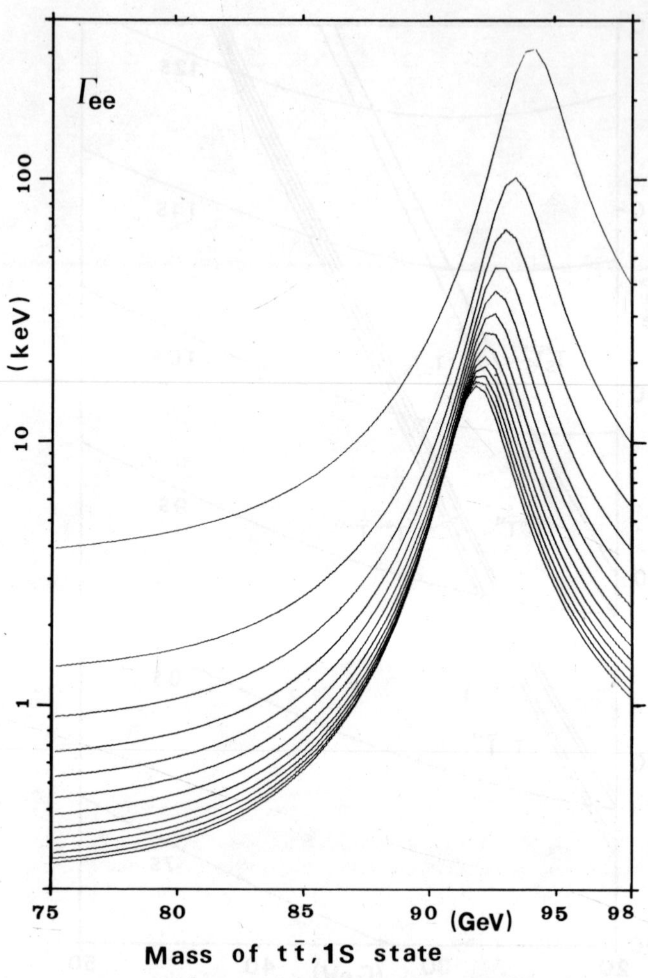

Fig.3 Γ_{ee}(nS) as a function of the mass of $t\bar{t}$, 1S state. The Z^0 pole effect and the QCD corrections are included.

CAN WE REALLY SEE DIQUARKS IN HIGH ENERGY PARTICLE COLLISIONS?*)

T. KOBAYASHI and T. HIROSE

Department of Physics, Faculty of Science
Tokyo Metropolitan University
Setagays-ku, Tokyo 158
JAPAN

ABSTRACT

Role of diquarks as a non-perturbative QCD effect is emphasized in the context of particle reactions at high energies. It is shown that recent EHS (European Hybrid Spectrometer) data would encourage our quark-diquark picture of nucleon combined with fragmentation mechanism a la Field-Feynman and Lund string.

1 Not only from non-perturbative aspects of Quantum Chromodynamics (QCD), but from beyond the standard model diquarks are becoming an interesting topic in contemporary high energy phenomenology.[1-8] It is expected that they have a potential to produce considerable future information on the dynamics of subhadronic system and confinement. However, diquarks have become the target of criticism. -- Why do you need them? We are satisfied by quarks and gluons as far as the constituent of hadrons are concerned. -- Perturbatively you are right. It is just non-perturbative aspects that ascribed to diquarks. QCD effects connected with bound two-quark system are nothing but non-perturbative ones. Let us recall superconductivity. It is regarded as a non-perturbative aspect of Quantum Electrodynamics (QED). The idea of Cooper pairs, a non-perturbatively formed substructure led us to clarify this long standing phenomenon. Confinement in QCD is just the corresponding object. It is out of perturbative QCD. Such substructure as bound diquark system would shed light on dynamics of confinement and eventually on basic understanding of reaction mechanism

*) Dedicated contribution to "Essays in Honour of the 60th Birthday of Professor Yoshio Yamaguchi"

of particle collisions.

Several experimental observations have encouraged us to make a step forward the diquark picture. For instance, (a) deep inelastic lepton scattering on nucleons, (b) Drell-Yan process, (c) large p_T hadron production in hadronic collisions and (d) hadron production in e^+e^- annihilations could naturally better understood by introducing diquarks. Particularly, in (d) one has obtained larger production cross sections of baryons than what one expects from ordinary models without diquarks. A good agreement with experimental data on e^+e^- could be obtained by assuming 14% of a diquark pair directly coupled to the virtual photon. As far as reactions (a) to (c) are concerned, the existence of a diquark system has been revealed through experimental data involving hard processes as their subprocesses in a nucleon (hadron). A quark gets "knocked out" of a nucleon through hard interactions and a diquark system is left behind. Subsequently the diquark system would fragment into hadrons due to colour confinement force. For those hard processes perturbative QCD would be helpful to some extent.

2 On the other hand, in soft processes, i.e. in low p_T processes quarks or diquarks in a nucleon should be regarded as spatially extended object of two quarks, with the binding force possibly being due to the colour interaction given by QCD. That is, they are so-called dressed valence quark, in which gluons and sea quarks are effectively taken into account.

There is, however, a broad spectrum as to how to treat diquarks. "Hello diquark, goodbye gluon!" may be most extreme.[9] Some people regard a diquark as an elementary point-like constituent just like quark or gluon. Others proposed a spatially extended diquark with its radius (0.1 - 0.2) fm. There are also varieties as to whether it is a correlated subunit or it is an uncorrelated object. Of course heavy diquark systems like (uc) or (tb) may be possible in addition to lighter one like (ud) or (dd). Our standpoint is rather ascetic. We do not assign any more specific property to diquark systems than extended two quark system via strong colour forces. Our diquark, however, is assumed not to split into quarks throughout interactions. The colour magnetic force binding suggests that spin-isospin singlet diquarks as $(ud)_0$ are less massive than those of spin-isospin triplet counterparts $(ud)_1$, $(uu)_1$ and $(dd)_1$.

The purpose of the present article is how our diquark idea has successfully been tested through soft hadron production process in pp collisions.[10] Figure 1 illustrates schematically the reaction mechanism. We assume that one of dressed quark (dressed quark) "i" is

knocked out of the nucleon "a" and then fragments into hadrons "c".

Referring to Fig. 1 one can write the inclusive distribution of a hadron "c" produced in a + b→c + X as

$$\frac{1}{\sigma}\frac{d\sigma}{dx} = \sum \int dx' G_{a\to i}(x') \left(\frac{x'}{x}\right) D_{i\to c}\left(\frac{x}{x'}\right) \tag{1}$$

where $G_{a\to i}(x')$ stands for the distribution of dressed quark "i" inside the hadron "a" and $D_{i\to c}(z)$ the fragmentation function of the dressed quark "i" into hadron "c". x denotes the Feynman variable representing the fraction of longitudinal momentum of the parent hadron "a" carried by the quark "i", and z = x/x' corresponds to the same relation between the quark "i" and the daughter hadron "c".

Experimental data on p + p→Λ+ X show the remarkable enhancement at large x. To understand that, first of all, $G_{a\to i}(x)$ should be enhanced at large x. Of course, the larger the average value of x would be the more favourable. Thus the two-body system of quark and diquark composing a nucleon instead of three independent quark system seems to be a natural choice. We may parametrize the momentum sharing function for the quark (or diquark) in a nucleon as

$$g_{QD}(x,y) \propto x^{\alpha_q} y^{\alpha_{qq}} \delta(1-x-y)$$

where α_q and α_{qq} denote dynamical parameters and x(y) stands for the momentum fraction of the quarks (diquarks). The quark distribution function appearing in (1) can be written in the quark-diquark picture as follows:

$$G_{QD}(x) = \frac{1}{B(\alpha_q+1, \alpha_{qq}+1)} \int_0^1 dy\, g_{QD}(x,y)$$

$$= \frac{1}{B(\alpha_q+1, \alpha_{qq}+1)} x^{\alpha_q}(1-x)^{\alpha_{qq}} \tag{2}$$

where B(x, y) is the Euler's beta function. As far as α_q and α_{qq} are dynamical parameters, they have to be determined from the first principle. Unfortunately we don't know it yet. So we should have recourse to a model. Relying upon the dual Regge model is our working hypothesis, and we regard α_q and α_{qq} as the intercepts of Regge trajectories of mesons and baryonia, respectively. $\alpha_q = -1/2$ for u and d quarks from the ρ-ω trajectory and $1/2 < \alpha_{qq} < 2$ have been assumed. For the latter, however, we have not reliable experimental information. Fortunately there is little difference for the choice of $\alpha_{qq} = 1/2$, 1 and 2. As expected the distribution (2) for the quark-diquark configuration attributes to the quark smaller momenta, and consequently leading to higher momenta of the diquark than those given by the three independent quark system.

3 The ingenious combination of quark-diquark structure of a nucleon

with fragmentation models is our success of the diquark approach to soft hadronic reactions unexplored so far. We adopt the familiar Field-Feynman[11)] and the Lund string[12)] models for fragmentation processes which are installed in the Monte Carlo program EPOCS[13)] developed originally for the reaction e^+e^- hadrons at TRISTAN. The QCD elementary processes $e^+e^- \rightarrow q\bar{q}$, $q\bar{q}g$, q, $\bar{q}$ and g being quarks, antiquarks and gluons, respectively, and their fragmentation into hadrons a la Field-Feynman or Lund string are components of the EPOCS.

The TASSO data[14)] on $e^+e^- \rightarrow$ hadrons at $\sqrt{s}$ = 34 GeV have successfully enabled us to determine following parameters common to both models: (a) the QCD scale parameter Λ, (b) vector to pseudoscalar meson ratio (similarly decuplet to octet baryon ratio), (c) the suppression factor of a strange quark, (d) the transverse momentum of the quark with respect to the parent one and (e) the diquark pick up rate from the sea. There are still a few parameters specific to each model. Fragmentation functions, of course, have another parameter:

$$D(z) = \begin{cases} 1 - a_F + 3a_F(1-z)^2 & \text{for Field-Feynman} \\ (1+c)(1-z)^c & \text{for Lund string} \end{cases} \qquad (3)$$

It should be emphasized here that we have extensively examined the consistency between the model predictions and various TASSO data in order to fix these parameters.

4 Armed with these machineries we can calculate the inclusive distributions of a hadron produced in pp collisions. From them we shall illustrate the $p + p \rightarrow \Lambda + X$ as a typical example in which the diquark effects are most clearly exhibited. The is mainly created

from the valence diquark (ud) with spin zero combined with an s quark in sea. The large diquark fraction with high momentum plays an essential role for Λ production in the large x region. This is just what we have expected diquarks. On the contrary the disagreement of three independent quark model with experimental data is particularly striking for this case.

5 The idea of diquark is not new, but its important role has not fully been appreciated so far. In fact, the word "diquark" seems to differently be used by different authors. Still no overall unanimous view has emerged yet as to what is diquark and as to whether it is a non-perturbative counterpart of perturbative QCD where only quarks and gluons play a role. It is our feeling that we should sooner or later confront with a composite structure of quarks and leptons. Diquarks seem to suggest us a prototype of the composite structure of

constituent objects. Our present attempts are still premature. Future experimental information, for instance, baryon correlations expected soon from TRISTAN gives important clue to have a deeper insight to diquark systems. It is expected that we can open a window to clarify a more general composite structure of fundamental objects with the help of experimental data will be obtained from TEVATRON, LEP, HERA or SSC.

The content is based upon a part of EHS colaboration work mainly done together with T. Tsurugai.

References

1) Fredriksson, S., Jandel, M. and Larsson, T.I., Z. Phys.s C14, 35 (1982); C19, 53 (1983); Phys. Rev. D28, 257 (1983); Phys. Rev. Lett. 51, 2179 (1984).
2) Fredriksson, S., Phys. Rev. Lett. 52, 724 (1984).
3) Fredriksson, S. and Larsson, T.I., Phys. Rev. D28, 255 (1983).
4) Kinoshita, K., Noda, H. and Tashiro, T., Prog. Theor. Phys. 68, 1699 (1982); 2086.
5) Noda, H. and Tashiro T., Prog. Theor. Phys. 73, 158 (1985).
6) Kawabe, T., Phys. Lett. 114B, 263 (1982); Z. Phys. C16, 367 (1983).
7) Ito, I., Matsui, S., Toyoda, M. and Tsai, S.Y., preprint NUP-A-86-1.
8) Sukhatme, U.P., Lassila, K.E. and Orava, R., Phys. Rev. D25, 2975 (1982).
9) Fredriksson, S., Talk at the XIX-th Recontre de Moriond, La Plagne, Savoie, France, March 1984.
10) Hirose, T., Kobayashi, T. and Tsurugai, T., preprint TMUP-8407. Bailly, J.L. et al. EHS-RCBC Collaboration, preprint CERN/EP 86-32 (1986).
11) Field, R.D. and Feynman, R.P., Phys. Rev. D15, 2590 (1977); Nucl. Phys. B136, 1 (1978).
12) Anderson, B., Gustafson, G. and Soderberg, B., Z. phys. C20, 317 (1983); Sjostrand, T., Computer Phys. Comm. 28, 229 (1983).
13) Kato, K. and Munehisa, T., KEK report 84-18.
14) Althoff, M. et al., Z. Phys. C22, 307 (1984).

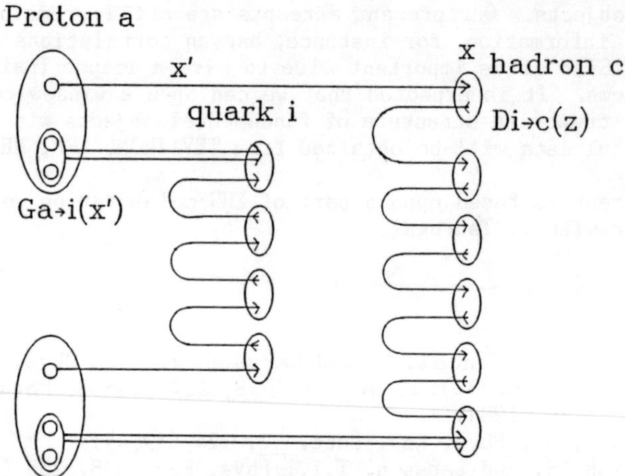

Fig. 1 Schematic diagram of proton structure and quark (diquark) fragmentation

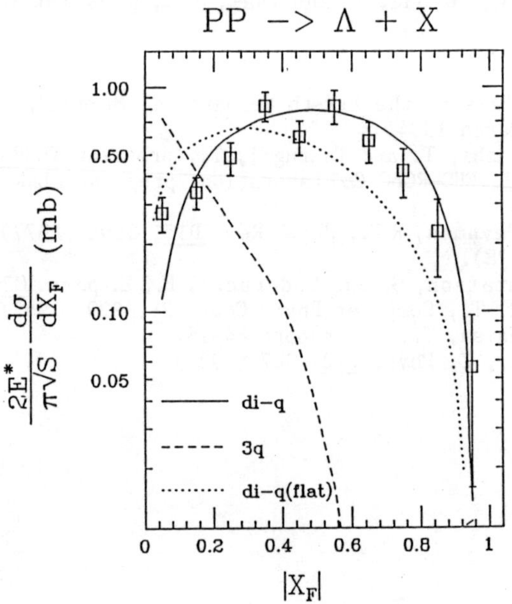

Fig. 2 Feynman x distribution for $p + p \to \Lambda + X$ at 360 GeV/c

DOUBLE LOGARITHMS IN QED RADIATIVE CORRECTIONS TO THE REACTION $e^+e^- \to \nu_\ell \bar{\nu}_\ell \gamma$

Masataka Igarashi
Department of Physics, Tokai University, Hiratsuka City, Kanagawa 259-12, JAPAN

and

Nobuya Nakazawa
Department of Physics, Kogakuin University, Shinjuku-ku, Tokyo 160, JAPAN

ABSTRACT

The QED virtual corrections to the neutrino counting reaction are summarized. By including the correction due to the soft photon emission, it is shown that, in the region where C.M. energy is comparable to M_Z, the perturbation calculation of the corrections can be used in spite of the appearance of double logarithms.

1. INTRODUCTION

Recently we have obtained the QED radiative corrections to the neutrino counting reaction[1)]

$$e^+(p_+) + e^-(p_-) \to \bar{\nu}_\ell(q_+) + \nu_\ell(q_-) + \gamma(k) \tag{1}$$

under the assumptions: (i) neutrinos are massless, (ii) the reaction is mediated only by Z-boson and (iii) the counterterm to the axial vector coupling is the same as that to the vector coupling.[2)] The second assumption is valid above $\sqrt{s} > M_Z$ where the cross-section is sizable and the reaction can be used as a realistic method of counting the neutrino number.[3)]

Meanwhile, the QED corrections to $e^+e^- \to \gamma Z$ reaction are obtained by another group under the same treatment of the axial coupling as ours.[4] They claim that the magnitude of the corrections takes so large negative value that the perturbation expansion may be broken down. Its origin is attributed to the appearance of double logarithms, which is known to exist in the QED corrections to a similar type of reaction, i.e., $e^+e^- \to \gamma\gamma$.[5] In this note we summarize our result and discuss the validity of perturbation expansion in these reactions.

2. SUMMARY OF QED VIRTUAL CORRECTIONS

The lowest order cross-section of reaction (1) mediated by Z-boson is well-known.[3]

$$\frac{d^2\sigma^0}{dxdy} = \alpha^2 \frac{f^2 + g^2}{(M_Z^2 - Q^2)^2 + M_Z^2\Gamma_Z^2} \frac{\Gamma_\nu}{M_Z} \times Q^2 \frac{s - Q^2}{s} [\frac{\tau}{\kappa} + \frac{\kappa}{\tau} + 2\frac{sQ^2}{\kappa\tau}] \,, \tag{2}$$

where

$$\begin{aligned} x &= 2k_0/\sqrt{s} \,, \\ y &= \cos\theta \,, \\ \kappa &\cong sx(1 + y)/2 \,, \\ \tau &\cong sx(1 - y)/2 \,, \\ Q^2 &= (q_+ + q_-)^2 = s(1 - x) \,, \\ f &= \frac{M_Z}{4\sqrt{M_Z^2 - M_W^2}} \frac{M_Z}{M_W} (3 - 4\frac{M_W^2}{M_Z^2}) \,, \\ g &= F = G = \frac{M_Z}{4\sqrt{M_Z^2 - M_W^2}} \frac{M_Z}{M_W} \,. \end{aligned} \tag{3}$$

and $\Gamma_\nu = N_\nu \alpha (F^2 + G^2)M_Z/3$ is the partial decay width of Z-boson into N_ν neutrino-anti-neutrino pairs. Since the momenta of neutrinos are not observed or integrated out, both the lowest order and the corrected cross-sections are free from the parity violating terms and symmetric with respect to the interchange $\kappa \leftrightarrow \tau$. If one integrated Eq.(2) over x under the narrow width approximation, the result is the cross-section for $e^+e^- \to \gamma Z$[6] multiplied by a factor of Γ_ν/Γ_Z:

$$\int \frac{d^2\sigma^0}{dxdy} dx \xrightarrow[Q^2 \to M_Z^2]{} \frac{d\sigma^0}{dy} \times \frac{\Gamma_\nu}{\Gamma_Z} \,, \tag{4}$$

$$\frac{d\sigma^0}{dy} = 2\pi\alpha^2(f^2 + g^2)\,\frac{(s + M_Z^2)^2 + (s + M_Z^2)^2 y^2}{(s - M_Z^2)s^2(1 - y^2)}\,. \tag{5}$$

Furthermore, if one sets $f = 1$, $g = 0$ and $M_Z = 0$ in Eq.(5), it becomes the cross-section for $e^+e^- \to \gamma\gamma$. Thus the magnitudes of corrected cross-sections relative to the lowest order one should become at $Q^2 = 0$ those in $e^+e^- \to \gamma\gamma$ and at $Q^2 = M_Z^2$ those in $e^+e^- \to \gamma Z$ independent of the narrow width approximation.

The cross-section with QED virtual corrections can be written as

$$\frac{d^2\sigma^{Virt}}{dxdy} = \frac{d^2\sigma^0}{dxdy}\left[\frac{\tau}{\kappa} + \frac{\kappa}{\tau} + 2\frac{sQ^2}{\kappa\tau}\right]^{-1}$$

$$\times\ \frac{\alpha}{\pi}\left[\mathscr{C}^{SE}(\kappa) + \mathscr{C}^{VTX}(\kappa) + \mathscr{C}^{BOX}(\kappa) + (\kappa \leftrightarrow \tau)\right]\,, \tag{6}$$

where the superscripts SE, VTX and BOX mean the electron self energy, the vertex correction and the box contribution, respectively. Each corrections are given as follows:

$$\mathscr{C}^{SE}(\kappa) = \left(\frac{\tau}{\kappa} + \frac{sQ^2}{\kappa\tau}\right)\left[-\ln\left(\frac{m^2}{\lambda^2}\right) + \frac{3}{2} + \frac{1}{2}\ln\left(\frac{\kappa}{m^2}\right)\right]\,, \tag{7}$$

where m is the electron mass and λ is a fictitious mass of the photon to regulate the infrared divergence.

$$\mathscr{C}^{VTX}(\kappa) = \left(\frac{\tau}{\kappa} + \frac{sQ^2}{\kappa\tau}\right)\left[2\ln\left(\frac{m^2}{\lambda^2}\right) - 3 + \frac{Q^2}{2(s - \tau)}\ln\left(\frac{\kappa}{Q^2}\right)\right]$$
$$+ \left[1 - \frac{s^2Q^2}{\tau(s - \tau)^2}\right]\left[\frac{s - \tau}{\kappa}Q^2G + 2\frac{Q^2}{\kappa}\ln\left(\frac{\kappa}{Q^2}\right)\right]$$
$$+ \frac{1}{2}\frac{sQ^2}{(s - \tau)^2}\ln\left(\frac{\kappa}{Q^2}\right) - \left(\frac{s}{s - \tau} + \frac{s - \tau}{\kappa}\right)\ln\left(\frac{\kappa}{m^2}\right)$$
$$+ \frac{s - \tau}{\kappa} - \frac{1}{2}\frac{s^2}{\kappa\tau}\frac{Q^2}{s - \tau}\,. \tag{8}$$

The function G is a complicated function composed of double logarithms, but its explicit form is not necessary becuase it cancels the same function contained in the box contribution. In the region $4m^2 > Q^2 \geq 0$ where the reaction (1) cannot be used for its original purpose, the function $\ln(Q^2/m^2)$ in Eq.(8) should be replaced by its analytic continuation arctangent; however, we left them as they stand since they are always multiplied by the factor Q^2. The box contribution is given by

$$\mathscr{C}^{\rm BOX}(\kappa) = \left(\frac{\tau}{\kappa} + \frac{sQ^2}{\kappa\tau}\right)\left[\frac{1}{2}\ln^2\left(\frac{s}{m^2}\right) - \ln\left(\frac{s}{\lambda^2}\right)\ln\left(\frac{s}{m^2}\right) + \frac{2}{3}\pi^2\right]$$
$$+ \frac{s^2 + (s-\tau)^2}{2\kappa\tau}\left[\ln^2\left(\frac{s}{\kappa}\right) + \ln^2\left(\frac{\kappa + Q^2}{Q^2}\right) - \ln^2\left(\frac{\kappa}{Q^2}\right)\right.$$
$$\left. + 2\,{\rm Sp}\left(\frac{Q^2}{\kappa + Q^2}\right) + 2\,{\rm Sp}\left(1 - \frac{Q^2}{s}\right) - \frac{\pi^2}{3}\right]$$
$$- \frac{Q^2}{\kappa}\left(1 - \frac{s^2Q^2}{\tau(s-\tau)^2}\right)(s-\tau)G + \left[\frac{s}{s-\tau} + \frac{s^2}{\kappa\tau} - \frac{s}{\tau}\right]\ln\left(\frac{\kappa}{m^2}\right)$$
$$+ \ln\left(\frac{\kappa}{s}\right) - \frac{Q^2}{s-\tau}\left[2\frac{s^2}{\tau(s-\tau)} - \frac{s+\tau}{2}\right]\ln\left(\frac{\kappa}{Q^2}\right)$$
$$+ \left(\frac{Q^2}{s - Q^2}\right)^2\ln\left(\frac{s}{Q^2}\right) - \frac{Q^2}{s - Q^2} - 1\,, \qquad (9)$$

where Sp(x) is the Spence function defined in ref.7) as $Li_2(x)$.

In the $Q^2 \to 0$ limit, these three formulas (7) to (9) coincide with the corresponding ones in $e^+e^- \to \gamma\gamma$, respectively.[8] The infrared divergence in these formulas must be cancelled by that in the correction due to the soft photon emission

$$\mathscr{C}^{\rm SOFT}(\kappa)$$
$$= \left(\frac{\tau}{\kappa} + \frac{sQ^2}{\kappa\tau}\right)\left[\left(\ln\frac{s}{m^2} - 1\right)\ln\left(\frac{4\Delta_c^2}{\lambda^2}\right) - \frac{1}{2}\ln^2\left(\frac{s}{m^2}\right) + \ln\left(\frac{s}{m^2}\right) - \frac{\pi^2}{3}\right], \qquad (10)$$

where Δ_c is the maximum energy of undetectable photon. Finally we have

$$\mathscr{C}^{\rm ALL}(\kappa) = \left(\frac{\tau}{\kappa} + \frac{sQ^2}{\kappa\tau}\right)\left[-\left(\ln\left(\frac{s}{m^2}\right) - 1\right)\ln\left(\frac{4\Delta_c^2}{s}\right) + \frac{3}{2}\ln\left(\frac{s}{m^2}\right) + \frac{\pi^2}{3} - 2\right]$$
$$+ \frac{s^2 + (s-\tau)^2}{2\kappa\tau}\left[\ln^2\left(\frac{s}{\kappa}\right) + \ln^2\left(\frac{\kappa + Q^2}{Q^2}\right) - \ln^2\left(\frac{\kappa}{Q^2}\right)\right.$$
$$\left. + 2\,{\rm Sp}\left(\frac{Q^2}{\kappa + Q^2}\right) + 2\,{\rm Sp}\left(1 - \frac{Q^2}{s}\right) - \frac{\pi^2}{3}\right]$$
$$+ \frac{Q^2}{s-\tau}\left[\frac{s}{2(s-\tau)} + \frac{1}{2} - \frac{3}{2}\frac{\tau}{\kappa}\right]\ln\left(\frac{\kappa}{Q^2}\right)$$
$$+ \left(\frac{Q^2}{s - Q^2}\right)^2\ln\left(\frac{s}{Q^2}\right) + \left(1 + \frac{3}{2}\frac{\tau}{\kappa}\right)\ln\left(\frac{\kappa}{s}\right)$$
$$- \frac{s}{s - Q^2} + \frac{s}{2(s-\tau)} - \frac{\tau}{2\kappa}\,. \qquad (11)$$

The sum of virtual corrections contained in the above expression is in accordance with that given in ref.9). It should be noted that the electron mass-singularity appears only in the term multiplicative to the lowest order cross-section.

3. DOUBLE LOGARITHMIC APPROXIMATION

If only the double logarithmic terms are kept in Eq.(11), the total correction becomes

$$\mathscr{C}^{ALL}_{DLA}(\kappa) = (\frac{\tau}{\kappa} + \frac{sQ^2}{\kappa\tau})f_1(s,m,\Delta_c) + \frac{s^2 + (s - \tau)^2}{2\kappa\tau} f_2(s,\kappa,Q^2) \qquad (12)$$

with

$$f_1(s,m,\Delta_c) = \ln \frac{s}{m^2} \ln \frac{4\Delta_c^2}{s} , \qquad (13)$$

$$f_2(s,\kappa,Q^2) = \ln \frac{s}{\kappa} + \ln (\frac{\kappa + Q^2}{Q^2}) - \ln (\frac{\kappa}{Q^2}) . \qquad (14)$$

The function $f_1(s,m,\Delta_c)$ depends on the choice of Δ_c. If one takes a large value for Δ_c, the result may be considered to simulate at least partially the effect of hard photon emission. In an extreme case, if Δ_c is taken to be close to its maximum $\sqrt{s}/2$, the first term in Eq.(12) can be neglected.

The function $f_2(s,\kappa,Q^2)$ clearly shows the existence of double logarithm $\ln^2(s/\kappa)$ in the Q^2 zero limit or in the reaction $e^+e^- \to \gamma\gamma$. The latter function can be rewritten in other forms:

$$f_2(s,\kappa,Q^2) = \ln^2\frac{s}{Q^2} + 2 \ln \frac{s}{Q^2} \ln \frac{Q^2}{\kappa} + \ln^2\frac{\kappa + Q^2}{Q^2}$$

$$= \ln^2\frac{s}{\kappa + Q^2} + 2 \ln \frac{\kappa + Q^2}{\kappa} \ln \frac{s}{Q^2} . \qquad (15)$$

The first expression means that, if $s \sim Q^2$ and $Q^2 \gg \kappa$, there is no double logarithms, while the second expression shows that, if $s > \kappa \gg Q^2$, the function $f_2(s,\kappa,Q^2)$ becomes rather similar to that in the Q^2 zero limit. In other words, the reaction $e^+e^- \to \gamma Z$ does not show large double logarithmic behavior in the energy region available in near future; but it behaves just as the reaction $e^+e^- \to \gamma\gamma$ in extremely high energy region ($s \gg M_Z^2$). This is a physically reasonable conclusion, which is first formulated in ref.1) and explicitly stated in ref.9).

Finally, we make a comment on the numerical values of the total correction. We set W = 105 GeV and Δ_c = 0.2 GeV. The formula in Eq.(12) under the double logarithmic approximation and the exact one in Eq.(11) give -63% and -52%, respectively, both for $e^+e^- \to \gamma\gamma$ and γZ almost independent of the scattering angles. That is, the correction is dominated by the function $f_1(s,m,\Delta_c)$ in Eq.(12) and this means that the inclusion of hard photon emission is indispensable. The latter is known to give the correction of the magnitude more than a few ten percent.[8] Therefore the QED radiative corrections as a whole are of the magnitude at most a few ten percent with minus sign and can be treated perturbatively at least in the energy region of the colliders of next generation.

REFERENCES

1) Igarashi, M. and Nakazawa, N., preprint TKU-HEP 86/01 and KUDP-86/02 (1986).

2) Berends, F. A., Kleiss, R. and Jadach, S., Nucl. Phys. B202, 63 (1982); Bohm, M. and Hollik, W., Nucl. Phys. B204, 45 (1982).

3) Ma, E. and Okada, J., Phys. Rev. Lett. 41, 287 (1978) and ibid, 41, 1759 (1978); Gaemers, K. J. F., Gastmans, R. and Renard, F. M., Phys. Rev. D19, 1605 (1979); Barbiellini, G., Richter, B. and Siegrist, J. L., Phys. Lett. B106, 414 (1981).

4) Dombey, N., Cole, J. and Boudjema, F., Phys. Lett. 167B, 108 (1986).

5) Sudakov, V. V., Zh, Eksp. Teor. Fiz. 30, 87 (1957) [Sov. Phys. JETP 3, 65 (1956)].

6) Hinchliffe, I., Proc. of the '82 DPF Summer Study, Snowmass, Colorado (1982).

7) Lewin, L., Dilogarithm and Associated Functions, Macdonald, London (1968).

8) Fujimoto, J., Igarashi, M. and Shimizu, Y., KEK preprint (1986).

9) Berends, F. A., Burgers, G. J. H. and van Neerven, W. L., Institut-Lorentz preprint (1986).

SEARCH FOR SLEPTONS AT THE Z° PEAK IN $e^{-}e^{+}$ COLLIDER

Keisho Hidaka

Tokyo Gakugei University, Department of Physics
Koganei-Shi, Tokyo 184
JAPAN

ABSTRACT

We investigate single scalar lepton production process $e^{-}e^{+} \to Z^{\circ} \to \tilde{\ell}^{\mp}\tilde{\gamma}\ell^{\pm}$ at the Z° peak. We find that a detectable number of these scalars should be produced at the SLC and LEP-I colliders even if their masses substantially exceed the beam energy $E=m_Z/2 \simeq 46GeV$.

1. INTRODUCTION

Selectrons $\tilde{e}$ (the supersymmetric partners of electron) could be pair-produced in $e^{-}e^{+}$ annihilation if the beam energy were to exceed their mass. As pointed out by several authors[1,2,3] the selectrons with masses larger than the beam energy could be produced singly in the process

$$e^{-}e^{+} \to \tilde{e}^{\mp}\tilde{\gamma}e^{\pm} \tag{1}$$

where $\tilde{\gamma}$ is the photino which is assumed to be almost massless. In this article assuming light photino we point out that a detectable number of sleptons ($\tilde{\ell}^{\mp}$) should be singly produced via a process such as

$$e^{-}e^{+} \to Z^{\circ} \to \tilde{e}^{\mp}\tilde{\gamma}e^{\pm} \tag{2}$$

in the run at the Z° peak at the SLC and LEP-I colliders even if their masses substantially exceed the beam energy $E=m_Z/2 \simeq 46GeV$[4].

2. EVALUATION OF THE TOTAL CROSS SECTION

Several diagrams can contribute to the process (1)[2]. As the Z° formation dominates the total $e^{-}e^{+}$ cross section at the Z° peak ($E=m_Z/2$), the contribution (2) dominates the process (1) at the peak. The following two diagrams contribute to the process (2) :

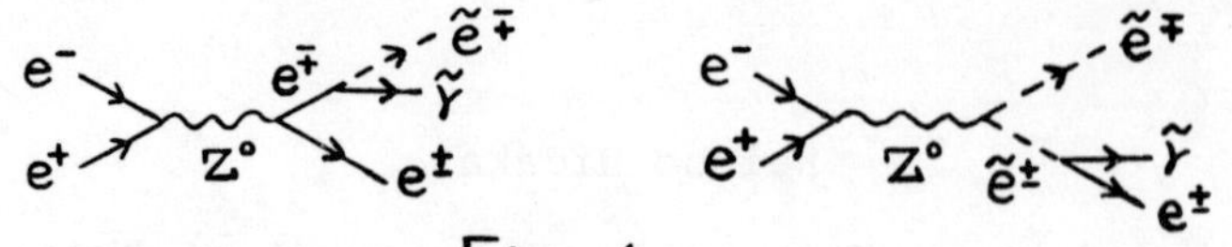

Fig. 1

The total cross section for the process (1) at the Z° peak in this approximation is then given by

$$\sigma_{TOT}(e^{-}e^{+} \to \tilde{e}^{\mp}\tilde{\gamma}e^{\pm}) = 1.26\times10^{-37}(1.68\times10^{-38})\ \mathrm{cm}^{2}$$

for $m_{\tilde{e}}$=50GeV(60GeV)$\gg m_{\tilde{\gamma}}$. With the design luminosities of SLC and LEP-I ($L=10^{31}$ and $10^{32}\ \mathrm{cm}^{-2}\mathrm{sec}^{-1}$, respectively) one has the expected number of events for the single $\tilde{e}$ production at the Z° peak in a 3-month run as follows:
9.8(1.3) events for $m_{\tilde{e}}$=50(60)GeV$\gg m_{\tilde{\gamma}}$ at SLC,
98 (13) events for $m_{\tilde{e}}$=50(60)GeV$\gg m_{\tilde{\gamma}}$ at LEP-I.
The same holds for the single $\tilde{\mu}$ and $\tilde{\tau}$ production. Hence it should be possible to detect the charged sleptons($\tilde{e}$,$\tilde{\mu}$ and $\tilde{\tau}$) in the run at $\sqrt{s}=m_{Z}$ at SLC and LEP-I if their masses are less than 50GeV and 60GeV, respectively. We find that considerably higher(by about 8 GeV) slepton masses can be reached by increasing (luminosity)x(running time) by a factor 5.

Our approach can also be applied to the single $\tilde{e}$ production in $\bar{p}p$ collision (via the Z° production and its subsequent decay into $\tilde{e}\tilde{\gamma}e$, i.e., $\bar{p}p \to Z^{\circ}X \to \tilde{e}^{\mp}\tilde{\gamma}e^{\pm}X$) at the SPS-collider and the TEVATRON collider. Unfortunately we find that with the expected parameters of the colliders $\tilde{e}$ masses only up to $\sim m_{Z}/2 \simeq 46$GeV can be reached in such experiment due to insufficient luminosity.

3. EXPERIMENTATION OF SLEPTON SEARCH

The signal and backgrounds for the $\tilde{e}^{\mp}\tilde{\gamma}e^{\pm}$ event are as follows:
(i) Signal; The produced $\tilde{e}^{\mp}$ decays immediately into $e^{\mp}$ and $\tilde{\gamma}$ and the final state becomes $e^{-}e^{+}\tilde{\gamma}\tilde{\gamma}$. The $\tilde{\gamma}$ behaves like a neutrino and escapes detection. Hence the signal for this event is an acoplanar $e^{-}e^{+}$ pair with large missing transverse momentum p_{T} and missing energy.
(ii) Backgrounds; The $\tau^{-}\tau^{+}$ pair background($\tau^{\mp} \to e^{\mp}\nu\bar{\nu}$) should not be serious, since the e^{-} and e^{+} from $\tau^{\mp}$ decay are almost back to back and hence coplanar. Backgrounds from pair production of $L^{-}L^{+}$(heavy lepton), $W^{\pm}W^{\mp}_{virtual}$, $\tilde{W}^{+}\tilde{W}^{-}$(wino), $H^{+}H^{-}$(higgs) and $\tilde{H}^{+}\tilde{H}^{-}$(higgsino) where each particle decays into "$e^{\mp}$+ missing neutrals(ν and $\tilde{\gamma}$)" could be discriminated from the single slepton production events ($\tilde{l}^{\mp}\tilde{\gamma}l^{\pm}$) on the basis that

they lead also to acoplanar $e^{\mp}\mu^{\pm}$ events with $N(e^{+}\mu^{-}) + N(e^{-}\mu^{+}) = 2\,N(e^{-}e^{+})$ in remarkable contrast to the $\tilde{l}^{\mp}\tilde{\gamma}l^{\pm}$ events. Moreover $\sigma_{TOT}(e^{-}e^{+} \to W^{\pm}\text{"}W^{\mp}\text{"} \to e^{\pm}e^{\mp}\nu\bar{\nu})$ is negligibly small at the Z^{o} peak. Backgrounds from two-photon events, $e^{-}e^{+}\gamma$, $\tau^{-}\tau^{+}\gamma$ events and so on[3] should not be serious, since such events will tend to have p_T balance and/or coplanarity.

It is to be noted that in actual experiment one introduces various cuts on the momenta of the e^{-} and e^{+} in the final state, such as missing p_T and acoplanarity angle cuts, to eliminate these background events and that such cuts reduce the number of signal events somehow. Appropriate cuts in this case would be as follows:

(a) Production(or polar) angle cut (to be introduced according to detector acceptance),

$$\theta_{min} < (\theta_{-} \text{ and } \theta_{+}) < 180^{o} - \theta_{min},$$

where $\theta_{\mp}$ is the angle between the e^{-} beam and the produced $e^{\mp}$.

(b) Energy cut (to ensure detection and identification of the $e^{\mp}$ in the detector),

$$(E_{-} \text{ and } E_{+}) > E_{min},$$

where $E_{\mp}$ is the energy of the produced $e^{\mp}$.

(c) Energy sum cut (to eliminate the two-photon background),

$$E_{-} + E_{+} > C\sqrt{s} = C\, m_Z.$$

(d) Acoplanarity-angle (ϕ) cut (to eliminate the backgrounds from the $\tau^{-}\tau^{+}$ pair, two-photon, $e^{-}e^{+}\gamma$ and $\tau^{-}\tau^{+}\gamma$ events),

$$\phi > \phi_{min},$$

where $\phi \equiv 180^{o} - \cos^{-1}(\not{p}_{-T}\cdot\not{p}_{+T} / |\not{p}_{-T}||\not{p}_{+T}|)$ and $\not{p}_{\mp T}$ is the transverse momentum vector of the $e^{\mp}$.

(e) Missing-p_T ($\not{p}_T$) cut (to eliminate the backgrounds from the two-photon ($e^{-}e^{+} \to e^{-}e^{+}e^{-}e^{+}$) and $e^{-}e^{+}\gamma$ events),

$$\not{p}_T > \not{p}_{T\,min},$$

where $\not{p}_T = |\not{p}_{-T} + \not{p}_{+T}|$.

For a detector with polar angle acceptance of $\theta_{min} = 15^{o}$, we can take these cut parameters typically as follows:

$$\theta_{min} = 15^{o},\ E_{min} = 2\text{ GeV},\ C = 0.2,\ \phi_{min} = 30^{o},\ \not{p}_{T\,min} = 18\text{ GeV/c}. \qquad (3)$$

By combining the isotropic $\tilde{e}^{\mp}$ decay with the $\tilde{e}^{\mp}$ production cross section evaluated with the diagrams of Fig.1, we calculate the total cross section $\sigma_{TOT}(e^{-}e^{+} \to Z^{o} \to \tilde{e}^{\mp}\tilde{\gamma}e^{\pm} \to e^{-}e^{+}\tilde{\gamma}\tilde{\gamma})$ with simultaneous cut conditions (a)-(e) for various values of $m_{\tilde{e}}$ and $m_{\tilde{\gamma}}$ at $\sqrt{s} = m_Z = 93$ GeV. In Fig.2 is presented the result obtained by Monte Carlo calculation.

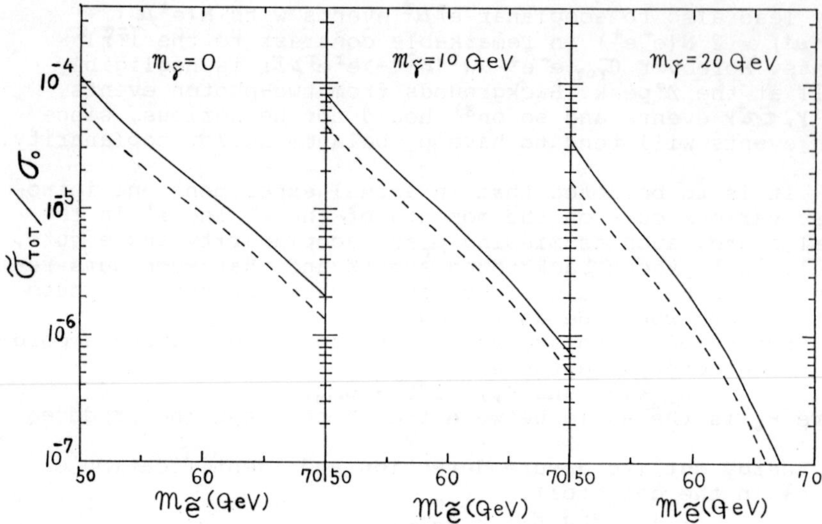

Fig.2 Total production cross section $\tilde{\sigma}_{TOT} \equiv \sigma_{TOT}(e^-e^+ \to Z^0 \to \tilde{e}^{\mp}\tilde{\gamma}e^{\pm} \to e^-e^+\tilde{\gamma}\tilde{\gamma})$ divided by $\sigma_0 \equiv \sigma_{TOT}(e^-e^+ \to Z^0 \to e^-e^+)$ $=1.52\times10^{-33}\mathrm{cm}^2$ plotted against $m_{\tilde{e}}$ for various values of $m_{\tilde{\gamma}}$ at $\sqrt{s} = m_Z$. The solid curves are for the $\tilde{\sigma}_{TOT}$ with no cut. The dashed curves are for the $\tilde{\sigma}_{TOT}$ with simultaneous cut conditions (a)-(e) with eq.(3).

Once the events of an e^-e^+ pair with missing momentum which satisfy all the cut conditions (a)-(e) were observed in an actual experiment, by investigating the event rate and the distributions of the observed events we could determine the $m_{\tilde{e}}$ and $m_{\tilde{\gamma}}$ with the assumption that all the observed events were from the process $Z^0 \to \tilde{e}^{\mp}\tilde{\gamma}e^{\pm} \to e^-e^+\tilde{\gamma}\tilde{\gamma}$. The investigation of the event rate and distributions of the events observed in actual experiment which satisfy all the cut conditions (a)-(e) could also play an important role in discriminating the possible origines of the observed events, such as

(A) $e^-e^+ \to Z^0 \to \tilde{e}^{\mp}\tilde{\gamma}e^{\pm} \to e^-e^+\tilde{\gamma}\tilde{\gamma}$ (with $m_{\tilde{e}} \geq \sqrt{s}/2 = m_Z/2$),
(B) $e^-e^+ \to \tilde{e}^-\tilde{e}^+ \to e^-e^+\tilde{\gamma}\tilde{\gamma}$ (with $m_{\tilde{e}} < \sqrt{s}/2 = m_Z/2$) and
(c) $e^-e^+ \to \tilde{Z}^0\tilde{\gamma} \to e^-e^+\tilde{\gamma}\tilde{\gamma}$ (with $m_{\tilde{Z}} + m_{\tilde{\gamma}} < \sqrt{s} = m_Z$)[5].

Here $\tilde{Z}^0$ is the superpartner of the Z^0. The analysis of the event rate and distributions (,such as angular-energy, acoplanarity-angle and missing-p_T distributions,) at $\sqrt{s} = m_Z$ of the $e^-e^+\tilde{\gamma}\tilde{\gamma}$ events originating from the processes (A)~

(C) will be presented elsewhere[6]. Here it is to be noted that $\sqrt{s}$ dependences of the cross sections of the processes (A)-(B) and (C) are quite different: The cross sections of (A) and (B) have a Z° resonance peak whereas that of (C) does not.

From Fig.2 we see that the simultaneous cuts (a)-(e) with the cut parameters of eq.(3) reduce the number of the signal events for the process $e^- e^+ \to Z^\circ \to \tilde{e}^\mp \tilde{\gamma} e^\pm \to e^- e^+ \tilde{\gamma}\tilde{\gamma}$ typically by about 30-40 %. This reduction is mainly due to the cuts (d) and (e). Using Fig.2 and the expected luminosities of SLC and LEP-I, one obtains the expected number of the $e^- e^+ \tilde{\gamma}\tilde{\gamma}$ events from the Z° decay which satisfy all the cut conditions (a)-(e) with eq.(3) at the Z° peak in a 3-month run as follows (for $m_{\tilde{e}_R} = m_{\tilde{e}_L}$):
5 (0.5) (SLC), 50 (5) (LEP-I) events
for $m_{\tilde{e}}$= 52 (64) GeV and $m_{\tilde{\gamma}}$= 0 GeV,
for $m_{\tilde{e}}$= 50 (61.5) GeV and $m_{\tilde{\gamma}}$= 10 GeV,
for $m_{\tilde{e}}$= 48 (55) GeV and $m_{\tilde{\gamma}}$= 20 GeV and
for $m_{\tilde{e}}$= 46.5 (48) GeV and $m_{\tilde{\gamma}}$= 30 GeV.
The same holds for the single $\tilde{\mu}^\mp$ production via the Z° formation. Hence in case $m_{\tilde{\gamma}}$= 0 GeV it should really be possible to detect the charged slepton $\tilde{l}$(= $\tilde{e}$ and $\tilde{\mu}$) production in the run at $\sqrt{s} = m_Z$ at SLC (LEP-I) if $m_{\tilde{l}}$< 52 (64) GeV. In case $m_{\tilde{\gamma}}$= 10,20 and 30 GeV these numbers are reduced to $m_{\tilde{l}}$< 50 (61.5), 48 (55) and 46.5 (48) GeV,respectively.

4. DISCUSSION AND CONCLUSION

Our approach can also be applied to the production of single scalar quark, "the squark"($\tilde{q}$), at the Z° peak in the $e^- e^+$ collision, such as the process $e^- e^+ \to Z^\circ \to \tilde{q}\tilde{\gamma}\bar{q}$[4,7]. A similar analysis for this process will be presented elsewhere[6].

We conclude that a detectable number of sleptons should be singly produced at the Z° peak at the SLC and LEP-I even if their masses substantially exceed the beam energy.

References

1) Gaillard, M.K. et al., Phys. Lett. 116B, 279 (1982).
2) Kuroda, M. et al., Phys. Lett. 127B, 467 (1983).
3) Gladney, L. et al., Phys. Rev. Lett. 51, 2253 (1983) and Fernandez, E. et al., Phys. Rev. Lett. 52, 22 (1984).
4) Hidaka, K., Komatsu, H. and Ratcliffe, P., Phys. Lett. 150B, 399 (1985), addendum 156B, 451 (1985); See also Johnson, J.E. and Rudaz, S., Phys. Rev. D30, 1590 (1984).
5) Dicus, D.A. et al., Phys. Rev. D29, 1317 (1984).
6) Hidaka, K., Komatsu, H. and Ratcliffe, P., in preparation.
7) Atwood, D. et al., Phys. Rev. D33, 1260 (1986).

CHIRAL GOLDSTINOS DUE TO DYNAMICAL SUSY BREAKING

Masakuni IDA
Department of Physics, Kobe University
Kobe 657

In this note we discuss possibilities in which the origin of matter is ascribed not to the compactification of space-time, but to dynamical breaking of SUSY induced by the first breakdown of the Yang-Mills symmetry. By matter we mean quarks and leptons (q/ℓ) or chiral fermions to compose them at a lower energy scale.

Let us take the D = 4, N = 1 supergravity coupled to a super-Yang-Mills theory with gauge group G. We first consider a dynamical breaking $G \longrightarrow G'$ in the limit of $\kappa \longrightarrow 0$ and $g \longrightarrow 0$, where κ and g are the gravitational and the YM coupling constant. Let G be a simply-connected, compact, semi-simple Lie group and T be a toral subgroup of it. A subgroup G'of G is called a C-subgroup if $C_G(T) \supseteq G' \supseteq \{C_G(T)\}_{ss}$ for some T. Here $C_G(T)$ is the centralizer of T in G and $\{C_G(T)\}_{ss}$ ($\equiv H$) represents its semi-simple part. Hereafter we restrict M, M = G/G', to C-spaces(simply-connected, compact, homogeneous, complex manifolds). Then G' should be a C-subgroup of G.

For $G' = C_G(T)$ (= H x T), M is Kaehlerian and we can construct on M a SUSY nonlinear Lagrangian[1] $\mathcal{L}_M(\phi, \partial_\mu\phi)$, where $\phi = (A,\psi,F)$ are the components of chiral superfields $\Phi_\pm$. The Lagrangian generally gives rise to U(1) anomalies, which forbid gaugeization of T and, hence, G. We thus take $G' = H$ to avoid U(1) anomalies. But $M = G/H$ is non-Kaehler because $b_2(M) = 0$, and we have explicit SUSY breaking.[2)]

We appeal to nonlinear realization of SUSY in order to restore it. The Goldstone fermion field $\lambda(x)$ of Volkov and Akulov super-transforms as

$$\delta\lambda = \varepsilon/a + \xi^\mu \partial_\mu \lambda, \qquad \xi^\mu \equiv -(i/2)a\bar{\varepsilon}\gamma^\mu\lambda,$$

where a is a dimensional constant characterizing dynamical breaking of SUSY. We define $\tilde{\phi}$ as functions of ϕ and λ so that $\delta\tilde{\phi} = \xi^\mu\partial_\mu\tilde{\phi}$ and replace $\mathcal{L}_M(\phi, \partial_\mu\phi)$ by $|W|\mathcal{L}_M(\tilde{\phi}, \nabla_\mu\tilde{\phi})$, where[3)]

$$|W| \equiv \det(\delta_\mu^\nu - (i/2)a^2\bar{\lambda}\gamma_\mu\partial^\nu\lambda\ ),$$

$$\nabla_\mu \equiv (W^{-1})_\mu{}^\nu\partial_\nu.$$

It can be seen that λ couples to ϕ so as to restore SUSY.

On the basis of these considerations we state our first basic assumption: The Goldstone bosons emerging by the first spontaneous breaking $G \longrightarrow H$ are accompanied by physical goldstinos. In other words, we assume that the physical degrees of freedom are equal to those given by linear SUSY. Then H is a semi-simple C-subgroup of G, and M (=G/H) requires dynamical breaking of SUSY because of its non-Kaehlerian nature. Our second basic assumption is: The goldstino representations under H contain chiral ones.

We now turn g and κ on, which cause the Higgs and the super-Higgs mechanism. The simultaneity of the two dynamical breakings produces massless chiral goldstinos, while the graviton and the gaugeons corresponding to H remain massless. We are led to a new effective theory denoted symbolically by EG + YM(H) + matter, where matter means here chiral goldstinos.

In our scheme the scale for the emergence of matter (Λ_{mat}) is equal to that of dynamical SUSY breaking (Λ_{SUSY}), which will lie below that of the compactification to D = 4 (Λ_{comp}):

$$\Lambda_{mat} = \Lambda_{SUSY} \; (< \Lambda_{comp}).$$

This is in sharp contrast with the widely accepted picture

$$\Lambda_{q/\ell} = \Lambda_{comp} \; (>> \Lambda_{SUSY}).$$

Our scheme presumes Λ_{SUSY} to be not much below Λ_{comp} ($\simeq M_{pl}$). Matter existe only in the world of nonlinear SUSY and has no physical superpartners.

We should fix G to build a realistic model. For this purpose we consider the D = 10 heterotic strings with G = Ss(32) or $E_8 \times E_8'$.[4] Since the compactification of this theory to D = 4 is not well understood we follow the arguments of Candelas, Horowitz, Strominger and Witten,[5] but choose the compact space K to be a torus (T^6) in place of a Calabi-Yau manifold. Since K is flat, G remains unbroken. Moreover, there are no chiral fermions because of $\chi(K) = 0$. Six U(1) gaugeons of Kaluza-Klein type will emerge corresponding to the isometries of T^6. We assume them to become massive without violating N = 1 SUSY in D = 4.

We set up our starting point in D = 4 to be N = 1 SG + SYM(G) with G = $E_8 \times E_8'$, for no chiral goldstinos emerge if G = Ss(32). As discrete symmetries it has S as well as C and P, where S represents the interchange $E_8 \longleftrightarrow E_8'$. In the simultaneous breakings of G and SUSY, we require that $\tilde{C}$(=CS) and $\tilde{P}$(=PS) are preserved although C, P and S are not and that G/H contains chiral H representations with chiral goldstinos being anomaly-free with respect to H.

We get two cases. In the first one chiral goldstinos have family degrees of freedom and can be identified with q/ℓ.[6] We find H to be $(E_6)^2$, $(D_5)^2$ or $(A_5)^2$, the family number being 3, 4 or 3. We are led to non-SUSY GUTs with shadow matter, but they inherit the difficulty of ordinary GUTs, the problem of q/ℓ mass hierarchy. There seems to be little hope for this case to be realistic.

In the second case, in which H is $(D_7)^2$ or $(A_7)^2$, chiral goldstinos have no family degrees of freedom. Moreover, they cannot be interpreted as preons because they behave as $\underset{\sim}{1}$ for one of the two factor groups. But they may produce preons through a phase transition of $H \rightarrow A_6 \times A_6'$,[2] in which chiral preons are given by $\{(\underset{\sim}{21},\underset{\sim}{7}) + (\underset{\sim}{7}^*, \underset{\sim}{21}^*)\}_L$. This coincides with the starting point of the SU(7) x SU(7) unified preon model.[7]

We thus interpret one of the A_6's to be the hypercolor gauge group responsible for the confinement of hypercolored preons, which leads to 2 + 2 families of q/ℓ. $\Lambda_{q/\ell}$ will lie midway between Λ_{SUSY} and Λ_{ew}, the scale of the electroweak symmetry breaking. This time we have no shadow matter. Small q/ℓ masses could be understood by a BCS-like mechanism [7] due to the ordinary color interaction in cooperation with strong, short-range forces resulting from the preon confinement. We think that this latter case is worth further studies.

In the string theory fundamental particles are not point-like, and all of them lie on Regge trajectories just like hadrons do. Is there any new definition of elementarity which can distinguish them from hadrons? Let us try to define elementary particles to be those which exist already in the linear-SUSY phase. According to this criterion gauge particles (the graviton and the gaugeons) together with their superpartners are elementary, but hadrons are not because the hadronic scale is certainly below Λ_{SUSY}.

The elementarity of q/ℓ depends on the scheme we adopt. They are elementary in the usual picture, in which $\Lambda_{q/\ell} >> \Lambda_{SUSY}$. On the contrary, our scheme requires $\Lambda_{mat} = \Lambda_{SUSY}$, so all the matter particles are to be regarded as composite. They live in the nonlinear-SUSY phase, without having physical superpartners.

This note is dedicated to Professor Yoshio Yamaguchi, who ingeneously introduced the unity to the complicated world of hadrons.

REFERENCES

1) Zumino B., Phys. Lett. 87B, 203(1979).
2) Ida M., Prog. Theor. Phys. 75, 105(1986).
3) Ivanov E. A. and Kapustnikov A. A., J. Phys. A11, 2375 (1978).
4) Gross D., Harvey J., Martinec E. and Rohm R., Nucl. Phys. B256, 253(1985).
5) Candelas P., Horowitz G., Strominger A. and Witten E., Nucl. Phys. B258, 75(1985).
6) Buchmüller W., Love S. T., Peccei R. and Yanagida T., Phys. Lett. 115B, 233(1982).
7) Ida M., Zeits. Phys. C28, 529(1985); C29, 291(1985).

GENERATION SYMMETRY AND HIDDEN AXIONS IN NO-SCALE SUPERGRAVITY

Hikaru Sato

Department of Physics, Hyogo University of Education
Yashiro, Hyogo 673-14, JAPAN

ABSTRACT

We incorporate the generation symmetry and $U(1)_{PQ}$ into the hidden sector of no-scale supergravity models. The pattern of the generation symmetry breaking is determined by the radiative corrections. Invisible axions live in the gravity hidden sector.

The "no-scale" unified models[1] were proposed to solve the gauge hierarchy problem, where the potential of the hidden sector is absolutely flat with zero cosmological constant[2] and undetermined scale of local SUSY breaking. The electroweak symmetry breaking and the local SUSY breaking scales are fixed dynamically by radiative corrections.[3-6]

The models proposed so far have introduced the broken generation symmetry by hand by assuming the largest t-quark Yukawa coupling. When the t-quark coupling is large enough, non-gravitational radiative corrections may drive the $(mass)^2$ of one of the Higgs fields negative, triggering spontaneous electroweak symmetry breaking.[7,8]

Recently we have proposed to incorporate the generation symmetry into the hidden sector of the no-scale SUSY standard model in such a way that the pattern of the generation symmetry breaking is determined dynamically by radiative corrections.[9] In this paper we solve the strong CP problem by introducing the $U(1)_{PQ}$ symmetry[10] into the gravity hidden sector so that the axion lives in the hidden sector without conflict with the cosmological[11] and astrophysical[12] bounds.

We assume the global generation (or family) symmetry $SU(3)_F$ and the observed three families are assigned as $\underline{3}$ of $SU(3)_F$, so that we consider the supersymmetric $SU(2)_L \times U(1)_Y \times SU(3)_F$ model coupled with N = 1 supergravity. The hidden sector includes three sets of complex scalar fields $v^\alpha = \{x^\alpha, y^\alpha, z^\alpha\}$ which transform as $\underline{6}$ of $SU(3)_F$ but are $SU(2)_L \times U(1)_Y$ singlets. The index $\alpha = (i,j)$, $(i,j = 1,2,3)$ is represented by the pair of indices in the fundamental representation. The Kähler potential G_0 of the hidden sector, which yields a flat vanishing potential, has the following form:

$$G_0 = -3\,\ell n(\ell n\,\rho + C) + D\ , \quad \rho = x^\alpha x^*_\alpha + y^\alpha y^*_\alpha + z^\alpha z^*_\alpha = v^\alpha v^*_\alpha\ , \tag{1}$$

where C, D are numerical constants.

The observable sector includes the $SU(3)_F$ non-singlet matter superfields and two $SU(3)_F$ singlet Higgs fields:

$$w^a = \{\ \bar{U}^i,\ \bar{D}^i,\ \bar{E}^i,\ Q^i,\ L^i,\ H_1,\ H_2\}\ . \tag{2}$$

The observable and hidden sectors should be combined in such a way that Yukawa couplings have finite flat limit.[4] This is done by making the ansatz for the superpotential as

$$\begin{aligned} f(w) &= [z^*_{ij}\bar{U}^i Q^j H_2 + y^*_{ij}\bar{D}^i Q^j H_1 + x^*_{ij}\bar{E}^i L^j H_1 + m_4 H_1 H_2]\,\Phi(\rho) \\ &= v^*_\alpha f^\alpha(w)\,\Phi(\rho) + m_4\,H_1 H_2\ \Phi(\rho)\ , \end{aligned} \tag{3}$$

where $\Phi(\rho)$ is to be determined such that Yukawa couplings have a finite flat limit, $M = m_P/\sqrt{8\pi} \to \infty$ with fixed $m_{3/2}$,

$$m_{3/2} \simeq e^{G_0/2}\,M = (\ell n\,\rho + C)^{-3/2}\,e^{D/2}\,M\ . \tag{4}$$

The scalar potential in the flat limit reads

$$V = \left| m_{3/2} w^*_a + F(\rho) v^*_\alpha \frac{\partial f^\alpha}{\partial w^a} \right|^2 + \text{D terms} \quad , \tag{5}$$

where $F(\rho)$ is obtained by the limit, $3\, e^{G/2}(\Phi/G_0')' \to -F(\rho)$.

The effective Yukawa coupling constants in the observable sector amount to

$$h^{ij} = F(\rho)\, z^{ij}, \quad \tilde{h}^{ij} = F(\rho)\, y^{ij}, \quad f^{ij} = F(\rho)\, x^{ij} \ , \tag{6}$$

which obey the following constraint:

$$h^{ij} h^*_{ji} + \tilde{h}^{ij} \tilde{h}^*_{ji} + f^{ij} f^*_{ji} = |F(\rho)|^2 \rho \ . \tag{7}$$

If these Yukawa couplings can trigger the electroweak symmetry breaking dynamically and produce the nonvanishing vacuum expectation values (VEVs) for H_1 and H_2,[7,8] mass matrices of quarks and leptons are given by

$$M_U = F(\rho) \langle H_2 \rangle \{ z^{ij} \}, \quad M_D = F(\rho) \langle H_1 \rangle \{ y^{ij} \}, \quad M_L = F(\rho) \langle H_1 \rangle \{ x^{ij} \}. \tag{8}$$

Nonvanishing VEVs for x^{ij}, y^{ij}, z^{ij} break the local SUSY and generation symmetry at the hidden sector but the scale and pattern of the breaking are not determined at the tree level. Phenomenologically preferable mass matrices and the scale of SUSY breaking may be determined dynamically in terms of the interactions in the observable sector. In ref.[9] we have presented a simplified model which realize this scenario.

Now it is straightforward and natural to extend our generation symmetry to include the axial U(1) symmetry. The Kähler potential (1) and the superpotential (3) is invariant under the following U(1) transformation;

$$\{ \bar{U}^i, \bar{D}^i, \bar{E}^i, Q^i, L^i \} \to e^{i\alpha} \{ \bar{U}^i, \bar{D}^i, \bar{E}^i, Q^i, L^i \} \ , \tag{9a}$$

$$\{ x^{ij}, y^{ij}, z^{ij} \} \to e^{2i\alpha} \{ x^{ij}, y^{ij}, z^{ij} \} \ . \tag{9b}$$

We shall assign this U(1) to the $U(1)_{PQ}$ symmetry.[10] Note that the $U(1)_{PQ}$ defined by eqs.(9) does not change the Higgs superfield H_1, H_2. Thus the axion comes entirely from the hidden scalar fields and it decouples with the Higgs fields. The $U(1)_{PQ}$ is violated by the VEVs $\langle v^{\alpha}\rangle$ in the hidden sector which are determined by radiative corrections.

Under the $U(1)_{PQ}$, the QCD angle θ changes as $\theta \to \theta - 2N_f\alpha$, for N_f generations. QCD instantons, however, introduce a potential with minima at values of θ for which there is no strong CP-violation: $\bar{\theta} = \theta + \arg\det M = 0$, where M is the quark mass matrix. Keeping only the axion field a(x), the scalar fields in the hidden sector may be written as

$$v^{\alpha} = \langle v^{\alpha}\rangle e^{ia/f_a}\ , \quad v^{\alpha} = x^{\alpha},\ y^{\alpha},\ z^{\alpha}\ . \tag{10}$$

The $U(1)_{PQ}$ current is written as

$$\begin{aligned} J_{\mu}^{PQ} &= 2\, G^{\alpha}_{0\beta}\, v^{*}_{\alpha}\overleftrightarrow{\partial}_{\mu} v^{\beta} + \bar{u}\gamma_{\mu}\gamma_5 u + \bar{d}\gamma_{\mu}\gamma_5 d + \cdots \\ &= f_a(\rho)\ \partial_{\mu}a + \bar{u}\gamma_{\mu}\gamma_5 u + \bar{d}\gamma_{\mu}\gamma_5 d + \cdots\ , \end{aligned} \tag{11}$$

and the axion decay constant $f_a(\rho)$ is given by

$$f_a^2(\rho) = 2\,G^{\alpha}_{0\beta}\langle v^{*}_{\alpha}\rangle\langle v^{\beta}\rangle = \frac{6}{(\ln\rho + C)^2} M^2\ . \tag{12}$$

Since $M = m_P/\sqrt{8\pi} \simeq 10^{18}$ GeV, the numerical constant C must be adjusted so as to satisfy the cosmological and astrophysical bounds[11,12] for $f_a(\rho)$; $10^9\,\mathrm{GeV} \lesssim f_a(\rho) \lesssim 10^{12}\,\mathrm{GeV}$.

The couplings of the axion with ordinary matter are given by

$$\mathcal{L}_a = -\frac{1}{f_a}\left(M_U^{jk}\bar{u}_j i\gamma_5 u_k + M_D^{jk}\bar{d}_j i\gamma_5 d_k + M_L^{jk}\bar{\ell}_j i\gamma_5 \ell_k\right) a + \mathrm{h.c.} \tag{13}$$

The strength of the axion's couplings is suppressed by a factor, (quark-lepton mass)/M so that the axion decouples with ordinary matter in a flat limit, $M \to \infty$. In the gravity hidden sector, however, the axion may interact "strongly" with other supergravity multiplets. This gives another possible mechanism to dissipate the cosmological energy density stored by the axion field. In the usual axion models, axions are in the observable sector. Since their coupling to ordinary matter is very weak, the axion energy density of order of $f_a^2 m_a^2$ cannot be dissipated enough and exceeds the critical density needed to close the universe if $f_a > 10^{12}$ GeV.[11]

In our model, axions are in the hidden sector. They can communicate with ordinary matter only through gravitational interactions suppressed by M^{-1}. If there are zero-mass fields in the hidden sector which interact strongly with axions, axions can decay rapidly without suppressed by a factor f_a^{-1} and lose their energy density. In this case f_a could be as large as $M \simeq 10^{18}$ GeV and axions become truely invisible.

REFERENCES

[1] For a review see, Nills, H.P., Phys. Rep. 110C, 2 (1984).

[2] Cremmer, E., Ferrara, S., Kounnas, C. and Nanopoulos, D.V., Phys. Lett. 133B, 61 (1983).

[3] Ellis, J., Lahanas, A.B., Nanopoulos, D.V. and Tamvakis, K., Phys. Lett. 134B, 429 (1984).

[4] Ellis, J., Kounnas, C. and Nanopoulos, D.V., Nucl. Phys. B241, 406 (1984).

[5] Ellis, J., Kounnas, C. and Nanopoulos, D.V., Nucl. Phys. B247, 373 (1984); Phys. Lett. 143B, 410 (1984).

[6] Ellis, J., Enqvist, K. and Nanopoulos, D.V., Phys. Lett. 147B, 99 (1984); Phys. Lett. 151B, 357 (1984).

[7] Inoue, K., Kakuto, A., Komatsu, H. and Takeshita, S., Prog. Theor. Phys. 68, 927 (1983).

[8] Ellis, J., Nanopoulos, D.V. and Tamvakis, K., Phys. Lett. 121B, 123 (1983).

[9] Sato, H., Phys. Lett. 161B, 107 (1985).

[10] Peccei, R.D. and Quinn, H.R., Phys. Rev. Lett. 38, 1440 (1977); Phys. Rev. D16, 1791 (1977).

[11] Preskill, J., Wise, M.B. and Wilczek, F., Phys. Lett. 120B, 127 (1983);
Abbott, L.F. and Sikivie, P., Phys. Lett. 120B, 133 (1983);
Dine, M. and Fischler, W., Phys. Lett. 120B, 137 (1983).

[12] Dicus, D.A., Kolb, E.W., Teplitz, V.L. and Wagoner, R.V., Phys. Rev. D22, 839 (1980);
Fukugita, M., Watamura, S. and Yoshimura, M., Phys. Rev. D26, 1840 (1982).

DYNAMICAL ASPECTS OF SUPERSYMMETRIC COMPOSITE MODELS OF QUARKS AND LEPTONS

Masaki Yasuè

Institute for Nuclear Study, University of Tokyo
Midori-cho, Tanashi, Tokyo 188
JAPAN

ABSTRACT

Chiral symmetry breakings and spectrum of light composite particles including quarks and leptons are discussed in supersymmetric confining theories based on $SU(N)^{loc}_{sc}$ with M-flavors for $N \leq M$ and with an extra antisymmetric $\underline{6}_{sc}$ field for N=4.

The qualitative understanding of the observed mass hierarchies of quarks and leptons has been a challenging problem in particle physics. It is one of the motivations for considering quarks and leptons as composites of further fundamental constituents. The compositeness of quarks and leptons is also implied by analogy and similarities between quarks and leptons, which can be traced back to the symmetry between baryons (p, n, Λ) and leptons (e, μ, ν) found in weak interactions.[1] Some of the "algebraic" properties of the constituents are so determined to explain the observed quark-lepton quantum numbers. However, the quark-lepton mass hierarchies will only be explained by considering the "dynamical" properties of the constituents.

The present strategy for composite model building is to assume, as an underlying dynamics for composite quarks and leptons, a confining force based on a non-abelian gauge theory. To generate light composite fermions with masses<<inverse sizes of composites ($\gtrsim$1 TeV), the underlying dynamics should be equipped with a certain mass-protection mechanism based on a symmetry-principle. Mass-protection symmetries are chiral symmetry and supersymmetry (SUSY) and the mass-protection mechanisms arise for i) unbroken chiral symmetry,[2] ii)

unbroken SUSY with spontaneously broken chiral symmetry[3] and iii) spontaneously broken SUSY.[4]

Whether these mass-protection symmetries are unbroken or spontaneously broken is determined by the underlying dynamics. If the underlying dynamics is described by a QCD-like confining theory, in which anomalies are cancelled by the L- and R-handed fermions, the relevant symmetries, i.e. the chiral symmetries for L- and R-handed fermions, are dynamically broken once quarks and leptons are formed out of odd number of fermions or a fermion and boson in the confining phase;[2,5] therefore, the presence of light composite fermions is not guaranteed by the chiral symmetries for the constituents. In order to maintain (some of) chiral symmetries, the underlying dynamics should be described by either chiral gauge theories or gauge theories with scalars coupled to fermions in a L-R asymmetric way. The natural extension of the latter possibility of course leads to a supersymmetric gauge theory.

There exists a strong constraint on the existence of chiral symmetries, which comes from the anomaly-matching of the chiral symmetries.[2] The non-vanishing anomalies must be carried by the Nambu-Goldstone bosons or massless composite fermions in the confining phase.[6] This constraint is found to be naturally satisfied by invoking the notion of complementarity.[7] Complementarity uses the equivalence of two phases:[8] The Higgs phase and confining phase of the same theory and postulates that i) The breaking pattern of the chiral symmetries in the Higgs phase should be the same as that in the confining phase and ii) The spectrum of the massless particles in the Higgs phase should be the same as that in the confining phase. Since the anomaly-matching is trivial in the Higgs phase, it is guaranteed in the confining phase. Thus, if complementarity is applied to the models, the spectrum of massless particles in the confining phase is provided by examining the well-understood Higgs phase.

Realistic composite models can be formulated in supersymmetric gauge theories. The simplest underlying dynamics is supersymmetric quantum chromodynamics (SQCD) with N-colors and M-flavors of $(\Phi_L)^A_i$ and

$(\Phi^C_R)^i_A$ (A=1~N; i=1~M). Recent extensive studies of the dynamical properties of SQCD with the massless flavors have indicated the dynamical breakdown of the chiral symmetries for M-flavors and for the gluino, i.e. $SU(M)_L \times SU(M)_R \times U(1)_A$.[9-11] In case that SUSY is not spontaneously broken, SQCD provides the Nambu-Goldstone mechanism of generating massless fermions as supersymmetric partners of the Nambu-Goldstone bosons,[3] which are associated with the spontaneous symmetry breakdown. Since for N>M the vacuum instability seems to suggest the spontaneous breakdown of SUSY, this mechanism calls for $N \leq M (<3N)$.

The massless SQCD with $N \leq M$ can be examined by complementarity since SQCD involves scalars in the fundamental representations of $SU(N)^{loc}_{sc}$. Complementarity leads to $G = SU(M)_L \times SU(M)_R \times U(1)_V \times U(1)_A \to H = SU(M)_L \times SU(M)_R \times U(1)_A$ for N=M and $\to H = SU(N)_{L+R} \times SU(M-N)_L \times SU(M-N)_R \times U(1)'_V \times U(1)'_A$ for N<M.[10] All anomaly-matching conditions are satisfied by the composite fermions contained in the Nambu-Goldstone superfields associated with G→H that will describe the quark-lepton fields. However, in realistic models, SUSY must be explicitly broken to generate the mass splitting between bosons and fermions.

To evaluate SUSY-breaking effects, one can use the effective lagrangian with the superpotential W_{eff} for the composite superfields, which accommodates the anomalous U(1) transformation.[11] The form of W_{eff} is determined to be: $W_{eff} = S\{\ell n[(S/\Lambda^3)^{N-M} \det(T/\Lambda^2)] + f(Z)\}$ with $Z = \det(T^j_i)/Y_L T^{M-N} Y^C_R$,*) where T^j_i is made of Φ_{Li} and Φ^{Cj}_R; S of two chiral spinor gauge superfields; and Y_L (Y^C_R) of the N-antisymmetrized Φ_L (Φ^C_R). The SUSY-vacuum consistent with complementarity is provided by $f(Z) = -\rho \ell n Z$ $(0<\rho<1)$ for N=M[12] and $f(Z) = \rho \ell n(Z - \langle Z \rangle|_{\theta=0})$ $(2<\rho<N)$ for N<M.[13] One can show that all remaining chiral symmetries are spontaneously broken once SUSY is explicitly broken. If SUSY is broken by the gluino mass m(g), then for N=M[12] $\pi(S) \sim \Lambda^2 m(g)$ and for N<M[13] $\pi(S) \sim \Lambda^3 (m(g)/\Lambda)^{\rho/[\rho-(M-N)]}$ and $\pi(Ti) \sim \Lambda^2 (m(g)/\Lambda)^{(M-N)/[\rho-(M-N)]}$ (i=N+1~M) with $\rho > M-N$ (and $\rho > 2$), where $\pi(S)$ and $\pi(Ti)$ are the lowest components of $\langle S \rangle$ and $\langle T^i_i \rangle$; $\pi(S) = \langle S \rangle|_{\theta=0}$ and $\pi(Ti) = \langle T^i_i \rangle|_{\theta=0}$.

---

*) $Y_L T^{M-N} Y^C_R = \sum \varepsilon^{i1\cdots iM} \varepsilon_{j1\cdots jM} (Y_L)_{i1\cdots iN} T^{jN+1}_{iN+1} \cdots T^{jM}_{iM} (Y^C_R)^{j1\cdots jN}$.

Composite quarks and leptons in one generation can be included in the Nambu-Goldstone superfields for N=4,6 with M=6 (describing up- and down-weak isospins of $SU(2)_W$ and four extended colors of $SU(4)_C$) and for N=4 with M=8 (describing two sets of the isospins and the four colors). The electroweak $[SU(2)_W]_L^{loc}$ symmetry is spontaneously broken by either the confining force or the SUSY-breaking terms. For N=M=6 and for N=4 with M=8, $SU(2)_W$ is included in the vectorial $SU(N)_{L+R}$, requiring $\Lambda=O(G_F^{-1/2})\sim1$ TeV while for N=4 with M=6, $SU(2)_W$ remains as a chiral symmetry, $SU(M-N)_L\times SU(M-N)_R$, and will be broken by the SUSY-breaking effect, requiring $(M_{SS}/\Lambda)^\alpha=O(G_F^{-1/2})$, where M_{SS} is the SUSY-breaking scale and $\alpha=1/(\rho-2)$ in the case of $M_{SS}=m(g)$. If the models are carefully examined, phenomenological difficulties manifest themselves: For N=4,6 with M=6, the models contain light charged lepto-quarks and $SU(3)_C^{loc}$-octet fermions because G→H includes $[SU(4)_C]_L\times[SU(4)_C]_R\to SU(4)_C$ and their masses (m) can be estimated from W_{eff} as $m/m_{q,\ell}\sim1$ for N=M=6 and $\sim(M_{SS}/\Lambda)^\alpha<1$ for N=4 and M=6; and for N=4 with M=8, the model contains the mirror states of quarks and leptons such as "$(w_L w_L w_L)_i c_R^C$" that mix with the ordinary quark-lepton states such as "$w_{Ri}^C c_R^C$" through $\langle T_i^i\rangle|_{\theta=0}\neq0$ (i=1~4) and $\langle Y_{L1\cdots4}\rangle|_{\theta=0}\neq0$, where "$w_L$" ("$w_R^C$") stand for two $\underline{2}$'s (four $\underline{1}$'s) of $[SU(2)_W]_L^{loc}$ and "c" for $\underline{4}$ of $SU(4)_C$. The elimination of these charged exotics in the light fermion spectrum is only possible outside SQCD.[12)]

A composite model without the light charged exotic fermions, consistent with complementarity, can be constructed in the model based on $SU(4)_{sc}^{loc}$,[14)] which contains one antisymmetric superfield $\hat{a}^{[AB]}$ (A,B=1~4) in addition to the M-flavors ($M\leq10$ from the asymptotic freedom). The global symmetry is $SU(M)_L\times SU(M)_R\times[U(1)]^3$ that undergoes spontaneous breakdown to $SU(2)_W\times SU(M-2)_L\times SU(M-2)_R\times U(1)_w\times U(1)_c\times U(1)_A$, where $U(1)_{w,c,A}$ stand for the number conservation of $\Phi_{i=1,2}$, $\Phi_{i=3\sim M}$ and the gluino, respectively. The breaking is induced by scalar condensates in the confining phase; $\langle\tilde{w}_{Li}\tilde{w}_R^{Cj}\rangle=O(\Lambda^2)\delta_i^j$ (i,j=1,2) and $\langle\tilde{a}\tilde{a}\rangle=O(\Lambda^2)$, corresponding to $\langle\tilde{w}_{Li}^A\rangle=O(\Lambda)\delta_i^A$, $\langle\tilde{w}_{RA}^{Ci}\rangle=O(\Lambda)\delta_A^i$ (i,A=1,2) and $\langle\tilde{a}^{[AB]}\rangle=O(\Lambda)G^{AB}$ ($G^{AB}=-G^{BA}$ with $G^{14}=G^{32}=1$) in the Higgs phase, where $\tilde{w}_{Li}=\Phi_{Li}|_{\theta=0}$ and $\tilde{w}_R^{Ci}=\Phi_R^{Ci}|_{\theta=0}$ (i=1,2).

Four generations of quarks and leptons are contained in the model with ten flavors (M=10); $\hat{c}_{i=1\sim4}$ and $\hat{s}_{i=1\sim4}$ transforming as $\underline{4}$'s of $SU(4)_C$ and $\hat{w}_{Li=1,2}$ ($\hat{w}_{Ri=1,2}$) as $\underline{2}$ ($\underline{1}$'s) of $[SU(2)_W]_L^{loc}$, and are included in the Nambu-Goldstone superfields, ($\hat{w}_L\hat{c}_L$, $\hat{w}_R^C\hat{c}_R^C$) and ($\hat{w}_L\hat{s}_L$, $\hat{w}_R^C\hat{s}_R^C$), and in the chiral superfields, ($\hat{w}_L^*\hat{a}\hat{c}_L$, $\hat{w}_R^{C*}\hat{a}c_R^C$) and ($\hat{w}_L^*\hat{a}\hat{s}_L$, $\hat{w}_R^{C*}\hat{a}s_R^C$), where $\hat{w}_L^*$ and $\hat{w}_R^{C*}$ are the chiral L-handed superfields, the complex conjugates of $\hat{w}_L$ and $\hat{w}_R^C$ surrounded by the gauge superfields. Exotics are all neutral and described by the chiral superfields ($\hat{w}_L\hat{a}\hat{w}_L$, $\hat{w}_R^C\hat{a}\hat{w}_R^C$) and by the Nambu-Goldstone superfields ($\hat{w}_L\hat{w}_R^C$, $\hat{a}\hat{a}$). The $SU(2)_W$-triplet $\hat{w}_L\hat{w}_R^C$ associated with $\langle\tilde{w}_{Li}\tilde{w}_R^{Cj}\rangle\neq0$ will be eliminated by the $[SU(2)_W]_L^{loc}$ flavor gauge superfield. Dangerous flavor-changing interactions of $K^+\to\pi^+e\mu$ and $K_L\to e\mu$ will be suppressed if the 1st and 2nd generations consist of $\hat{w}\hat{c}$ and $\hat{w}\hat{s}$, respectively.[15] However, the order-of-magnitude estimation of masses of quarks and leptons suggests $\hat{w}\hat{c}$ for the 1st generation but $\hat{w}^*\hat{a}\hat{c}$ for the 2nd one and the suppression is not realized. Other processes such as $\mu\to e\gamma$ and K_0-$\bar{K}_0$ are well suppressed by the (approximate) $U(1)_W$ symmetry and proton decays[16] are all forbidden by the $U(1)_c$ symmetry.

We have discussed dynamical aspects of supersymmetric composite models. Since dynamical models of quarks and leptons are at least equipped with approximate chiral symmetries, it is a rather general phenomenon that light quarks and leptons accompany light exotic fermions in order to meet the consistency of the unbroken chiral symmetry. So far the present quark-lepton phenomenology has indicated no observation of the exotic fermions (and bosons) and requires the large mass splittings between the exotics and the quarks and leptons. However, even in the attempt in constructing one-generation-models, the simple-minded SQCD is suffering from the existence of the light charged fermions or from the appreciable mixing of the quark-lepton states with their mirror states. In the case of $SU(4)_{sc}^{loc}$, one possible cure will be the inclusion of the antisymmetric $\underline{6}_{sc}$ state.

REFERENCES

1) Yamaguchi, Y., Prog. Theor. Phys. Suppl. 11, 1 (1959); Gamba, A., Marshak, R.E. and Okubo, S., Proc. Nat. Acad. of Sci. 45, 881 (1959).
2) 't Hooft, G., in Recent Developments in Gauge Theories, Cargèse, 1979, edited by G. 't Hooft et al. (Plenum Press, New York, 1980), p.135.
3) Buchmüller, W., Love, S.T., Peccei, R.D. and Yanagida, T., Phys. Lett. 115B, 223 (1982); Buchmüller, W., Peccei, R.D. and Yanagida, T., Phys. Lett. 124B, 67 (1983); Barbieri, R., Masiero, A. and Veneziano, G., Phys. Lett. 128B, 179 (1983); Greenberg, O.W., Mohapatra, R.N. and Yasuè, M., Phys. Lett. 128B, 65 (1983). See also Aoki, K.-I., Bando, M., Kugo, T. and Uehara, S., Prog. Theor. Phys. 67, 1193 (1982).
4) Terazawa, H., Prog. Theor. Phys. 64, 1763 (1980); Bardeen, W.A., and Visnjić, V., Nucl. Phys. B194, 422 (1982); Miyazawa, H., in Proc. 4th INS Winter Seminar on Structure of Quarks and Leptons, Tokyo, 1981, edited by M. Yasuè (INS, Univ. of Tokyo, 1982), p.71.
5) See for example, Weingarten, D., Phys. Rev. Lett. 51, 1830 (1983); Witten, E., Phys. Rev. Lett. 51, 2351 (1983); Nussinov, S., Phys. Rev. Lett. 51, 2081 (1983); 52, 966 (1984); Phys. Lett. 136B, 93 (1984); Phys. Lett. 139B, 203 (1984); Vafa, C. and Witten, E., Nucl. Phys. B234, 173 (1984).
6) Frishman, Y., Schwimmer, A., Banks, T. and Yankielowicz, S., Nucl. Phys. B177, 157 (1981); Coleman, S. and Grossman, B., Nucl. Phys. B203, 205 (1982).
7) Dimopoulos, S., Raby, S. and Susskind, L., Nucl. Phys. B173, 208 (1980); Matsumoto, T., Phys. Lett. 97B, 131 (1980).
8) Fradkin, E. and Shenker, S.H., Phys. Rev. D19, 3682 (1979); Banks, T. and Rabinovici, E., Nucl. Phys. B160, 349 (1979).
9) See for example, Amati, D., Rossi, G.C. and Veneziano, G., Nucl. Phys. B249, 1 (1985) and references therein.
10) Taylor, T.R., Phys. Lett. 128B, 403 (1983); Gérard, J.-M. and Nilles, H.P., Phys. Lett. 129B, 243 (1983).
11) Taylor, T.R., Veneziano, G. and Yankielowicz, S., Nucl. Phys. B218, 493 (1983).
12) Masiero, A., Pettorino, R., Roncadelli, M. and Veneziano, G., Nucl. Phys. B261, 633 (1985).
13) Yasuè, M., preprint INS-Rep.-589 (INS, Univ. of Tokyo), June, 1986.
14) Yamada, H. and Yasuè, M., preprint INS-Rep.-581 (INS, Univ. of Tokyo), April, 1986, to be published in Phys. Lett. B.
15) Bars, I., in Proc. 1984 DPF Summer Study on Design and Utilization of the SSC, Snowmass, June 23-July 13, 1984, edited by R. Donaldson and J.G.Morfin (AIP, New York, 1985), p.832; Greenberg, O.W., Mohapatra, R.N. and Nussinov, S., Phys. Lett. B148, 465 (1984).
16) Yamaguchi, Y., Prog. Theor. Phys. 22, 373 (1959); Sakharov, A.D., Pis'ma ZhETF 5, 32 (1967) [JETP Lett. 5, 24 (1967)]; Pati, J.C. and Salam, A., Phys. Rev. Lett. 31, 661 (1973).

III. QUANTUM FIELD THEORIES

VARIATIONS ON A THEME OF JULIAN SCHWINGER

—— On and Around Anticommuting Numbers ——

S. Kamefuchi
Institute of Physics, University of Tsukuba
Ibaraki 305, Japan

ABSTRACT

A short review is given of our recent works related in varied degrees to anticommuting (Grassmann) numbers which were first introduced into quantum field theory by Schwinger.

0. THEME: Anticommuting Numbers

To the best of my knowledge, Schwinger was the first, in the realm of quantum field theory, to employ anticommuting numbers or what are nowadays called Grassmann numbers. In his 1951 paper[1)], entitled "On the Green's Functions of Quantized Fields", he introduced anticommuting numbers as external sources of Fermi fields. In fact I can, even now, vividly recall the surprise I received when encountered for the first time such unusual objects in physics: the power as well as the simplicity of his method impressed me enormously.

In the above-mentioned paper Schwinger based his arguments on the variation principle for quantized fields which he himself had formulated elsewhere[2)]. Referring to the transition amplitude $<\sigma_1|\sigma_2>$ between a pair of states $|\sigma_j>$ given on spacelike surfaces σ_j (j=1,2), he expressed the principle in the form:

$$\delta\langle\sigma_1|\sigma_2\rangle = i\langle\sigma_1|\delta\int_{\sigma_2}^{\sigma_1}(dx)L|\sigma_2\rangle \quad . \tag{0.1}$$

Here L, the Lagrangian (density) of the system, contains among others the term $([\bar{\psi},\eta] + [\bar{\eta},\psi])/2$, in which η's are assumed to anticommute with themselves as well as with the Fermi field ψ, and thus play the role of external sources of ψ and $\bar{\psi}$. As shown by him, various useful relations can be derived by performing the variations $\delta_\eta, \delta_{\bar{\eta}}$ with respect to η, $\bar{\eta}$. In recent years, on the other hand, anticommuting numbers have found another application in the problem of the so-called supersymmetry. As is well known, by means of such numbers symmetry transformations can be so generalized that Bose and Fermi fields are related in larger multiplets.

Now the above idea of Schwinger's -- of playing with anticommuting numbers η -- is the main theme of the present essay. Starting from this I will be, so to speak, composing a series of variations on this theme, that is, discussing, one after another, various applications and generalizations of anticommuting numbers. In so doing I will take material mostly from my recent works done in collaboration with Y. Ohnuki and M. Omote.

1. VARIATION I: Anticommuting Sources

Instead of a Fermi field ψ let us consider for simplicity a system comprising just a single Fermi oscillator. In terms of the usual annihilation-creation operators a and $a^\dagger$ the most general Hamiltonian of the Schwinger type is

$$H = \omega a^\dagger a + f a^\dagger \eta + f^* \eta^* a + \omega_0 \quad . \tag{1.1}$$

Here (and in the following) η and its *-conjugate η^* are assumed to satisfy $[\eta,\hat{\eta}]_+ = [\eta,\hat{a}]_+ = 0$ with $\hat{\eta} \equiv \eta$ or η^*, $\hat{a} \equiv a$ or $a^\dagger$, $[A,B]_\pm \equiv AB \pm BA$ and ω, ω_0 and f are constants. Thus $\hat{\eta}$ plays the role of an external source of Fermi quanta. Now the procedure commonly adopted in dealing with such a system is first to evaluate those expectation values <···> of observables that are, in general, η-dependent and then to attach physical meanings only to their limiting values $\partial/\partial\hat{\eta} \langle\cdots\rangle|_{\hat{\eta}=0}$. In our opinion, however, this kind of procedure is not quite satisfactory in the following respects.

Mathematically, we cannot assign any numerical values to Grassmann numbers $\eta, \eta', \ldots$, because they are defined by $[\eta,\eta]_+ = [\eta,\eta']_+ = \cdots = 0$ and nothing else. In fact, all formulae for Grassmann numbers should remain invariant under the scale change $\eta \rightarrow \text{const.}\eta$. Physically, this means that in (1.1) the coupling strength f is not defined at all. Moreover, expectation values of observables are in general η-dependent and hence insusceptible of physical interpretation. That is to say, the theory with $\eta \neq 0$ is not entitled to be a physical theory. Thus a question arises as to whether it is a legitimate procedure to derive physical results from a non-physical theory. At any rate such an attitude does not suit my personal taste: I would prefer a theory, if possible, which is physical at every stage and in every respect.

In this connection let us make two more remarks on the Hamiltonian (1.1). First, it can easily be diagonalized: even after diagonalization, however, there remains the problem of how to interpret the term $-|f|^2\eta^*\eta/\omega$ thereby produced. Second, H has a technical deficiency such that all higher-order effects due to the repetion of $(fa^\dagger\eta f^*\eta^* a)$ identically vanish. [Note, however, that the problem does not arise in field theory, for the measure of such effects is zero.]

In view of the above situation let us now propose an alternative formulation of anticommuting sources[3]. Thus, instead of (1.1) we adopt the Hamiltonian of the form

$$H = H_1 + \eta H_2 + H_3\partial_\eta + \eta H_4\partial_\eta \tag{1.2}$$

with $H_1 = \omega_0$, $H_2 = -fa^\dagger$, $H_3 = -f^*a$ and $H_4 = \omega a^\dagger a$. Here η is assumed to be self-*-conjugate, i.e., $\eta^* = \eta$, and the left-differentiation ∂_η with respect to η is subject to $\partial_\eta^{\,2} = 0$ and $\partial_\eta^* = \partial_\eta$. Our Hamiltonian then has two merits. First, the scale of f is fixed in (1.2); second, the higher-order effects of interactions as mentioned above survive here and take the right form.

On the other hand, any observables, to be denoted hereafter by F, should consist of those terms which are even with respect to $\hat{a}$ and $\hat{\eta}$. For definiteness let us therefore restrict ourselves to, say, the even sector of the state-vector space, and write general state vectors as $\Phi(\eta) = \Sigma_{j=1,2}\, c_j \Phi_j(\eta)$ where c_j's are complex numbers, $\Phi_0(\eta) \equiv |0\rangle$ and $\Phi_1(\eta) \equiv \eta a^\dagger |0\rangle$. Norms, or more generally, inner products of $\Phi(\eta)$'s are defined by

$$(\Phi,\Phi') \equiv \int d\eta\ \Phi(\eta)^* \Omega \Phi'(\eta) \tag{1.3}$$

with $\Omega \equiv \eta - \partial_\eta$. On the other hand, matrix elements of F, which is always written as $F = F_1 + \eta F_2 + F_3 \partial_\eta + \eta F_4 \partial_\eta$, are defined by

$$(\Phi,F\Phi') \equiv \int d\eta\ \Phi(\eta)^* \Omega F \Phi'(\eta) \quad . \tag{1.4}$$

Reality of $(\Phi,F\Phi)$ or the relation $(\Phi,F\Phi')^* = (\Phi',F\Phi)$ is guaranteed, provided F is self-conjugate: $F = \tilde{F} \equiv \Omega F \Omega^{-1}$, i.e., $F_1^\dagger = F_1$, $F_2^\dagger = F_3$ and $F_4^\dagger = F_4$. Further, the completeness condition of the basis states reads: $\Sigma_j \Phi_j(\eta)\Phi_j(\eta')^* \Omega' = \delta(\eta-\eta')$. [In the present case it is convenient to put $\int d\eta = 0$ and $\int d\eta\eta = 1$; then the delta function $\delta(\eta-\eta')$ to satisfy $\int d\eta' \delta(\eta-\eta') f(\eta') = f(\eta)$ is given as $\delta(\eta-\eta') = \eta'-\eta$.]

When the Schrödinger equation is assumed in the usual form $i\dot{\Phi}(\eta,t) = H\Phi(\eta,t)$, all (Φ,Φ')'s are kept constant in time. Integrating the equation, for example, from $t = -\infty$ to $+\infty$, we then obtain the S-matrix of the usual Dyson form.

Summarizing, the present formulation removes all the deficiencies inherent in the usual procedure, and thus provides a satisfactory physical theory. It is expected that more realistic systems can be dealt with in a similar manner. Lastly, let us add that the criticism made above mostly applies to the usual formulation of supersymmetry, where again the problem of physical interpretation is not seriously considered.

2. VARIATION II: Introduction of G-numbers

Given a system of Bose and Fermi fields ϕ_j $(j=1,2, \cdots, n)$, it is customary in quantum field theory to adopt anticommutation relations for any pairs of different Fermi fields and commutation relations for any other pairs of different fields. It is known, however, that there is a certain amount of freedom in setting up commutation relations among different fields. In fact, the above-mentioned choice is usually called the normal case, and all other choices in which some of the commutators $[\ ,\ ]_{\mp}$ of the normal case are replaced by $[\ ,\ ]_{\pm}$ are called anomalous cases. It is also known that by means of the so-called Klein transformations we can pass from one case to another, so far as the locality condition is kept intact (Lüders' theorem)[4]. Thus we can write the most general form of equal-time commutation relations as

$$[\phi_i(x),\phi_j^{\dagger}(y)]_{-(ij)} = \delta_{ij}\delta(x-y)\ ,\quad [\phi_i(x),\phi_j(y)]_{-(ij)} = 0\ , \tag{2.1}$$

where $(ij) = (ji) = +$ or $-$ $(i,j=1,2,\cdots,n)$ are called relative signatures. [For the case of relativistic fields (2.1) is not general enough and needs a slight modification. The same remark also applies to the discussion in variation V.] For Bose (Fermi) fields ϕ_j, $(jj) = +(-)$, and let us assume hereafter that $(jj) = +(-)$ for $j=1,2,\cdots, b$ $(j=b+1,b+2,\cdots,b+f \equiv n)$.

Grassmann numbers may be regarded as a kind of classical counterparts of quantal Fermi fields. Thus, corresponding to the general case (2.1) we may similarly introduce a set of n kinds of numbers ξ_j as the classical counterparts of ϕ_j's, their "commutation relations" being

$$[\xi_i\ ,\ \xi_j]_{-(ij)} = 0\ . \tag{2.2}$$

Let us call ξ_j's with $(jj) = +(-)$ *generalized Bose* or *gB-numbers* (*generalized Fermi* or *gF-numbers*) and call both kinds *generalized* or *g-numbers*[5]. For gB-numbers it is further assumed that $\xi_j^N \neq 0$

for any positive integer N. In the normal case, where $(ij) = -$ only for $i,j = b+1, b+2, \cdots, b+f$, gB- and gF-numbers are reduced to complex and Grassmann numbers, respectively.

It is worth emphasizing here that many of algebraic or analytic formulae for ordinary complex or Grassmann numbers can be generalized smoothly to the case of g-numbers. Here let us mention only a few of them. Quantities $T_{ij\cdots}$ may be called g-tensors when their commutation properties are the same as $\xi_i \xi_j \cdots$. Given a g-tensor of second rank M_{ij} we can construct a g-matrix $M = \|M_{ij}\|$ and define the corresponding determinant by $\mathrm{Det}\, M \equiv \exp[\mathrm{Tr}\log M]$, where $\mathrm{Tr}\, M \equiv \Sigma^n_{j=1}(jj)M_{jj}$. Such Det M is a g-scalar, or commutes with any other quantities.

It is often convenient to introduce *-conjugates ξ^*_j of ξ_j: under *-conjugation all g-numbers undergo $\xi_j \leftrightarrow \xi^*_j$ while the order of $\hat{\xi}_j$ ($\equiv \xi_j$ or ξ^*_j) is reversed in each monomial. For g-functions $f(\xi, \xi^*)$ in $\hat{\xi}_j$'s we can define analytical operations such as differentiation, integration, Fourier transformation, etc. The most important among these is the following integral formula:

$$\int \xi_1^{\,r_1} \xi_1^{*\,s_1} \xi_2^{\,r_2} \xi_2^{*\,s_2} \cdots \xi_n^{\,r_n} \xi_n^{*\,s_n} \exp[-\xi^*\cdot\xi](d\xi d\xi^*) = \prod_{j=1}^{n} (\delta_{r_j s_j} r_j!) \quad , \tag{2.3}$$

where $r_j, s_j = 0,1,2,\cdots (r_j, s_j = 0,1)$ for $j = 1,2,\cdots,b$ $(j = b+1, b+2, \cdots, b+f)$, $\xi^*\cdot\xi \equiv \Sigma^n_{j=1}\xi^*_j\xi_j$ and $(d\xi d\xi^*) \equiv d\xi_1 d\xi^*_1 d\xi_2 d\xi^*_2 \cdots d\xi_n d\xi^*_n$. For the case of gB-numbers, in particular, $\int \cdots d\xi d\xi^*$ should be interpreted as $\iint_{-\infty}^{\infty} \cdots dudv/\pi$ where $\xi^* = u+iv$ and $\xi^* = u-iv$.

Some physical application of g-numbers will be given as the next variation.

3. VARIATION III: Bose-Fermi Similarity

For the sake of simplicity let us replace each field ϕ_j in the previous discussion with a single oscillator, to be described by $\hat{a}_j$ $(j=1,2,\cdots,n)$, and consider a system consisting of n oscillators.

With the j-th oscillator we then associate a pair of g-numbers ξ_j and ξ_j^* and assume the following commutation relations:

$$[a_i, a_j^\dagger]_{-(ij)} = \delta_{ij} \quad , \quad [a_i, a_j]_{-(ij)} = 0 \quad ;$$
$$[\xi_i, \hat{\xi}_j]_{-(ij)} = 0 \quad , \quad [\xi_i, \hat{a}_j]_{-(ij)} = 0 \quad . \tag{3.1}$$

Denoting the ground state of the system by $|0\rangle$ we also assume, for completeness, formal relations such as $\hat{\xi}_j|0\rangle = |0\rangle\hat{\xi}_j$ and $\langle 0|\hat{\xi}_j = \hat{\xi}_j\langle 0|$. It should be understood here that under Hermitian conjugation the order of factors in each monomial is reversed, and c-, g- and q-numbers undergo, respectively, complex-, *- and Hermitian conjugations. With these preliminaries we now proceed to show that in terms of g-numbers, or in the g-representation, Bose and Fermi oscillators can be described in a unified manner[5].

To this end we first introduce coherent states by

$$|\xi\rangle \equiv \exp[a^\dagger \cdot \xi]|0\rangle \quad , \quad \langle \xi^*| \equiv \langle 0| \exp[\xi^* \cdot a] \tag{3.2}$$

with scalar products defined by $A \cdot B \equiv \Sigma_{j=1}^n A_j B_j$. As usual, these are eigenstates of $\hat{a}_j$: $a_j|\xi\rangle = \xi_j|\xi\rangle$ and $\langle\xi^*|a_j^\dagger = \langle\xi^*|\xi_j^*$, and their inner products and the completeness relation are given by $\langle\xi^*|\xi\rangle = \exp[\xi^* \cdot \xi]$ and $\int|\xi\rangle\langle\xi^*|\exp[-\xi^* \cdot \xi](d\xi d\xi^*) = 1$, respectively. Next, for an arbitrary state vector $|\ \rangle$ we define the g-wavefunction $\psi(\xi^*)$ and its dual $\bar{\psi}(\xi)$ by

$$\psi(\xi^*) \equiv \langle\xi^*|\ \rangle \ , \quad \bar{\psi}(\xi) \equiv \langle\ |\xi\rangle \quad , \tag{3.3}$$

the normalization condition being $\int\bar{\psi}(\xi)\psi(\xi^*)\exp[-\xi^* \cdot \xi](d\xi d\xi^*) = 1$. For observables $F(a^\dagger, a)$ given as normal-ordered polynomials in $\hat{a}_j$'s we find $\langle\xi^*|F(a^\dagger, a)|\xi\rangle = F(\xi^*, \xi)\exp[\xi^* \cdot \xi]$. When the Hamiltonian $H(\hat{a}^\dagger, a)$ also is normal-ordered, the Schrödinger equation reads:

$$i\frac{\partial\psi(\xi^*, t)}{\partial t} = \int H(\xi^*, \xi')\psi(\xi'^*, t)\exp[\xi^* \cdot \xi' - \xi'^* \cdot \xi'](d\xi' d\xi'^*) \quad . \tag{3.4}$$

Now, what is most remarkable in the g-representation is that integration of (3.4) or related calculations can be carried out, to a considerable extent, without explicitly specifying the individual relative signatures. As an example, let us quote just one result, that is, the path-integral expression for transition amplitudes. For the transition $|I\rangle$ at $t=t_I \to |F\rangle$ at $t=t_F$ it is given by

$$\langle F|U(t_F,t_I)|I\rangle$$
$$= \lim_{N\to\infty} \iint\cdots\int \bar{\psi}_F(\xi_{(N)})\exp[\sum_{k=0}^{N} i\Delta t L_{(k)}]\psi_I(\xi^*_{(0)})\prod_{k=0}^{N}(d\xi_{(k)}d\xi^*_{(k)}) \tag{3.5}$$

in terms of the quantities: $U(t,t')\equiv\exp[-iH(t-t')]$, $\Delta t\equiv(t_F-t_I)/N$, $L_{(k)}\equiv i(\xi^*_{(k)}\cdot(\xi_{(k)}-\xi_{(k-1)})/\Delta t) - H(\xi^*_{(k)},\xi_{(k-1)})$, $\xi_{(-1)} = H(\xi^*_{(0)},\xi_{(-1)}) = 0$ and $L_{(0)} = i(\xi^*_{(0)}\cdot\xi_{(0)})/\Delta t$.

Summarizing, we can say that in the g-representation Bosons and Fermions behave very similarly, so much so that no separate treatments are needed: Bose-Fermi similarity is most explicitly manifested here. We also note that just as "g" is placed in between "c" and "q" in the alphabet, g-numbers play a role which is something in between the roles of c- and q-numbers. The transformation theory of quantum mechanics is extended in this way to the g-representation as well.

4. VARIATION IV: Hypersymmetry

Symmetry transformations are usually formulated on the basis of Lie groups where group parameters to specify individual group elements are taken to be complex numbers. It is also well known that when Grassmann numbers are used as group parameters, symmetry transformations are generalized to the so-called supersymmetry transformations; here it is customary to attach a prefix "super" to every word connected therewith. As may naturally be imagined, the generalization along this line can be pushed further by adopting our g-numbers as group parameters: in this case let us employ a new prefix "*hyper*" for the sake of distinction.

From the definition of g-numbers it is evident that hypersymmetry includes symmetry or supersymmetry as a special case. We may therefore expect that the introduction of this new concept will enable us to deal with those symmetries which have so far remained unnoticed or undefinable. That this in fact is the case will be shown below.

Let us consider a Lie group G whose elements are specified by a set of g-numbers ξ_j $(j=1,2,\cdots,n)$. Denoting by L_j the generator corresponding to the parameter ξ_j we can write an arbitrary element of G as $\exp[\Sigma_j \xi_j L_j]$. A formal generalization of the usual Lie group arguments then leads us to the following relations[6]:

$$[L_i, L_j]_{-(ij)} = (ij)C^k_{ij}L_k \quad ,$$

$$(ki)[L_i,[L_j,L_k]_{-(jk)}]_{-(ki)(ij)} + (ij)[L_j,[L_k,L_i]_{-(ki)}]_{-(ij)(jk)}$$
$$+(jk)[L_k,[L_i,L_j]_{-(ij)}]_{-(jk)(ki)} = 0 \,, \qquad (4.1)$$

where the structure constants C^k_{ij} are g-tensors subject to

$$C^k_{ij} = -(ij)C^k_{ji} \quad ,$$
$$(\ell k)C^m_{jk}C^i_{\ell m} + (\ell j)C^m_{k\ell}C^i_{jm} + (jk)C^m_{\ell j}C^i_{km} = 0 \qquad (4.2)$$

(with summation convention over m being understood).

As a simple example of hypersymmetry let us consider QED of the anomalous case as characterized by the following: the electromagnetic field $A_\mu(x)$ is in interaction with a pair of Majorana fields $\psi_j(x)$ $(j=1,2)$ for which $\psi_j(x) = \psi^c_j(x) = \psi^\dagger_j(x)$ in the Majorana representation of γ-matrices; their commutation relations are of the anomalous type, i.e., $[\hat{\psi}_1(x),\hat{\psi}_2(y)] = [\hat{\psi}_j(x),A_\mu(y)]_+ = 0$ for x-y being spacelike; nevertheless the Lagrangian L is assumed to be of the usual form:

$$L = -\frac{1}{2}\sum_{j=1}^{2}\bar{\psi}_j(\gamma\partial+m)\psi_j + \frac{ie}{2}(\bar{\psi}_1\gamma_\mu\psi_2 + \bar{\psi}_2\gamma_\mu\psi_1)A_\mu + L_0(A_\mu) \qquad (4.3)$$

with L_0 being the free Lagrangian of A_μ.

Now the field equations derived from (4.3) leads us, as usual, to the conservation law $\partial_\mu j_\mu = 0$ of the current $j_\mu \equiv ie(\bar{\psi}_1\gamma_\mu\psi_2 + \bar{\psi}_2\gamma_\mu\psi_1)$. We should therefore expect that there exist underlying gauge transformations, and moreover that such transformations for the charged fields are expressible as a kind of rotations in the charge space with axes j=1,2. It is easy to see, however, that if the ordinary angle variable θ is used here, the commutation relations of $\hat{\psi}_j$'s will not be maintained by the transformations. And, in fact, it is only by invoking g-numbers that the required transformations can be consistently formulated.

Let $\theta(x)$ be a self-*-conjugate gB-number such that $[\theta(x),\theta(x')] = [\theta(x),A_\mu(y)] = [\theta(x),\hat{\psi}_j(y)]_+ = 0$. In terms of such $\theta(x)$ our hypergauge transformations will then be given as[7)]

$$
\begin{aligned}
&\psi_1(x) \to \cos\theta(x)\cdot\psi_1(x) + i\,\sin\theta(x)\cdot\psi_2(x) \quad , \\
&\psi_2(x) \to i\,\sin\theta(x)\cdot\psi_1(x) + \cos\theta(x)\cdot\psi_2(x) \quad , \qquad (4.4)\\
&A_\mu(x) \to A_\mu(x) - \frac{1}{e}\,\partial_\mu\theta(x) \quad .
\end{aligned}
$$

It is also easy to confirm that the current j_μ defined above is in fact the Noether current for the transformations (4.4).

5. VARIATION V: Parafields and Paranumbers

In this variation let us start, as in variation II, by considering a set of n fields ϕ_j (j=1,2,···,n). Here, however, all ϕ_j's are assumed to be Bose (Fermi) fields and to anticommute (commute) with each other, that is, all (jj) = +(−) and all (ij) = −(+) with i,j=1,2, ···,n; i≠j. Out of these we next construct a new field $\hat{\psi} \equiv \Sigma_{j=1}^{n}\hat{\phi}_j$ (Green's representation[8)]). The use of commutation relations (2.1) will then lead us to the following trilinear relations for equal-time quantities:

$$
[\psi(x),\ [\psi^\dagger(y),\psi(z)]_\pm] = 2\delta(x-y)\psi(z) \quad ,
$$

$$[\psi(x),\ [\psi^\dagger(y),\psi^\dagger(z)]_\pm] = 2\delta(x-y)\psi^\dagger(z) \pm 2\delta(x-z)\psi^\dagger(y) \quad ,$$

$$[\psi(x),\ [\psi(y),\psi(z)]_\pm] = 0 \quad , \qquad (5.1)$$

where the upper (lower) signs apply to the case $(jj) = +(-)$. Further, for the vacuum state $|0\rangle$ satisfying $\psi(x)|0\rangle = 0$, we have

$$\psi(x)\psi^\dagger(y)|0\rangle = n\delta(x-y)|0\rangle \quad . \qquad (5.2)$$

The relations (5.1) with upper (lower) signs are none other than the basic commutation relations for a para-Bose (para-Fermi) field. On the other hand, (5.2) indicates that the order of parastatistics for this case is n: that is to say, up to n particles we can construct, in the para-Bose (para-Fermi) case, completely antisymmetric (symmetric) states with non-vanishing norms.

Next, let us consider n kinds of g-numbers and their *-conjugates $\hat{\xi}_j$ $(j=1,2,\cdots,n)$ such that $[\xi_i,\hat{\xi}_j]_{-(ij)} = [\xi_i,\hat{\phi}_j(x)]_{-(ij)} = 0$ with the same (ij)'s as given above for ϕ_j's. Together with $\hat{\xi}_j$ we also consider $\hat{\xi}_j{}',\hat{\xi}_j{}'',\cdots$ of the same kind as $\hat{\xi}_j$: $[\xi_i,\hat{\xi}_j{}']_{-(ij)} = [\xi_i{}',\hat{\xi}_j{}'']_{-(ij)} = [\xi_i{}',\hat{\phi}_j(x)]_{-(ij)} = \cdots = 0$. With such numbers let us construct a new set of numbers by $\hat{\xi} \equiv \Sigma_{j=1}^n \hat{\xi}_j$, $\hat{\xi}' \equiv \Sigma_{j=1}^n \hat{\xi}_j{}',\cdots$, and call them *para--Bose* (*para-Fermi*) *numbers* of order n. The following relations then hold:

$$[\hat{\xi},[\hat{\xi}',\hat{\xi}'']_\pm] = 0 \quad , \quad [\hat{\xi},\hat{\xi}',\cdots,\hat{\xi}^{(m)}]_\mp = 0 \quad \text{for } m \geqslant n \ , \qquad (5.3)$$

where the upper (lower) signs apply to para-Bose (para-Fermi) numbers, and $[\ ,\ ,\cdots]_\mp$ denotes completely antisymmetrized (symmetrized) products[9].

Para-Bose and para-Fermi numbers, which we generally call *para-numbers*, may be regarded as classical counterparts of para-Bose and para-Fermi operators, respectively. Thus take, for example, the annihilation-creation operators $\hat{a}$ for an arbitrary mode of the para-Bose (para-Fermi) field $\psi(x)$. The coherent states for $\hat{a}$ are then given in the same form as (3.2) but with a new definition of scalar products $A \cdot B \equiv [A,B]_{\pm}/2$, where again the upper (lower) sign applies to the para-Bose (para-Fermi) case.

Having defined coherent states, we can then proceed in the same way as in variation III: by introducing a special representation corresponding to the g-representation thereof we can describe the system of paraquanta solely in terms of paranumbers. Thus, for example, by integrating the Schrödinger equation for the wavefunction $\psi(\xi^*,t) \equiv \langle\xi^*|t\rangle$ we can derive the path-integral formula for transition amplitudes which has a form similar to (3.5).

In the above we have been treating the simplest case of only one pair of a and $a^\dagger$. Evidently, it is a straightforward matter to generalize the above consideration to the case of many pairs.

0'. CODA: Superselection and Duality

Following the conventions in music I had better conclude the present variations by going back to the original theme. This time let me adopt, however, a different key to discuss a different application of the anticommuting number η in the context of Schwinger's variation principle.

As is well known, the superselection rule for Fermion number N_F sharply separates the state-vector space into the even and odd sectors where N_F=even and odd, respectively, and transitions are absolutely forbidden between the two sectors. In view of this I now rewrite the basic relation (0.1) in a form such as

$$\delta\langle\sigma_1,\pm|\ \sigma_2,\pm\rangle = i\,\langle\sigma_1\ \pm|\ \delta\int_{\sigma_2}^{\sigma_1}(dx)L^{(\pm)}|\ \sigma_2,\pm\rangle\ , \qquad (0'.1)$$

where $|\sigma,\pm\rangle$ denote states in the even/odd sectors, and assume moreover that the Lagrangians $L^{(\pm)}$ for the even/odd sectors can, in general, take different forms. Now, as regards those variations $\delta\hat{\psi}$ of a Fermi field $\hat{\psi}$ which vanish on the boundary surfaces σ_1 and σ_2, the right-hand side of (0'.1) may generally be written as

$$i\langle\sigma_1,\pm|\int_{\sigma_2}^{\sigma_1}(dx)(\delta\hat{\psi}\,\hat{A}^{(\pm)} + \hat{B}^{(\pm)}\delta\hat{\psi})|\sigma_2,\pm\rangle \quad . \qquad (0'.2)$$

If the variations are such that left-variations $\delta\hat{\psi}\,\hat{A}^{(\pm)}$ cannot be rewritten in the form of right-variations $\hat{B}^{(\pm)}\delta\hat{\psi}$ and vice versa, the two terms in the parentheses must be regarded as variationally independent. In such a case, $L^{(+)}$ will provide the field equations: $\hat{A}^{(+)} = 0$ and $\hat{B}^{(+)} = 0$ for $\hat{\psi}$'s acting on $|\sigma,+\rangle$ and $|\sigma,-\rangle$ respectively, whereas $L^{(-)}$ will provide $\hat{B}^{(-)} = 0$ and $\hat{A}^{(-)} = 0$ for $\hat{\psi}$'s acting on $|\sigma,+\rangle$ and $|\sigma,-\rangle$ respectively. Here, for consistency I must require the conditions $\hat{A}^{(\pm)} \propto \hat{B}^{(\mp)}$.

As discussed in detail elsewhere[10)], the requirements mentioned above are all satisfied by para-Fermi fields of even order. Let me consider, for example, a para-Fermi field $\psi(x)$ of order 2 and assume that

$$L^{(\pm)} = -\frac{1}{2}[\bar{\psi},(\gamma\partial + m)\psi] \mp \frac{\kappa}{2}[\bar{\psi},\psi]_+ + \frac{ie}{2}[\bar{\psi},\gamma_\mu\psi]A_\mu + L_0(A_\mu) \quad . \qquad (0'.3)$$

Then the field equations become $(\gamma\partial + m \pm \kappa)\psi(x) = ie\gamma_\mu A_\mu(x)\psi(x)$ for $\psi(x)$'s acting on states $|\sigma,\pm\rangle$. This dual nature of the ψ may be utilized, for example, for explaining the dual nature of leptons. In such a model, particles with mass $m+\kappa$ and $m-\kappa$, both obeying Fermi statistics, are identified with muons and electrons, respectively, and the μ-e symmetry of lepton interactions will then come out as a natural consequence.

There still remains for me to say something about the anti-commuting number η. This I will do now by showing that with the aid of η the two equations in (0'.1) can be brought to a unified form. In so doing I follow the procedure employed in variation I and write: $L \equiv L^{(+)} + \eta(L^{(-)} - L^{(+)})\partial_\eta$, $|\sigma\rangle \equiv |\sigma,+\rangle + \eta|\sigma,-\rangle$, and $\langle\langle\sigma|F|\sigma'\rangle\rangle' \equiv \int d\eta\,\langle\sigma|\Omega F|\sigma'\rangle'$ for an arbitrary observable F. Then, the variation principle (0'.1) can be cast in the form:

$$\delta \ll \sigma_1 | \sigma_2 \gg \;=\; i \ll \sigma_1 | \delta \int_{\sigma_2}^{\sigma_1} (dx)\, L\, | \sigma_2 \gg \quad , \tag{0'.4}$$

when $\eta|\sigma,-\rangle = -|\sigma,-\rangle\eta$ is noted. I have thus come back to the equation of the same form as the original (0.1).

I would like to dedicate this essay to Professor Yoshio Yamaguchi on the occasion of his sixtieth birthday. His friendship has always been a source of pleasure, advice and encouragement for me. I am very grateful for this, and hope that he will continue to be happy and creative in many more years to come. Thanks are also due to my collaborators Y. Ohnuki and M. Omote for permitting me to include here some of our unpublished results.

REFERENCES

1) Schwinger, J., Proc. Nat. Acad. Sci. 37, 452, 455 (1951).
2) Schwinger, J., Phys. Rev. 82, 914 (1951).
3) Omote, M. and Kamefuchi, S., to be published.
4) See, for example, Ohnuki, Y. and Kamefuchi, S., *Quantum Field Theory and Parastatistics*, Univ. of Tokyo Press / Springer Verlag (1982), Appendix F.
5) Ohnuki, Y. and Kamefuchi, S., Nuovo Cimento 70A, 435 (1982); 73A, 328 (1983); 77A, 99 (1983); 83A, 275 (1984).
6) Omote, M. and Kamefuchi, S., Nuovo Cimento 50A, 21 (1979); See also Ref.4), Appendix G.
7) Ohnuki, Y. and Kamefuchi, S., to be published.
8) Green, H. S., Phys. Rev. 90, 270 (1953).
9) Ohnuki, Y. and Kamefuchi, S., J. Math. Phys. 21, 609 (1980); See also Ref.4), Chap.27.
10) Ohnuki, Y. and Kamefuchi, S., Ref.4), Chap.21.

NONLINEAR REALIZATION OF NONCOMPACT GROUPS

Y. Ohnuki

Department of Physics, Nagoya University

Nagoya 464, Japan

ABSTRACT

Under certain conditions a proof is given of the positivity of effective quantum Hamiltonian for Goldstone fields which are generated by spontaneous symmetry breakdown of noncompact groups.

I.

In 1967 Schwinger[1] proposed a new form of chiral invariant Lagrangian to describe a system containing Goldstone pions, where field variables were required to obey nonlinear transformations of the chiral SU(2) group. At that time I was at Institute for Nuclear Study, University of Tokyo, as a visiter. Y. Yamaguchi and I then tried to generalize[2] this formalism together with some other members of the institute. Our main results were published in the Supplement of Progress of Theoretical Physics Extra Number (1968)[3], where we explicitly constructed possible forms of nonlinear transformations for Goldstone and also matter fields within a general framework of chiral groups. Especially, the matrix $\sqrt{M}$ introduced therein was shown to play a role of "boost" in the chiral space. In fact, there exist analogous features between representations of the nonlinear chiral SU(2) and the Poincaré group in many respects as seen in the following: pion field ↔ 3-momentum of massive particle, isospin group ↔ little group of the Poincaré group, chiral transformation for matter field of isospin j ↔ Wigner rotation of massive field of spin j, etc.. Accordingly we can attain the nonlinear realization of the chiral group for an arbitrary isospin (unitary spin) field with the aid of the induced representation technique employed in obtaining irreducible unitary representations of the Poincaré group.[4)]

A further generalization of such approach was made by Coleman, Wess and Zumino[5)]. They discussed a general type of spontaneous symmetry breakdown G → H, where G is a semisimple compact Lie group containing H as a subgroup.

In this note we shall apply their arguments to examine a spontaneous symmetry breakdown of semisimple noncompact groups. For this purpose we assume that they are obtained from compact groups by using Weyl's unitary trick.[6)] Most of physically interesting noncompact groups seem to belong to this case.

We denote the noncompact group as $\bar{G}$, which is related to a compact group G through the unitary trick. The symmetry of the group $\bar{G}$ is assumed to spontanously break down into its maximum compact subgroup H. Under these conditions it will be shown in an elementary way that the quantum Hamiltonian for Goldstone fields with the lowest order of derivatives becomes a positive definite operator.

II.

Let us denote the hermitian generators of the compact group G as $T_1, T_2, \cdots, T_m, K_1, K_2, \cdots, K_n$, where $T_1, T_2, \cdots, T_m$ are the generators of the subgroup H. Following Coleman, Wess and Zumino[5)] we write g_0 (ϵ G) as

$$g_0 = \exp(i\phi\cdot K)\cdot h \tag{1}$$

with

$$\phi\cdot K \equiv \sum_{j=1}^{n} \phi_j K_j \ , \tag{2}$$

where h is an element of H and ϕ_j's real parameters. For any $g \in G$ we can also write the group element gg_0 in the form

$$gg_0 = \exp(i\phi'\cdot K)\cdot h' \ , \tag{3}$$

where $h'(\in H)$ is, in general, a function of g and ϕ_j's. If ϕ_j's are functions of space-time point x, they can be regarded as Goldstone fields[5] generated by the spontaneous breakdown $G \to H$. The transformation

$$\phi_j(x) \underset{g}{\to} \phi_j'(x) \tag{4}$$

forms a nonlinear realization of G. In order to make it possible to apply the unitary trick we further assume that the coset space G/H is symmetric, i.e., the generators of G is assumed to obey a Lie algebra of the following type:

$$\begin{aligned} [T_a, T_b] &= i \sum_{c=1}^{m} \gamma_{ab}^{c}\, T_c \ , \\ [T_a, K_j] &= i \sum_{k=1}^{n} \gamma_{aj}^{k}\, K_k \ , \\ [K_j, K_k] &= i \sum_{a=1}^{m} \gamma_{jk}^{a}\, T_a \ , \end{aligned} \tag{5}$$

where γ's are real structure constants. The algebra (5) admits the automorphism $K_j \to -K_j$, $T_a \to T_a$ $(j=1,2,\cdot\cdot,n,\ a=1,2,\cdot\cdot,m)$. Then it can be shown[7] that corresponding to the transformation (3) the quantity M defined by

$$M \equiv \exp(2i\phi(x)\cdot K) \tag{6}$$

undergoes linear transformations such that

$$M \underset{g}{\to} \begin{cases} g\,M\,g^{-1} & \text{for } g \in H\,, \\ g\,M\,g & \text{for } g = \exp(i\lambda\cdot K) \end{cases} \tag{7}$$

Here λ_j $(j=1,2,\cdot\cdot,n)$ are real parameters. Hence we are immediately led to the chiral invariant Lagrangian of the well known form $L \propto \mathrm{tr}(\partial^\mu M^\dagger \partial_\mu M)$, since M is represented by a unitary matrix with finite dimensions.

On the basis of these arguments we now proceed to study the spontaneous symmetry breakdown of $\bar{G}$. As mentioned already its Lie algebra is derived from (5) by using the unitary trick, which means to replace the generators K_j by $\bar{K}_j \equiv iK_j$ $(j=1,2,\cdot\cdot,n)$ with T_a $(a=1,2,\cdot\cdot,m)$ kept unchanged. Consequently we have a noncompact (real) Lie algebra

$$\begin{aligned} [T_a, T_b] &= i \sum_{c=1}^{m} \gamma^c_{ab}\, T_c\,, \\ [T_a, \bar{K}_j] &= i \sum_{k=1}^{n} \gamma^k_{aj}\, \bar{K}_k\,, \\ [\bar{K}_j, \bar{K}_k] &= -i \sum_{a=1}^{m} \gamma^a_{jk}\, T_a\,. \end{aligned} \tag{8}$$

Then Eq.(1) may be rewritten as

$$g_0 = \exp(i\bar{\phi}(x)\cdot\bar{K})\cdot h \qquad (g_0 \in \bar{G}\,, \quad h \in H) \tag{9}$$

with real Goldstone fields $\bar{\phi}_j(x)$. However unitary representations of noncompact groups are known to be of infinite dimensions except for a trivial one-dimensional case. Therefore, in order to avoid unnecessary mathematical complications we shall use the finite-dimensional representation with antihermitian $\bar{K}_j$ ($=iK_j$), though it does not form a unitary representation of $\bar{G}$. Then the Eq.(9) turns out to be

$$g_0 = \exp(-\bar{\phi}(x)\cdot K)\cdot h \tag{10}$$

with $K_j = K_j^\dagger$ ($j=1,2,\cdots,n$) which satisfy the algebra (5) and are of course finite-dimensional. From $gg_0=\exp(-\bar{\phi}'(x)\cdot K)\cdot h'$, we again obtain the nonlinear transformation $\bar{\phi}_j(x) \underset{g}{\rightarrow} \bar{\phi}_j'(x)$ for Goldstone fields. In the same way as done in the compact group G with symmetric coset space G/H we are now able to introduce the matrix

$$\bar{M} \equiv \exp(-2\bar{\phi}(x)\cdot K) , \tag{11}$$

whose transformations under $\bar{G}$ are given by

$$\bar{M} \underset{g}{\rightarrow} \begin{cases} g\,\bar{M}\,g^{-1} & \text{for } g \in H , \\ g\,\bar{M}\,g & \text{for } g = \exp[-\lambda\cdot K] . \end{cases} \tag{12}$$

(λ_j: real parameter)

Note here that the matrix $\bar{M}$ is hermitian. Hence the invariant Lagrangian with minimal derivatives has to take the following form:

$$L \propto \mathrm{tr}\,(\partial^\mu \bar{M}^{-1} \partial_\mu \bar{M}) . \tag{13}$$

III.

The consistent Hamiltonian density invariant under $\bar{G}$ is known to be given by the formula[8)]

$$H(x) = \sum_j [\dot{\bar{\phi}}_j\, \partial/\partial\, \dot{\bar{\phi}}_j]L - L . \tag{14}$$

The term $[\dot{\bar{\phi}}_j \partial/\partial\dot{\bar{\phi}}_j]L$ is defined by expanding L into power series in $\bar{\phi}_j$'s and $\dot{\bar{\phi}}_j$'s, and applying the following rules: (i) $[\dot{\bar{\phi}}_j\partial/\partial\dot{\bar{\phi}}_j](\Sigma_\alpha X_\alpha)= \Sigma_\alpha[\dot{\bar{\phi}}_j\partial/\partial\dot{\bar{\phi}}_j]X$, (ii) $[\dot{\bar{\phi}}_j\partial/\partial\dot{\bar{\phi}}_j](XY)=[\dot{\bar{\phi}}_j\partial/\partial\dot{\bar{\phi}}_j]X\cdot Y+X\cdot[\dot{\bar{\phi}}_j\partial/\partial\dot{\bar{\phi}}_j]Y$, (iii) $[\dot{\bar{\phi}}_j\partial/\partial\dot{\bar{\phi}}_j]\dot{\bar{\phi}}_k=\delta_{jk}\dot{\bar{\phi}}_j$, $[\dot{\bar{\phi}}_j\partial/\partial\dot{\bar{\phi}}_j]\bar{\phi}_k=0$. Thus we find the Hamiltonian is of the form

$$H(x) = -N\ \mathrm{tr}(\partial_t\bar{M}^{-1}\partial_t\bar{M} + \nabla\bar{M}^{-1}\nabla\bar{M}) \tag{15}$$

with an appropriate constant N.

To make our story simple we introduce here a hermitian matrix

$$\Phi \equiv \tanh(\bar{\phi}\cdot K) \tag{16}$$

and rewrite the matrix $\bar{M}$ as

$$\bar{M} = \frac{1-\Phi}{1+\Phi}\ , \tag{17}$$

which leads us to the relations

$$\begin{aligned}\partial_t\ \bar{M}^{-1}\partial_t\ \bar{M} &= -4\ \frac{1}{1-\Phi}\ \partial_t\Phi\ \frac{1}{1-\Phi^2}\ \partial_t\Phi\ \frac{1}{1+\Phi}\ ,\\ \nabla\ \bar{M}^{-1}\ \nabla\ \bar{M} &= -4\ \frac{1}{1-\Phi}\nabla\ \Phi\ \frac{1}{1-\Phi^2}\nabla\ \Phi\ \frac{1}{1+\Phi}\ .\end{aligned} \tag{18}$$

Thus we easily see the Hamiltonian (15) can be written as

$$H(x) = 4N\ \mathrm{tr}\left\{\left(\frac{1}{\sqrt{1-\Phi^2}}\ \partial_t\Phi\ \frac{1}{\sqrt{1-\Phi^2}}\right)^2 + \left(\frac{1}{\sqrt{1-\Phi^2}}\nabla\ \Phi\ \frac{1}{\sqrt{1-\Phi^2}}\right)^2\right\}\ . \tag{19}$$

Since $(1-\Phi^2)$ is a positive matrix and hence $(1-\Phi^2)^{-1/2}\partial_t\Phi(1-\Phi^2)^{-1/2}$ and $(1-\Phi^2)^{-1/2}\nabla\Phi(1-\Phi^2)^{-1/2}$ are both hermitian, the right hand side of (19) with N>0 is evidently positive definite.

Summarizing the above we now arrive at the following theorem:

Theorem

In the spontaneous symmetry breakdown of semisimple noncompact group $\bar{G}$ into a compact maximum subgroup H the Hamiltonian for Goldstone fields with minimal derivatives is always positive definite, provided that the coset space $\bar{G}/H$ is symmetric and $\bar{G}$ is made compact by the unitary trick.

The theorem is of course applicable, for example, to $SO(p,q)/SO(p)\times SO(q)$ and $SU(p,q)/SU(p)\times SU(q)\times U(1)$, special cases of which have been discussed by several authors.[9)]

ACKNOWLEDGEMENTS

It is a great pleasure to dedicate this article to Professor Yoshio Yamaguchi on the occasion of his sixtieth birthday. Thanks are also due to T. Morozumi for useful discussions.

REFERENCES

1) Schwinger, J., Phys. Letters 18, 923 (1967); 24B, 473 (1967).

2) Yamaguchi, Y., "Fundamental Physics — 1967 Tokyo Summer Lecutures in Theoretical Physics —", ed. G. Takeda and Y. Hara: p.24, Syōkabō, Tokyo and W.A. Benjamin Inc., New York.
Ohnuki, Y. and Yamaguchi, Y., "Chiral Dynamics" (in Japanese) INS-TH YY-67-1 (1967).

3) Hiida, K., Ohnuki, Y. and Yamaguchi, Y., Prog. Theor. Phys. Suppl. Extra number, 337 (1968).

4) Wigner, E.P., Ann. Math. 40, 149 (1939).

5) Coleman, S., Wess, J. and Zumino, B., Phys. Rev. 177, 2239 (1969).

6) For example, see Hermann, R., "Lie Groups for Physicits", W.A. Benjamin Inc. (1966).

7) Zumino, B., B127, 189 (1977).

8) Kamefuchi, S., O'Raifeartaih, S.L. and Salam, A., Nucl. Phys. 28, 529 (1961).

9) Cohen, Y. and Ravinovici, E., Phys. Letters 124B, 371 (1983).
Davis, A.C., Macfarlane, A.J. and van Holten, J.W., Phys. Letters 125B, 151 (1983); Nucl. Phys. B232, 473 (1984).
van Holten, J.W., Phys. Letters 135B, 427 (1983).
Amit, D.J. and Davis, A.C., Nucl. Phys. B225 [FS9], 221 (1983).
Gomes, M. and Ha, Y.K., Phys. Letters 145B, 235 (1985).
Davis, A.C., Freeman, M.D. and Macfarlane A.J., Nucl. Phys. B258, 393 (1985).
Ha, Y.K., Nucl. Phys. B256, 687 (1985).
Morozumi, T. and Nojiri, S., Prog. Theor. Phys. 75, 677 (1986).

SOME RELATIONS OF PARAMETERS IN QUANTUM FIELD THEORY

K. Ishikawa

Department of Physics, Hokkaido University,
Sapporo, 060 Japan

ABSTRACT

Two schemes of parameter relations, linear relation and non-linear relation are discussed. The linear relation of coupling constants is derived directly from an underlying symmetry of the classical theory and is preserved usually in the quantum theory. The non-linear relation is not derived by a same manner but is derived by more involved way which is intrinsically connected with quantum theory. An underlying symmetry which leads the linear relation is shown to be essential in the non-linear relation too. Some extension is also discussed.

I. INTRODUCTION

A theory is tested by comparing theoretical calculations with observed values. However in most cases there is a difficulty in a comparison. An exact calculation is hardly achieved and some approximation is involved in most cases in theoretical side. In experimental side, experimental error in observed values would always exist. Some quantities are even unobservable. Thus difficulties in the test of the theory exist in many occasions.

A rigorous relation among physical quantities, if exist, helps a lot in the comparison. These relations might reflect characteristic features of the theory. It may be useful to find and study these relations from both the theoretical side and from the experimental side. A purpose of the present essay is to survey some known these relations. Although there is a few original result, the author hopes that this may help toward better understanding and new extension. He dedicates this essay to Professor Yoshio Yamaguchi, who is one of the inventors of applying higher internal symmetry to high energy physics, in the occasion of his sixtieth birthday.

Linear relations among coupling constants may be the simplest example. An invariance of a theory under some internal symmetry group enforce the coupling constants to be connected together by linear relations. If these coupling constants are observable, this may be one of the best place to look in testing the theory. Many internal symmetries of high energy physics have been found in this manner.

Apart from the linear relations, several non-linear relations are also known. We mean with this non-linear relation a simple relation among independent quantities such as

$$A = BC \tag{1}$$

but not the one such as

$$A = BC(1 + d_1C + d_2C^2 + d_3C^3 + \cdots\cdots) \tag{2}$$

The Eq.(1) is highly non-trivial but the Eq.(2) is not so since we can regard that the A is calculated by a perturbative expansion with a parameter C. Apparently such kind of non-linear relation as Eq.(1) is not derived from a simple symmetry argument although the underlying symmetry may reflect this result. Since this kind of non-linear relation is highly non-trivial, it is quite interesting and challenging for theorists to find and give a proof for it.

If a relation such as Eq.(1) and its stability against small perturbation is found in a new theory, we may be able to test the theory by this. Since no approximation is involved, this process can be a good way of probing the theory.

II. LINEAR RELATIONS OF COUPLING CONSTANTS

(1) Electric charges.

Probably the most well known relation are the equivalence of electric charge of electron, of proton, and of singly charged ion:

$$|e_{e\ell}| = |e_{pr.}| = |e_{ion}| \tag{3}$$

These values may have been different due to the differences in the interactions of three different objects. Field theoretical proof of Eq.(3) was given based on Ward-Takahashi identity[2], which is derived from U(1) gauge invariance of the theory.

In this example we can see another interesting thing. Namely the ion which is a composite object of the electron and the nucleus has the same unit of charge as the constituent itself. Usually the dynamics of the composite state is described by a new theory which is derived from but is independent from that of the constituent. It is sufficient to know the theory which describes the dynamics of the composite particle. In the present case, the interaction between ions is more complicated than that of electrons because ion has spatial structure. The total charge of ion is, however, integer multiple of the one of the electrons. This is trivial since electric charge is additive quantity. Stability of the charge against an additional interaction is shown by Ward-Takahashi identity derived from

$$\begin{gathered} \partial^{\mu} J_{\mu} = 0 \\ [J_0(x),\ \Psi(y)]\delta(x_0 - y_0) = -e\Psi(y) + \text{higher derivatives.} \end{gathered} \tag{4}$$

The second term in the right hand side of the above equation vanishes for the electron case. Regardless of this term, vertex part is related to propagator by

$$\Lambda_\mu(p,p) = e \frac{\partial}{\partial p_\mu} s^{-1}(p) \tag{5}$$

This shows an universality of the charge.

(2) Hadron internal symmetry. (flavor symmetry)

Meson-baryon coupling constants are derived from the following interaction Lagrangian which has a parameter α:

$$\begin{aligned}\frac{1}{g}\mathcal{L}_1 = {} & \bar{N}\vec{\tau}N\vec{\pi} + (1-2\alpha)\bar{\Xi}\vec{\tau}\Xi\vec{\pi} + \frac{3-4\alpha}{\sqrt{3}}\bar{N}Ny^0 \\ & - \frac{3-2\alpha}{\sqrt{3}}\bar{\Xi}\Xi y - (1-2\alpha)\bar{N}\vec{\tau}K\cdot\vec{\Sigma} - \bar{\Xi}\tau\cdot\bar{K}^c\cdot\vec{\Sigma} \\ & - (1-2\alpha)\vec{\Sigma}\cdot\vec{K}^{+}\vec{\tau}N - \bar{\vec{\Sigma}}\cdot K^{c+}\tau^{+}\Xi - \frac{3-2\alpha}{\sqrt{3}}\bar{N}\Lambda K \\ & - \frac{3-2\alpha}{\sqrt{3}}K^{+}\bar{\Lambda}N + \frac{3-4\alpha}{\sqrt{3}}\bar{\Xi}\Lambda K^c + \frac{3-4\alpha}{\sqrt{3}}K^{c+}\bar{\Lambda}\Xi \\ & + i(1-\alpha)\bar{\vec{\Sigma}} \times \vec{\Sigma}\cdot\vec{\pi} + \frac{2\alpha}{\sqrt{3}}\bar{\vec{\Sigma}}\Lambda\cdot\vec{\pi} + \frac{2\alpha}{\sqrt{3}}\cdot\bar{\Lambda}\vec{\Sigma}\cdot\vec{\pi} \\ & + \frac{2\alpha}{\sqrt{3}}\bar{\vec{\Sigma}}\cdot\vec{\Sigma}y_0 - \frac{2\alpha}{\sqrt{3}}\bar{\Lambda}\Lambda y_0 \quad .\end{aligned} \tag{6}$$

The coupling constants follow linear relation when α is eliminated. These relations are those of cores of flavor SU(3) symmetry, which was invented partly by Prof. Y. Yamaguchi.[1]

The current understanding of the flavor symmetry is based upon the existence of three light quarks with almost degenerate masses and the flavor invariant theory which describes their dynamics, quantum chromodynamics (QCD).

(3) Weak interaction universality.

The β-decay strength of μ-on and of neutron are almost the same despite the fact that the neutron is affected by strong interaction. This is understand from CVC and PCAC hypothesis.

Some examples of relations on weak interaction's parameters are

$$\Gamma(\tau \to \mu\nu_\mu\nu_\tau)/\Gamma(\tau \to e\nu_e\nu_\tau) = 1.03\pm0.05$$

$$\Gamma(\tau \to e\nu_e\nu_\tau)/\Gamma(\mu \to e\nu_e\nu_\mu)(\frac{m\tau}{m\mu})^5 = 0.99\pm0.07 \tag{7}$$

$$\frac{M_W^2}{M_Z^2\cos^2\theta w} = \underset{(UA1)}{1.018\pm0.041\pm0.021}, \quad \underset{(UA2)}{0.996\pm0.033\pm0.009} \ .$$

The first two relations mean lepton's universality. Namely the charged current strength of the electron, of the muon, and of the τ-lepton agree. This is understood from SU(2)×U(1) structure of the weak interaction easily.

The last relation in Eq.(7) is explained from weak $\Delta I=1/2$ rule, which means that the symmetry breaking occurs through $I=1/2$ Higgs scalar.

III. NON-LINEAR RELATIONS AMONG PHYSICAL PARAMETERS

The relations of the coupling constants discussed in the previous section are essentially linear, although some non-linearity appears, as can be seen in an example of Eq.(7), due to the mixing of particles. The linear relations among coupling constants are understandable easily, since ordinary internal symmetry is linear with respect to field operators.

Non-linear relations among parameters are a little more complicated. One way to obtain a non-linear relation is an extension of the previous method to a non-linear realization of symmetry, such as a non-linear σ-model. Whether the non-linear relation in the classical theory holds also in the quantum theory is not clear sometimes. We do not discuss this kind of nonlinear realization of the symmetry but discuss another way of obtaining non-linear relation which is purely quantum mechanical one.

Some well known observed examples are the following:

(1) A magnetic flux in superconductor is quantized as the integer multiple of unit flux Φ_0[3], which is given by

$$\Phi_0 = \frac{ch}{e} \quad \text{(flux quantization).} \tag{8}$$

(2) A frequency of an alternating current of Josephson junction with potential difference V[4] is given by

$$\nu = \frac{en}{h} \quad \text{(ac Josephson effect).} \tag{9}$$

(3) A Hall conductance of the two dimensional metal[5] is given by

$$\sigma_{xy} = \frac{e^2}{h} \times \text{integer} \quad \text{(integer quantum Hall effect).} \tag{10}$$

at the plateau region.

All three relations have electric charge e and Planck constant h and they are phenomena of Quantum Electrodynamics (QED). Apparently they are not understandable from such a simple symmetry argument that was used in the previous section.

First two relations are those which occur in superconductor, where a role of the phase of the order parameter is essential. The third one has no connection to the phase of the order parameter, since this phenomenon is observed in semi-conductor. Instead of order parameter, dimension of space and added magnetic field play important role. A topology shall be shown to be an important factor.

Since it is possible to measure the frequency and the conductance precisely, Josephson effect and quantum Hall effect are used in determing the fine structure constant α[6], if the relations Eq.(8) and Eq.(9) are exact. The precise value of α is needed in testing QED.

The validity of QED has been known up to some level but it may be broken at a next level. If this breakdown will really occur, we might be able to see some signal of the electrons's structure.

The exactness of the equations Eqs.(8),(9), and (10) must be examined carefully.

A naive application of underlying symmetry, U(1) gauge invariance in the present case, which is shown to lead a linear relation to the coupling constants, should play a critical role also in the present case. In fact in a general proof of the quantum Hall effect by the present author and his collaborator[7] Ward-Takahashi identity which is derived from current conservation and equal time commutation relation are the essential ingredients. From their work the Hall conductance is written as

$$\sigma_{xy} = \frac{e^2}{h}\frac{1}{24\pi^2}\int d^3p\, \varepsilon^{\mu\nu\rho}\, \mathrm{Tr}\left(\frac{\partial s(p)^{-}}{\partial p_\mu}s(p)\frac{\partial s(p)^{-}}{\partial p_\nu}s(p)\frac{\partial s(p)^{-}}{\partial p_\rho}s(p)\right) \tag{11}$$

, which is connected with topology of the mapping from the momentum space to the matrix space defined by the propagator s(p). The value becomes to the integer multiple of e^2/h if the propagator is analytic and single-valued in the gap region and in the localized state region.

The exactness of the Eq.(9) has been argued in the literature.[8] It seems, however, a more general proof may be desired.

The proof of the Eq.(8) was given by Byers and Yang[9] right after the discovery of the phenomenon. Their proof is based on the single-valuedness of the wave function. The Schrödinger equation becomes independent from the external vector potential when the suitable gauge transformation is made. The information of the vector potential is hidden in the boundary condition of the transformed wave function. It get a phase factor

$$e^{ie/c\hbar\Phi} \tag{12}$$

when a particle position is brought around a hole once, provided that the original wave function is single-valued. Hence the total energy is a periodic function of magnetic flux Φ with a period ch/e. At the maximum of free energy, the flux becomes the integer multiple of ch/e. In this proof, the single-valuedness of the wave function is essential.

The important role played by phase can be seen in their proof. A similar argument[10] has been given for a proof of Eq.(9) too. A role of topology is thus essential in Eqs.(8), (9), and (10).

IV. SUMMARY AND OUTLOOK

The known observed non-linear relations are related with QED, U(1) gauge theory. As was described in Sect.II, the unified electroweak theory uses non-Abelian gauge theory. Perhaps the non-Abelian nature would exist also in the condensed matter physics too. It would be interesting to find and to check non-linear relations in this kind of non-Abelian theory. As an example of this kind, let us describe briefly what would happen if weakly interacting fermion are put on a surface without asking its experimental observability.

Although the symmetry group of electroweak theory is SU(2)×U(1), only electrons can be confined into a surface. Hence we study here only diagonal elements.

Effective action in the lowest order of the external electroweak neutral fields is obtained as

$$
\begin{aligned}
s = {} & \frac{c_1}{2} \int d^3x\, \varepsilon^{\mu\nu\rho} A_\mu (\partial_\nu A_\rho) \\
& + c_2 \int d^3x\, \varepsilon^{\mu\nu\rho} A_\mu (\partial_\nu Z_\rho) \\
& + \frac{c_3}{2} \int d^3x\, \varepsilon^{\mu\nu\rho} Z_\mu (\partial_\nu Z_\rho)
\end{aligned}
\tag{13}
$$

The first term corresponds to the usual Hall effect described in the previous section. The coefficient c_1 is given by Eq.(11). The other two coefficients c_2 and c_3 are the function of Weinberg angle and are the same order of magnitude as c_1.

The second term in Eq.(13) leads the induced neutral current under the electric field. A feature of the electroweak theory is seen in this term. It might be interesting to find similar phenomena in different area, such as condenced matter physics.

ACKNOWLEDEMENT

The present work is partially supported by the Grands-in-Aid, Science and Culture, Japan. It is my great pleasure to dedicate this essay to Professor Yoshio Yamaguchi in the occation of his sixtieth birthday.

REFERENCE

1. Y. Yamaguchi, Prog. Theor. Phys. Suppl. No.11 (1959) 1 and 37.
2. J.C. Ward, Phys. Rev., 78, 1824 (1950); Y. Takahashi, Nuovo Cimento, 6, 370 (1957).
3. B.S. Deaver, Jr., and W.M. Fairbank, Phys. Rev. Letters 7, 43 (1961).
4. P.W. Anderson and J.M. Rowell, Phys. Rev. Letters 10 (1963) 230; W.H. Parker, B.N. Taylor, and D.N. Lagenberg, Phys. Rev. Letters 18 (1967), 287; J. Clarke; Phys. Rev. Letters. 21, (1968) 1566.
5. K.V. Klitzing, G. Dorda, and M. Pepper, Phys. Rev. Lett. 45 (1980) 494.
6. T. Kinoshita, in Proc. Inst. Symp. Foundations of Quantum Mechanics, Tokyo, edited by S. Kamefuchi et al., Phys. Soc. of Japan (1983) 357.
7. K. Ishikawa and T. Matsuyama, Hokkaido University preprints, EPHOU85 AUG012, SEP014, EPHOU86 APR007.
8. B.D. Josephson, Phys. Lett. 1 (1962) 251.
9. N. Byers and C.N. Yang, Phys. Rev. Letters, 7 (1961) 46.
10. F. Bloch, Phys. Rev. Letters, 21 (1968) 1241.

PATH-INTEGRAL FORMULATION OF ANOMALIES AND ZETA-FUNCTION REGULARIZATION

Ken-ichi Shizuya

Department of Physics, Tohoku University
Sendai 980, Japan

ABSTRACT

It is pointed out that the path-integral derivation of anomalies has natural foundation in zeta-function regularization of quantum theories. With this observation, the generality and limitations of the path-integral method are elucidated. The analyses of conformal and superconformal anomalies are reviewd to illustrate the general discussion.

Fujikawa's path-integral method provides a useful means of fomulating anomalies in field theory. It characterizes known anomalies (chiral[1]), dilatational[2]) and gravitational[3]) ones) as Jacobian factors in a universal manner.

Recently the path-integral framework has been extended to include conformal anomalies[4]) and superconformal (supercurrent) anomalies[5]). Until then this method was successfully applied only to theories with quadratic (bilinear) actions. It was known neither how to handle systems of nonlinear actions, nor how to treat higher-order anomalies. It has become clear from the analysis of these anomalies how to deal with these problems inherent to the path-integral method. The key observation is that the algorithm of the path-integral approach has natural foundation within the zeta-function regularization scheme: The path-integral method is thus a well-founded regularized procedure, which can be generelized to nonlinear systems as well as to higher orders in a systematic way.

The purpose of this paper is to review these recent studies and to elucidate the generality as well as limitations of the path-integral formulation of anomalies.

We shall mainly analyze a bosonic system for a generel discussion and survey some specific examples, conformal anomalies and superconformal anomalies, later.

Consider a quantum system described by the action $S[\Phi] = \int dx\, L(\Phi, \partial\Phi)$, which possesses some approximate (or, softly-broken) symmetry. Then, under the field transformation $\Phi \rightarrow \Phi' = \Phi + \delta^f\Phi$ with f parametrizing the infinitesimal symmetry operation, the Lagrangian undergoes the change

$$\delta L(\Phi, \partial\Phi) = \delta\Phi \frac{\partial L}{\partial\Phi} + (\partial_\mu \delta\Phi) \frac{\partial L}{\partial(\partial_\mu \Phi)} = \partial_\mu \Omega^\mu + \Delta . \tag{1}$$

Here the last line stands for the explicit variation of the Lagrangian and Δ denotes the effect of symmetry breaking. With the Noether current defind as

$$j^\mu = \delta^f\Phi \frac{\partial L}{\partial(\partial_\mu \Phi)} - \Omega^\mu \quad , \tag{2}$$

eq.(2) leads to the local current conservation law

$$\partial_\mu j^\mu = \Delta(x) - \delta^f\Phi(x)\ \delta S/\delta\Phi(x) \quad , \tag{3}$$

where $\delta S/\delta\Phi = \partial_\mu\{\partial L/\partial(\partial_\mu\Phi)\} - \partial L/\partial\Phi$.

Equation (3), being an operator equation, must be valid even when it is inserted in Green's functions. The source $J(x)$ for $\Phi(x)$ is easily introduced by the substitution

$$S \rightarrow S^J = S + \int dx\, J\, \Phi\ . \tag{4}$$

Then eq.(3) is equivalent to the Ward-Takahashi (WT) identity of the Green's function form:

$$\langle\ \partial_\mu j^\mu - \Delta + \delta^f\Phi(\delta S^J/\delta\Phi - J)\ \rangle = 0\ . \tag{5}$$

Naive use of the equation of motion $\langle\ \delta S^J/\delta\Phi\ \rangle = 0$ may suggests the vanishing of the corresponding term in eq.(5). Whether the Green's function of the composite operator $\delta^f\Phi(\delta S^J/\delta\Phi)$ vanishes or not, however, depends on how one regularizes its short-distance structure[1].

In the path-integral language, one starts with the generating functional of Green's functions

$$W[J] = \int[d\Phi]\ \exp(iS[\Phi] + i\int dx\, J\, \Phi\ )\ . \tag{6}$$

Let us here promote the field variation $\delta\Phi$ to the local variation

$$\delta_\xi\Phi(x) = \xi(x)\ \delta^f\Phi(x)\quad , \tag{7}$$

where $\xi(x)$ is an arbitrary function. Then the formal change of field variables $\Phi \rightarrow \Phi_\xi = \Phi + \delta_\xi\Phi$ in the path integral gives rise to the WT identity

$$\langle\ -\ \delta_\xi S[\Phi] - \int dx\, J\ \delta_\xi\Phi + i\ \ln A_\xi\ \rangle = 0\quad , \tag{8}$$

where $A_\xi[\Phi]$ stands for the Jacobian $A_\xi[\Phi] = \partial(\Phi_\xi)/\partial(\Phi)$. Equations (5) and (8) are equivalent, with the identification

$$\langle\delta^f\Phi(\delta S^J/\delta\Phi)\rangle \quad\leftrightarrow\quad i(\delta/\delta\xi)\langle\ln A_\xi\rangle\ . \tag{9}$$

So far eqs.(5), (8) and (9) are of formal meaning. They make sense only when one specifies how to regularize them. Thus, eq.(9) should be understood as meaning that regularizing the Jacobian in the path-integral approach corresponds to regularizing the composite operator $\delta^f \Phi(\delta S^J/\delta\Phi)$. In other words, the regularization prescription underlying the path-integral algorithm should be such as respects this correspondence. Zeta-function regularization is one such prescription, as argued below.

Introduction of Pauli-Villars regulator fields or the use of dimensional regularization modifies eqs.(5), (8) and (9); these prescriptions are not compatible with the path-integral algorithm. Accordingly, we avoid regulator fields and work in 4 dimensions.

Anomalies characterize quantum modification of classical symmetries. It is therefore natural to evaluate them in powers of $\hbar$ or by the loop expansion. Correspondingly, we perform the stationary-phase evaluation of the path-integral expression for the anomaly $\langle\delta^f \Phi(\delta S^J/\delta\Phi)\rangle$: We set $\Phi(x) = \phi(x) + \chi(x)$ and determine the classical field $\phi(x)$ so that the quantum field $\chi(x)$ has no expectation value, i.e., $\langle\Phi(x)\rangle = \phi(x)\langle 1\rangle$.

In particular, the one-loop structure of the theory is characterized by the quadratic quantum action

$$S_2[\phi;\chi] = \int dx\, \frac{1}{2}\, \chi(x)\ \Gamma[\phi]\ \chi(x)\ , \tag{10}$$

where

$$\langle x|\Gamma[\phi]|y\rangle = \delta^2 S^J[\phi]/\delta\phi(x)\delta\phi(y). \tag{11}$$

In general, $\Gamma[\phi]$ is an elliptic (invertible) differential operator. One can set up a quantum theory for this action by expanding $\chi(x)$ in terms of the (orthonormal) eigenmodes $\{p_k(x);\ \Gamma\ p_k(x) = \lambda_k\ p_k(x)\}$ of the operator Γ. In the bosonic case under consideration, the eigenvalues λ_k are negative in Euclidean space. The propagator $\langle\chi(x)\chi(x)\rangle$ contains a short-distance singularity, which may be regularized as follows:

$$\langle \chi(x)\chi(y)\rangle^{reg} = i \sum_k (1/\lambda_k)\, e^{\tau\lambda_k}\, p_k(x)\, p_k(y) \qquad (\tau > 0) \qquad (12)$$

$$= i \langle x |\, \frac{1}{\Gamma}\, e^{\tau\Gamma} | y\rangle \quad . \qquad (13)$$

The knowledge of this two-point function, combined with Wick's theorem, defines uniquely all other Green's functions and therefore the one-loop quantum theory.

The propagator (12) has a well-known parametric representation

$$\langle \chi(x)\chi(y)\rangle^{reg} = - i \int_\tau^\infty ds\, \langle x |\, e^{s\Gamma} | y\rangle \qquad (14)$$

in terms of the heat kernel $\langle x | e^{s\Gamma} | y\rangle$, which plays a fundamental role in zeta-function regularization[6)].

It is possible to verify that the above regularization prescription is compatible with the algorithm of the path-integral approach. Let us turn to the Green's function $\langle \delta^f \Phi(\delta S^J/\delta\Phi)\rangle$. For simplicity, we shall assume from now on that $\delta^f \Phi$ is linear in Φ and denote $\delta^f \Phi(x) = \Phi\, F = \int dy\, \Phi(y)\, \langle y | F | x\rangle$, where F may contain differential operators. The generelization to nonlinear cases is straightforward. The linear variation δ can be taken to act on Φ and χ separately. Then, at the one-loop level,

$$\langle \delta^f \Phi(\delta S^J/\delta\Phi)\rangle \quad \rightarrow \quad \langle\, \chi\, \frac{1}{2}(\, F\, \Gamma[\Phi] + \Gamma[\Phi]\, F^t)\, \chi\, \rangle \quad , \qquad (15)$$

where the operator $F\Gamma[\Phi]$ has been symmetrized; $\langle x | F^t | y\rangle = \langle y | F | x\rangle$ and $\Gamma = \Gamma^t$. We replace the propagator $\langle \chi \cdots \chi\rangle$ by the regularized one in eq.(13) to obtain the regularized expression for the anomaly

$$\langle \delta^f \Phi(x)\delta S^J/\delta\Phi(x)\rangle^{reg} = i \langle x | F_s e^{\tau\Gamma} | x\rangle = i \langle x | e^{\tau\Gamma}\, F_s | x\rangle \; , \qquad (16)$$

where $F_s \equiv \frac{1}{2}(F + F^t)$. Obviously, F_s can simply be replaced by F or F^t in eq.(16).

On the other hand, the Jacobian A_ξ associated with the change of the field $X \to X + \delta_\xi X$ with $\delta_\xi X \equiv \delta X(x)\xi(x) = X F \xi$ is given by

$$\ln A_\xi = \mathrm{Tr}[\ F\ \xi\ ]\ . \tag{17}$$

The quadratic form of the action (10) suggests us to use the operator Γ to regularize this Jacobian so that

$$\ln A_\xi^{reg} = \mathrm{Tr}[\ F\ \xi\ e^{\tau\Gamma}] \qquad (\tau > 0)\ . \tag{18}$$

Therefore,

$$i(\delta/\delta\xi(x))\ \ln A_\xi^{reg} = i\ \langle x|\ e^{\tau\Gamma}\ F\ |x\rangle \qquad (\tau > 0)\ . \tag{19}$$

One thus establishes the correspondence

$$\langle \delta^f\Phi(\delta S^J/\delta\Phi)\rangle^{reg} \quad \leftrightarrow \quad i(\delta/\delta\xi)\ \ln A_\xi^{reg} \tag{20}$$

at the one-loop level. The prescription for the regularized Jacobian (18) corresponds to the specific regularization of the propagator in eq.(13). In this sense, zeta-function regularization may be said to underlie the path-integral algorithm. This shows that the path-integral derivation of anomalies, or their identification as Jacobians, is a well-founded regularized procedure for general theories with quadratic (or, bilinear) actions. For such theories the one-loop teatment is exact.

In this connection, let us recall that chiral anomalies[1)] in gauge theories derive solely from the matter sector which is described by matter-field bilinear actions. Their expressions, written in terms of bare gauge fields, are therefore expected to be exact in conformity with the Adler-Bardeen theorm[7)]. This argument would be a heuristic way of understanding this theorem within the path-integral approach, although a careful analysis of the regularization of the gauge-field sector will be required for a rigorous proof.

For general nonlinear systems, the Jacobian interpretation of anomalies is correct at the one-loop level. It is not clear whether the same interpretation holds at higher orders, since the path-integral algorithm by itself does not specify how to perform higher-loop regularization. It will, however, be clear from our approach how to proceed to higher loops: Calculate $\langle \delta^f \Phi(\delta S^J/\delta\Phi) \rangle$ loop-wise by use of the regularized propagagator in eq.(13) and Wick's theorem, i.e., by means of zete-function regularization.

So far we have considered the bosonic case. Fermion theories can be treated in almost the same fashion: Let us denote the symmetry field variations as $\delta^f \bar{\psi} = \bar{\psi} F$ and $\delta^f \psi = G \psi$, and the bilinear action as

$$S_2 = \int dx \; \bar{\psi} \, \Gamma \, \psi \quad . \tag{21}$$

The anomaly $\langle \delta^f \Phi(\delta S^J/\delta\Phi) \rangle$ takes the form

$$\Xi = \langle \; \delta^f \bar{\psi}(\delta S^J/\delta\bar{\psi}) \; - \; (\delta S^J/\delta\psi)\delta^f \psi \; \rangle \quad , \tag{22}$$

which is rewritten as

$$\Xi = \langle \; \bar{\psi}(F\Gamma + \Gamma G)\psi \; \rangle \quad . \tag{23}$$

In accord with eq.(13), we regularize the propagator $\langle \psi\bar{\psi} \rangle$ as follows

$$\langle \; \psi(x) \; \bar{\psi}(y) \; \rangle^{reg} = \langle x | \; \frac{i}{\Gamma} \, e^{\tau\Gamma^2} \; | y \rangle \qquad (\tau > 0) \; . \tag{24}$$

This leads to the regularized anomaly

$$\Xi^{reg}(x) = -i \; \mathrm{tr} \; \langle x | (G + F^t) \; e^{\tau\Gamma^2} \; | x \rangle \quad , \tag{25}$$

where the minus sign comes from Fermi statistics. It is clear that this expression can be regarded as a regularized version of the naive Jacobian associated with the transformation $\psi \to \psi + \xi \, \delta^f \psi$ and

$\bar{\psi} \to \bar{\psi} + \xi\, \delta^f \bar{\psi}$. Our approach, in particular, clarifies that the Jacobian should be arranged as $-\mathrm{tr}(G + F^t)$ before regularization; the difference between F and F^t matters when F contains derivatives ∂_μ.

To substantiate the general argument given above, we shall finally survey some examples.

(I) Conformal anomaly in Φ^4 theory.

The conformal variation of a scalar field is written as

$$\delta^f \Phi = f^\mu \partial_\mu \Phi + \frac{1}{4}\,(\partial^\lambda f_\lambda)\Phi\ , \tag{26}$$

where $f^\mu(x)$ is the conformal Killing vector obeying $\partial_\mu f_\nu + \partial_\nu f_\mu = \frac{1}{2}(\partial^\lambda f_\lambda)\, g_{\mu\nu}$. Varying the Lagrangian

$$L[\Phi] = \frac{1}{2}\,(\partial_\mu \Phi)^2 - \frac{1}{2}\, m^2\, \Phi^2 - \frac{1}{4}\,\lambda\, \Phi^4 \tag{27}$$

gives rise to the conformal current

$$j^\mu = f_\nu\, \theta^{\mu\nu} \tag{28}$$

and the conformal breaking term

$$\Delta = \frac{1}{4}\,(\partial^\lambda f_\lambda)\, m^2 \Phi^2\ , \tag{29}$$

where $\theta^{\mu\nu}$ is the improved energy-momentum tensor

$$\theta^{\mu\nu} = (\partial^\mu \Phi)(\partial^\nu \Phi) - g^{\mu\nu}\, L[\Phi] + \frac{1}{6}(g^{\mu\nu}\partial^2 - \partial^\mu \partial^\nu)\, \Phi^2\ . \tag{30}$$

The current in eq.(28) is the improved current, which is obtained by discarding the contribution from an antisymmetric tensor $\Sigma^{\mu\nu}$ of the form $\partial_\nu \Sigma^{\mu\nu}$. In view of eq.(10), the one-loop quantum theory is characterized by the differential operator

$$\Gamma[\Phi] = -\,(\ \partial^2 + m^2 + 3\lambda\Phi^2\ ) \tag{31}$$

and the anomaly is calculated by use of eq.(18). It turns out that the resulting anomaly spoiles the local conservation law of the Poincare current $f^P_\nu \theta^{\mu\nu}$ with f^P_ν the Poincare Killing vector; global Poincare symmetry, however, is maintained[2]. As a matter of fact, it is possible to redefine the conformal current so that the (local) Poincare symmetry (or, energy-momentum conservation) is exact. With the new symmetric energy-momentum tensor

$$\tilde{T}^{\mu\nu} = \theta^{\mu\nu} + \frac{1}{2} g^{\mu\nu}\Phi \; \delta S/\delta\Phi \tag{32}$$

which satisfies $\partial_\mu \tilde{T}^{\mu\nu} = 0$, the new conformal current is given by

$$\hat{j}^\mu = f^P_\nu \, \tilde{T}^{\mu\nu} \quad . \tag{33}$$

The conformal anomaly associated with this current agrees with the usual trace anomaly.

Here we observe the phenomenon of anomaly trading. In general, one can redefine the energy-momentum tensor so that either its divergence or its trace has no anomaly. It is thus a matter of choice whether one preserves the Poincare symmetry or the dilatational one, although, on physics grounds, the former is usually taken to be exact. The path-integral method makes manifest the tradng of the anomaly between these symmetries.

Depending on how one regularizes the theory, the WT identities change their appearance. Their anomaly contents should, however, be the same. The matching of different regularization schemes can be made by trading the anomaly through the redefinition of currents in the WT identities.

(II) Superconformal anomalies.

Superconformal symmetry[8,9] plays an important role in characterizing the space-time symmetry contents and short-distance structure of supersymmetric theories. All the currents assiciated with this symmetry are embedded in a single superfield, called the supercurrent. The original derivation of the supercurrent, by Ferrara and Zumino[9], made use of Noether's variational procedure for component fields. The direct superspace version of the variational procedure for supercurrents has recently been constructed[5].

For definiteness, let us consider the Wess-Zumino model[8], describing the self-coupled chiral superfield $\Phi(z)$ and its conjugate $\bar{\Phi}(z)$, with the superspace action

$$S[\Phi,\bar{\Phi}] = \int d^8z\ \bar{\Phi}\Phi + \left\{ \int d^6z\ (\frac{1}{2}\, m\Phi^2 - \frac{1}{6}\, g\Phi^3) + \text{h.c.} \right\} , \tag{34}$$

where $d^8z \equiv d^4x\ d^2\theta\ d^2\bar{\theta}$, $d^6z \equiv d^4x\ d^2\theta$, etc.

Superconformal transformation of $\Phi(z)$ and $\bar{\Phi}(z)$ are written as

$$\delta\Phi(z) = -\frac{1}{4}\,\bar{D}^2[\Omega^\alpha D_\alpha + \frac{1}{3}\,(D^\alpha\Omega_\alpha)]\ \Phi(z) \quad ,$$

$$\delta\bar{\Phi}(z) = -\frac{1}{4}\,D^2[\bar{\Omega}_{\dot{\alpha}}\bar{D}^{\dot{\alpha}} + \frac{1}{3}\ \ \bar{D}_{\dot{\alpha}}\bar{\Omega}^{\dot{\alpha}}\ ]\ \bar{\Phi}(z) \quad , \tag{35}$$

where $\Omega(z)$ is an arbitrary spinor superfield and $\bar{\Omega}(z)$ its conjugate. They represent <u>local</u> superconformal transformation laws, appropriate for the Noether construction of the supercurrent. With Ω and $\bar{\Omega}$ properly constrained, they are reduced to the global transformation laws. The variation of the action gives rise to the supercurrent

$$R_{\alpha\dot{\alpha}} = (D_\alpha\Phi)(\bar{D}_{\dot{\alpha}}\bar{\Phi}) - \frac{1}{3}\,[D_\alpha\ ,\ \bar{D}_{\dot{\alpha}}](\bar{\Phi}\Phi) \tag{36}$$

and its conservation laws

$$\bar{D}^{\dot{\alpha}}R_{\alpha\dot{\alpha}} - \frac{1}{3}m\ D_\alpha\Phi^2 = -2\delta S/\delta\Omega^\alpha, \quad D^\alpha R_{\alpha\dot{\alpha}} - \frac{1}{3}m\ \bar{D}_{\dot{\alpha}}\bar{\Phi}^2 = 2\delta S/\delta\bar{\Omega}^{\dot{\alpha}}\ . \tag{37}$$

where $\delta S/\delta\Omega^\alpha = \delta/\delta\Omega^\alpha \int d^6z\ (\delta\Phi)\delta S/\delta\Phi$, etc.

The anomalies $\delta S/\delta\Omega$ and $\delta S/\delta\bar{\Omega}$ can be calculated by the path-integral algorithm. Here again we encounter the phenomenon of anomaly trading: The resulting anomalies make the conservation laws for $\bar{D}^{\dot{\alpha}}R_{\alpha\dot{\alpha}}$ $(D^\alpha R_{\alpha\dot{\alpha}})$ and $\partial^\mu R_\mu$ anomalous, where $R_\mu = \frac{1}{2}(\bar{\sigma}_\mu)^{\dot{\alpha}\alpha}R_{\alpha\dot{\alpha}}$. The latter $\partial^\mu R_\mu$, which contains energy-momentum conservation, can be made normal by a suitable redefinition of the supercurrent $R_{\alpha\dot{\alpha}} \to \hat{R}_{\alpha\dot{\alpha}}$. The new supercurrent then obeys the (anomalous) conservation law

$$\bar{D}^{\dot{\alpha}}\hat{R}_{\alpha\dot{\alpha}} = \frac{1}{3}\, m\, D_\alpha \Phi^2 - (g^2/192\pi^2)\, D_\alpha \bar{D}^2(\bar{\Phi}\Phi)\ ; \tag{38}$$

likewise, $D^\alpha \hat{R}_{\alpha\dot{\alpha}}$ has the anomaly $-(g^2/192\pi^2)\, \bar{D}_{\dot{\alpha}} D^2(\bar{\Phi}\Phi)$.

The present analysis developed for the simple models has also been applied to gauge theories (QCD and SQCD).

REFERENCES

1) Fujikawa, K., Phys. Rev. Lett. 42. 1195 (1979); Phys. Rev. D21, 2848 (1980).
2) Fujikawa, K., Phys. Rev. D23, 2262 (1981).
3) Alvarez-Gaume, L. and Witten, E., Nucl. Phys. B234, 269 (1984).
4) Shizuya, K. and Tsukahara, H., TU/86/296, Zeit. Phys. C to appear.
5) Shizuya, K., Tohoku Univ. preprint TU/86/299, June 1986.
6) Hawking, S., Comm. Math. Phys. 55, 133 (1977).
7) Adler, S.L. and Bardeen, W.A., Phys. Rev. 182, 1517 (1969).
8) Wess, J. and Zumino, B., Nucl. Phys. B70, 39 (1974).
9) Ferrara, S. and Zumino, B., Nucl. Phys. B87, 207 (1975).

AN INFINITE SET OF QUANTUM COMMUTING OPERATORS RELATED WITH SOLITON EQUATIONS AND THE VIRASORO ALGEBRA

Ryu SASAKI and Itaru YAMANAKA
Research Institute for Theoretical Physics
Hiroshima University, Takehara, Hiroshima 725, Japan

It is more than ten years since Professor Yoshio Yamaguchi gave a series of lectures on non-linear quantum field theories at the Univ. of Tokyo. As a graduate student at that time, one of the authors (R.S.) was very impressed by Prof. Yamaguchi's vigorous way of thinking and discussion and by his bold approach to the essential subject of particle physics. It is our great pleasure to dedicate the most recent results of our research in non-linear field theories to Professor Yoshio Yamaguchi as an essay in honor of his sixtieth birthday.

§1. Introduction

In this essay we report on one particular aspect of non-linear field theories, namely an infinite set of conserved quantities of a class of soliton equations and its quantum versions. Soliton theories in 1 + 1 dimensions are probably the best understood non-linear field theories and they provide an ideal theoretical laboratory for many new concepts in non-linear physics. One of the most characteristic features of soliton theories is the existence of an infinite set of conserved quantities[1)] and its involution properties under certain canonical structure or the Poisson bracket. Then a natural question arises whether such an infinite dimensional symmetry survives quantization. Namely, do we get an infinite set of quantum commuting operators which reduce to the classical one in the limit of $\hbar \to 0$, or do we not due to some anomalies caused by the high non-linearity of the interaction? Usually this problem is investigated system by system in terms of quantum inverse method[2)] or by transfer matrix approach based on a lattice.[3)]

We will answer this question in the affirmative for a class of soliton equations represented by the MKdV equation, eq. (2.1), which is also related with the sine(h)-Gordon equation. We also give a method of constructing the infinite set of commuting operators explicitly. Since the Poisson bracket characteristic for this class of soliton equations correspond exactly to the commutation relation of the two dimensional boson fields, naive quantization in the continuum theory is straightforward. As for the calculation of commutators among various operators, the techniques of the string theories and the conformal field theories are quite useful. We believe this explicit example will provide another interesting path for understanding the nature of quantum infinite dimensional symmetries.

This essay is organized as follows. In §2 we review briefly the classical theory of the infinite set of conserved quantities for the MKdV eq. In §3 the quantization procedure and the calculation techniques for commutators are explained. In §4 we show the method to determine the infinite set of quantum commuting operators. The final section is for summary and comments.

§2. Classical Theory of an Infinite Set of Conserved Quantities associated with the MKdV Eq.

In order to introduce notation and to be self-contained, we summarize some of the known results of the infinite set of conserved quantities associated with the Modified Korteweg de Vries equation (the MKdV eq)

$$u_t = u_{\sigma\sigma\sigma} - 6u^2 u_\sigma \ . \tag{2.1}$$

Here $u = u(t,\sigma)$ is a real (hermitian) field in 1 + 1 dimensions depending on time (t) and space (σ) variables and $u_t = \partial_t u$, $u_\sigma = \partial_\sigma u$,

$u_{\sigma\sigma} = \partial_\sigma^2 u$, ..., etc. Throughout this paper we impose the periodic boundary condition with a period 2π

$$u(t,\sigma+2\pi) = u(t,\sigma) \ , \quad 0 \le \sigma \le 2\pi \ . \tag{2.2}$$

The reason for this will become clear in later sections. It should be remarked, however, that the explicit forms of conserved densities to be discussed below are the same for the periodic boundary condition case and for the ordinary boundary condition case ($u(\sigma) \to 0$, as $|\sigma| \to \infty$, $-\infty < \sigma < \infty$).

An infinite set of conserved quantities[1),4)] is given as follows

$$I_n = -\frac{1}{2\pi}\int_0^{2\pi} d\sigma \ \frac{1}{2}u(\sigma)Y_{2n-1}(u) \ , \ n = 1, 2, 3, \ldots, \tag{2.3}$$

in which $Y_n(u)$ is a polynomial in u and its space derivatives u_σ, $u_{\sigma\sigma}$, ..., defined recursively

$$Y_{n+1} = \partial_\sigma Y_n + u\sum_{k=1}^{n-1} Y_k Y_{n-k} \ , \quad Y_1 = -u \ . \tag{2.4}$$

It should be remarked that in eqs. (2.3), (2.4) and hereafter the time t is always fixed and disregarded. It is easy to show that uY_{2n} is a total σ-derivative giving rise to a trivial conserved quantity. We define the order and the weight of a monomial in u, u_σ, $u_{\sigma\sigma}$,...,

$$M = u^{n_o} u_\sigma^{n_1} \ldots u_{m\sigma}^{n_m} \ ,$$

as follows

$$\text{order}(M) = \sum n_k \ ,$$

$$\text{weight}(M) = \sum (k+1) n_k \ . \tag{2.5}$$

Then we find that the polynomial uY_n has a definite weight

$$\text{weight}(uY_n) = n + 1 \,, \tag{2.6}$$

but its component monomials have different orders, all even,

$$2 \leq \text{order}(uY_{2n-1}),\quad \text{order}(uY_{2n}) \leq 2n \,. \tag{2.7}$$

A remarkable and highly nontrivial result of the soliton theory is that the infinite set of conserved quantities, I_n, $n = 1,2,3,\ldots,$ are in involution to each other with respect to the following Poisson bracket[5),6),7)]

$$\{u(\sigma),\ u(\sigma')\} = 2\pi\partial_\sigma\delta(\sigma-\sigma') \,, \tag{2.8}$$

$$\{I_n,\ I_m\} = 0 \,,\quad n,\ m = 1,2,\ldots, \,. \tag{2.9}$$

By adopting I_2 as a Hamiltonian, we get the MKdV eq. (2.1) as a canonical equation

$$\partial_t u(\sigma) = \{u(\sigma),\ I_2\} \,. \tag{2.10}$$

From the above involution property we find that the set $\{I_n\}$ is conserved for a large class of evolution equations

$$\partial_t u(\sigma) = \{u(\sigma), \mathcal{H}\} \,, \tag{2.11}$$

in which the Hamiltonian $\mathcal{H}$ is given by

$$\mathcal{H} = \sum_n h_n(t) I_n \,, \tag{2.12}$$

with arbitrary time dependent functions $h_n(t)$, $n = 1,2,\ldots,$. In fact we have

$$\partial_t I_m = \{I_m, \mathcal{H}\} = \sum_n h_n(t)\{I_m,\ I_n\} = 0 \,. \tag{2.13}$$

Thus we see that the infinite set of conserved quantities together with the Poisson bracket characterize the class of soliton equations.

§3. Quantization

In this section we discuss the quantization of the class of non-linear Hamiltonian dynamics introduced in the preceding section. Here we follow the naive canonical quantization procedure in continuum theory rather than applying the quantum inverse method or the transfer matrix approach. The canonical quantization is achieved by replacing the canonical Poisson bracket by the commutation relation

$$\{q_r, p_s\} = \delta_{r,s} \rightarrow [q_r, p_s] = i\hbar\delta_{r,s} .$$

Since we are interested in the relation between the classical and the quantum soliton theories all the $\hbar$ dependence are displayed explicitly. In the present case the fundamental commutation relation derived from the Poisson bracket (2.8) reads

$$[u(\sigma), u(\sigma')] = i\hbar 2\pi\partial_\sigma \delta(\sigma - \sigma') . \qquad (3.1)$$

As is well known in two dimensional field theories as well as in string theories, the above commutation relation is realized by (the derivative of) a hermitian boson field;

$$u(\sigma) = \sum_{n=-\infty}^{\infty} \alpha_n e^{-in\sigma} , \quad 0 \leq \sigma \leq 2\pi , \qquad (3.2)$$

$$[\alpha_m, \alpha_n] = \hbar m\delta_{m+n,o} , \quad (\alpha_m)^\dagger = \alpha_{-m} . \qquad (3.3)$$

We interpret α_m, $m > 0$ ($m < 0$) as creation (annihilation) operators and α_o as a zero mode. The vacuum $|0>$ is defined by

$$\alpha_m|0> = 0 , \quad m \geq 0 , \qquad (3.4)$$

and the Fock space is built by repeated application of creation operators on the vacuum.

In order to define the quantum version of the non-linear dynamics, eqs. (2.11), (2.12), we have to specify the operator ordering for I_n, n = 1, 2, 3,..., . Containing products of operators u, u_σ, $u_{\sigma\sigma}$, ..., at the same space-time point, I_n in quantum theory is full of divergences. The simplest way to get rid of the divergence is to adopt the normal product, which we denote by dots : : . So we denote by $\hat{I}_n$ the quantum operator corresponding to the classical quantity I_n. It has the following general structure

$$\hat{I}_n = :I_n: + \sum_{k=1} \hbar^k :I_n^{(k)}: , \quad n = 1, 2, 3, \ldots, \tag{3.5}$$

in which $I_n^{(k)}$ are yet unspecified but they should have the form

$$I_n^{(k)} = \frac{1}{2\pi}\int_0^{2\pi} d\sigma \mathcal{I}_n^{(k)} , \tag{3.6}$$

and $\mathcal{I}_n^{(k)}$ is a local polynomial in u, u_σ, $u_{\sigma\sigma}$,...,. The classical theory dictates only the $\hbar$-independent part, namely the first term on the r.h.s. of eq. (3.5).

We are now in a position to pose the question raised in the Introduction in a more definite way.

Question: Is it possible to define $I_n^{(k)}$ such that all $\hat{I}_n$ commute with each other ?

$$[\hat{I}_n, \hat{I}_m] = 0 , \quad n, m = 1, 2, \ldots, . \tag{3.7}$$

Since $\hbar$ is treated as a free parameter, eq. (3.7) demands that the coefficients of each power of $\hbar$ should vanish. We impose the following conditions on possible local counter terms $\mathcal{I}_n^{(k)}$;

(i) max. order$(\mathcal{I}_n^{(k)}) = 2(n - k)$, (3.8a)

implying $\mathcal{I}_n^{(k)} = 0$ for $k \geq n$, (3.8b)

max. weight$(\mathcal{I}_n^{(k)}) = 2n$, (3.8c)

$\mathcal{O}_n^{(k)}$ has even order, even weight terms only. (3.8d)

(ii) Irreducibility,

$I_n^{(k)}$ should not contain any of I_m as a part. (3.8e)

The necessity of the irreducibility condition is obvious. The meaning of the condition (3.8a∿c) is that the order and the weight of the added terms should not exceed those of the original term I_n. Here one should take into account of the fact that $\hbar$ has the order 2 due to the commutation relation (3.1). If one relaxes these conditions then the whole program eqs. (3.5), (3.7) would be uncontrollable.

In order to facilitate the calculation of commutators of normal ordered polynomials in field operators, we adopt the following technique which is rather common in string theories and in conformally invariant two dimensional field thoeries. Let us introduce a fictitious "time" variable τ and define a complex coordinate z by

$$z = e^{\tau + i\sigma} , \tag{3.9}$$

and consider the field theory defined on the complex plane

$$u(z) = \sum_n \alpha_n z^{-n} , \tag{3.10}$$

instead of the original theory defined on a circle S^1. It is obvious that by restricting the coordinate z to a unit circle we regain the original theory. For an arbitrary local polynomial p in the field variables u, u_σ, $u_{\sigma\sigma}$,..., we have

$$\begin{aligned} &\frac{1}{2\pi}\int_0^{2\pi} d\sigma\ p(u,u_\sigma,u_{\sigma\sigma},...,) \\ &= \frac{1}{2\pi i}\oint_0 \frac{dz}{z}\ p(u,Du,D^2u,...,) , \end{aligned} \tag{3.11}$$

in which

$$D = iz\partial_z \ . \tag{3.12}$$

The integration contour on the r.h.s. of eq. (3.11) should encircle the origin once and otherwise arbitrary due to the analyticity. Next we define the radial ordering[8)] R of two local operators $a(\zeta)$, $b(z)$ with complex coordinates ζ and z,

$$Ra(\zeta)b(z) = \begin{cases} a(\zeta)b(z) \ , & \text{if} \ |\zeta| > |z| \ , \\ \pm b(z)a(\zeta) \ , & \text{if} \ |\zeta| < |z| \ . \end{cases} \tag{3.13}$$

The minus sign is for fermionic operators a and b, which we do not consider in the present essay. The propagator of the u field is defined as the vacuum expectation value of the radial ordered product $Ru(\zeta)u(z)$,

$$\hbar\Delta(\zeta,z) = \langle 0|Ru(\zeta)u(z)|0\rangle = \hbar \frac{\zeta z}{(\zeta-z)^2} \ . \tag{3.14}$$

Let us denote by P the integral of a normal ordered polynomial p as is given in eq. (3.11),

$$P = \frac{1}{2\pi i}\oint_0 \frac{dz}{z} : p(z) : \ . \tag{3.15}$$

Similarly Q is defined by

$$Q = \frac{1}{2\pi i}\oint_0 \frac{dz}{z} : q(z) : \ .$$

Then by using the definition of the radial ordering and a simple deformation of integral contours, we get the following general formula for the commutator [P, Q],

$$[P,\ Q] = \frac{1}{(2\pi i)^2}\oint_0 \frac{dz}{z}\oint_{c_z} \frac{d\zeta}{\zeta} R : p(\zeta) : : q(z) : \ , \tag{3.16}$$

in which c_z is a small contour encircling the point z. By Wick's theorem we expand the radial ordered product $R{:}p(\zeta){:}{:}q(z){:}$ in terms of the propagator $\Delta(\zeta,z)$. The k-times contraction term contains a factor $\hbar^k \prod_i^k (D^{n_i} \Delta)$ and it corresponds to a (k-1) loop Feynman diagram. The first contraction term, k = 1, corresponding to a tree diagram and proportional to $\hbar$, gives the classical contribution, as is well known. The ζ-integration is the standard residue calculus. Therefore the resulting commutator, i.e., the r.h.s. of eq. (3.16) has the same structure as P and Q,

$$[P,Q] = \frac{1}{2\pi i} \oint_o \frac{dz}{z} : r(z) : , \qquad (3.17)$$

in which r is an appropriate polynomial in u, u_σ, $u_{\sigma\sigma}$,..., .

Before going into the details of the determination of the infinite set of quantum commuting operators in the next section, we give a simple and well known example of the calculation of commutators introduced in this section. The first member of the set of commuting operators is $:I_1:$, which has no additional term due to (3.8b),

$$\hat{I}_1 = : I_1 : = \frac{1}{2\pi i} \oint_o \frac{dz}{z} \frac{1}{2} : u(z)^2 : . \qquad (3.18)$$

By taking its moments, we define, ($L_o = \hat{I}_1$),

$$L_n = \frac{1}{2\pi i} \oint_o \frac{dz}{z} z^n \frac{1}{2} : u(z)^2 : , \quad n = 0, \pm 1, \pm 2, \ldots . \qquad (3.19)$$

Their commutation relations can be calculated in one line and we get the Virasoro algebra,

$$[L_n, L_m] = \hbar (n - m) L_{n+m} + \frac{\hbar^2}{12} n(n^2 - 1)\delta_{n+m,o} , \qquad (3.20)$$

in which the first term on the r.h.s. is the classical result due to a tree diagram while the second term, the central charge, comes from a one loop diagram and is a genuine quantum effect.

§4. Construction of an Infinite Set of Quantum Commuting Operators

In this section we show that the infinite set of quantum commuting operators $\hat{I}_n$,

$$[\hat{I}_n, \hat{I}_m] = 0 , \quad n, m = 1, 2,..., \tag{3.7}$$

does exist and that its members can be determined explicitly. As stated in the previous section we look for $\hat{I}_n$ in the following form

$$\hat{I}_n = : I_n : + \sum \hbar^k : I_n^{(k)} : , \tag{3.5}$$

in which $I_n^{(k)}$'s $(k \geq 1)$ satisfying the conditions (3.8) are to be determined. Let us write down some lower members explicitly in the notation of eq. (3.6);

$$\hat{I}_1 = : I_1 : , \quad \mathcal{I}_1 = \frac{1}{2} u^2 , \tag{4.1}$$

$$\hat{I}_2 = : I_2 : , \quad \mathcal{I}_2 = \frac{1}{2}(uD^2u - u^4) , \tag{4.2}$$

$$\hat{I}_3 = : I_3 : + \hbar : I_3^{(1)} : + \hbar^2 : I_3^{(2)} : ,$$

$$\mathcal{I}_3 = \frac{1}{2}uD^4u + 5u^2(Du)^2 + u^6 , \tag{4.3}$$

$$\mathcal{I}_3^{(1)} = a_1uD^4u + a_2u^2(Du)^2 + a_3uD^2u , \tag{4.3a}$$

$$\mathcal{I}_3^{(2)} = b_1uD^4u + b_2uD^2u , \tag{4.3b}$$

$$\hat{I}_4 = : I_4 : + \hbar : I_4^{(1)} : + \hbar^2 : I_4^{(2)} : + \hbar^3 : I_4^{(3)} : ,$$

$$\mathcal{I}_4 = \frac{1}{2}uD^6u - 7u^2(D^2u)^2 + \frac{7}{2}(Du)^4 - 35u^4(Du)^2 - \frac{5}{2}u^8 , \tag{4.4}$$

$$\mathcal{I}_4^{(1)} = c_1u^4(Du)^2 + c_2u^2(D^2u)^2 + c_3(Du)^4 + c_4u^2(Du)^2 + c_5uD^6u + c_6uD^4u + c_7uD^2u , \tag{4.4c}$$

$$\mathcal{L}_4^{(2)} = d_1u^2(D^2u)^2 + d_2(Du)^4 + d_3u^2(Du)^2$$

$$+ d_4uD^6u + d_5uD^4u + d_6uD^2u \ , \tag{4.4d}$$

$$\mathcal{L}_4^{(3)} = e_1uD^6u + e_2uD^4u + e_3uD^2u \ , \tag{4.4e}$$

. . . . , .

Here the coefficients a_1, a_2,..., e_3,..., are to be determined. A few remarks are in order. The individual monomials in $\mathcal{L}_n^{(k)}$ are linearly independent, namely no linear combination of them will lead to a total (σ) derivative. The choice of the set of monomials is not unique due to the freedom of addition of total (σ) derivative terms. $\hat{I}_2$ can have a term proportional to $\hbar$, i.e., $\mathcal{L}_2^{(1)} = \lambda uD^2u$, with an indefinite λ. However, the effect of this new term can always be absorbed to the original form of $\mathcal{L}_2$ by a suitable rescaling of $\hbar$ and of the field operator u. Therefore $\mathcal{L}_2^{(1)}$ is disregarded.

Since $\hat{I}_1 = L_o$, eq. (3.19), commutes with all of $:I_n^{(k)}:$ it does not give any information on the form of $\mathcal{L}_n^{(k)}$. Vanishing commutator with $\hat{I}_2$, on the other hand, gives severe restriction on $\mathcal{L}_n^{(k)}$ ($n \geq 3$) and all the undetermined coefficients are fixed uniquely by these conditions. Then for all these determined operators $\hat{I}_n$, $n = 3,4,5,\ldots$, vanishing of commutators $[\hat{I}_n, \hat{I}_m] = 0$, $n, m = 3,4,5,\ldots$, imposes an enormous (∞^2) set of non-trivial conditions. We have checked by explicit calculation with an aid of a formula manipulation program[9] that $[\hat{I}_3, \hat{I}_4] = 0$ actually holds. In fact, vanishing of this commutator is achieved by vanishing of coefficients of <u>dozens</u> of linearly independent monomials of field operators. The numerical values for the coefficients $a_1,\ldots,e_3$ are

$a_1 = 2/3$, $a_2 = 10$, $a_3 = -5/6$

$b_1 = -1/6$, $b_2 = 5/6$,

$$c_1 = -140\ ,\quad c_2 = -98/3\ ,\quad c_3 = 112/9\ ,\quad c_4 = -35/3\ ,$$

$$c_5 = 26/15\ ,\quad c_6 = -7/3\ ,\quad c_7 = -7/30\ ,$$

$$d_1 = -91/3\ ,\quad d_2 = 14/9\ ,\quad d_3 = 35/3\ ,\quad d_4 = 4/45$$

$$d_5 = -7/6\ ,\quad d_6 = 56/45$$

$$e_1 = 8/45\ ,\quad e_2 = 7/2\ ,\quad e_3 = -91/90\ . \tag{4.5}$$

Determination of higher operators $I_n^{(k)}$ is straightforward. Although further checking of vanishing of higher commutators is absolutely necessary and is currently pursued, there is little doubt that they are satisfied because of the highly nontrivial cancellation of $[\hat{I}_3,\hat{I}_4]$.

§5. Summary and Comments

By applying the techniques developed in the string theories and in the conformal field theories, we have almost established the existence of an infinite set of quantum commuting operators related to the MKdV equation as well as the Virasoro algebra. Our method gives an algorithm to determine the explicit forms of these operators. This set of operators may constitute the Cartan subalgebra of an infinite rank Lie algebra which might underlie the quantum integrable systems of some sort.

A lot of interesting things remain to be elucidated. We remark only a few of them here. Detailed checking of the miraculous cancellations of higher commutators and its clear and analytic understanding is vital for further steps.[10] The relationship between our naive continuum approach and the conventional quantum inverse method or the transfer matrix approach based on a lattice[3] is quite interesting.[11] The applicability of the present method to other types of soliton equations is certainly an interesting point.

References

1) See for example, A. C. Scott, F. Y. F. Chu and D. W. McLaughlin, Proc. IEEE 61 (1973) 1443.
R. K. Bullough and P. J. Caudrey, "Solitons" ed. R. K. Bullough and P. J. Caudrey, (Springer, Berlin, 1980).
2) L. D. Faddeev, Les Houches lecture notes, Session XXXIX 1982, Elsevier Sci. Publ.
3) J.-L. Gervais, Phys. Lett. 160B (1985) 277.
4) K. Konno, H. Sanuki and Y-H. Ichikawa, Prog. Theor. Phys. 52 (1974) 886,
M. Wadati, H. Sanuki and K. Konno, Prog. Theor. Phys. 53 (1975) 419.
5) H. Flaschka and A. C. Newell, in Lecture Notes in Physics 38, J. Moser, ed. (Springer, Heidelberg, 1975),
A. C. Newell, Proc. Roy. Soc. A365 (1979) 283.
6) Y. Kodama and M. Wadati, Prog. Theor. Phys. 56 (1976) 342, 1740.
7) J. Alberty, T. Koikawa and R. Sasaki, Physica 5D (1982) 43,
R. Sasaki, Proc. Roy. Soc. A373 (1980) 373.
8) S. Fubini, J. Hanson and R. Jackiw, Phys. Rev. D7 (1973) 1732.
9) I. Yamanaka, Hiroshima Preprint RRK 86, in preparation.
10) R. Sasaki and I. Yamanaka, Hiroshima Preprint RRK 86, in preparation.
11) M. Omote, R. Sasaki and I. Yamanaka, Hiroshima Preprint RRK 86, in preparation.

STRINGS AND KNOTS IN THE DISORDERED MEDIUM*)

Akio Sugamoto

National Laboratory for High Energy Physics (KEK)
Oho-machi, Tsukuba-gun, Ibaraki-ken, 305
JAPAN

ABSTRACT

Possibility to have Knots in the string theories is discussed. Topological study is reviewed on various types of knots. In aiming at obtaining the dynamics of knots, string motion in the dense, weakly-interacting scatterers is formulated in a self-consistent approximation.

I: INTRODUCTION

Recently, string theories were revived when M.Green and J.H.Schwarts[1] proved that only consistent string theories may be superstrings with SO(32) and E(8)xE(8) gauge groups. A candidate of phenomenologically favored string theory is a heterotic superstring[2] with E(8)xE(8) group, which is finite[3] and free from anomalies[4] at one loop level. Both phenomenological, and theoretical studies on these superstrigs are progressing very rapidly now.[5]

In these tendencies, it may not be a loss of time to consider what phenomena are expected in the string theories, which are, however, not realized in the ordinary theories of point particles. Among them, we will take up in this paper Knots of Strings, which are characteristic to closed strings such as heterotic ones. These amusing states of strings, knots, may play a crucial role in string theories in a simlar manner as monopoles and instantons play in the non-Abelian gauge theories.

Next section is devoted to a kinematical study of knots, which consists of discussions on knots group, torus knots, wild knot, and a relation between torus knots and singularities of algebraic curves which may be string world sheets.

---

*) Essay dedicated to professor Yoshio Yamaguchi in honor of his 60 th birthday.

In order to have stable states of knots, we need to suppress the flip-flop interation of strings by short distance repulsion and to have attraction at long distance which favors to fold strings inside a small region. We know that Anderson's localization[6] of electrons in the disordered medium can be studied by using path integral formulation[7] in a completely analogous way to the configulation problem of polymers located in a solvent[8]. These studies by S.F.Edwards are very beautiful and valuable, even though he has adopted self-consistent approximation methods. There is an indication of such favorable interation in these studies, repulsive at short distance and attractive at long distance.

Therefore, we will generalize Edwards' method to the case that strings are located in a disordered medium having dense, weakly-interacting, scatterers. This kind of disordered medium may be realized in the early stage of the universe and may have to do with the medium having condensation of vortices[9]. We will give in Sec.3, self-consistent equations (16) and (17) satisfied by the propagator of string under an effect of self-consistent potential U_{sc}, which reproduces the string motion in the disordered medium. We have an indication of long distant attraction in U_{sc}, but because of the crude approximation adopted, we cannot definitely elucidate the occurence of U_{sc} having desireable features discussed above. Therefore, our study of knots' dynamics is as yet at a primitive stage.

I wish to dedicate this paper to Professor Yoshio Yamaguchi in honor of his 60 th birthday. Recently I have looked over the preprints classified as YY-1 and YY-2, July(1968)[10] which were handed by him to his then students, Dr. Akitoshi Nishimura and myself. These preprints enthusiastically explain why high energy machines are necessary in Japan. His desire to have "Particle Physics Laboratory" was realized on April (1971) as KEK, where TRISTAN experiments will start from this autumn. In these preprints, he told that tradition of Japanese particle physics tend to favor theoretical or metaphysical studies not based on the experimental results, but we wellcome the fact that a few young researchers begin to perform original works in the regions directly related to experiments. We have now more of such young researchers. If the string fever spread all over the world might reduce the interest towards experiments, it would be a great pity for us. In this respect, I hope that this paper will not make him very sad.

2. KNOTS IN THE STRING THEORIES[11]

In our living space, we have many kinds of knots made up from closed strings. Since the living space is 3 dimensional, a closed string (1 dimensionally extended object), can afford us knotted states. In the space time with dimension geater than 4, however, every closed string cannot be knotted. This fact has been proved by E.C.Zeeman: If $p-q \geqslant 3$, q-th sphere S^q in the space of S^p is always unknotted.
Here "unknottedness" is equivalent to the fact that (S^p, S^q) are homeomorphic to $(\partial D^{p+1}, \partial D^{q+1})$, respectively, choosing certain discs (D^{p+1}, D^{q+1}) with dimensions greater by 1. Therefore, 9 or 25 dimensional space which is the space of superstrings or bosonic strings,

does not know the interesting knotted states. After compactification of extra 6 dimensions, we have strings living in 3 space-like and 1 time-like dimensions, where we can see the knotted states if we do not alter the configuration in this extra space. [Each configuration in the extra 6 dimensional space corresponds to one generation, so that except for the generation mixing, such configuration may be stable.]

Coming back to the 3 dimensional space, we know a criterion of telling knotted states from unknotted ones. C.P.Papakyriakopoulos has proved using Dehn's lemma the following fact: The necessary and sufficient condition for a knot K to be unknotted is

$$\pi_1(S^3 - K) \cong \mathbb{Z} \qquad (1)$$

Therefore if we calculate the knot group $\pi_1(S^3\text{-}K)$ and show that it is different from $\mathbb{Z}$, then the configulation K of closed string is knotted. An example is a trefoil knot whose knot group is a free Abelian group with 2 generators (a,b) satisfying a relation $a^2=b^3$, namely the group reads

$$G_{2,3} = \langle a, b \,;\, a^2 = b^3 \rangle \qquad (2)$$

which is not $\mathbb{Z}$ obviously. A generalization of a trefoil knot is (m,n) torus knot whose knot group is

$$G_{m,n} = \langle a, b \,;\, a^m = b^n \rangle \qquad (3)$$

where (m,n) is relatively prime.

String theories have sometimes the following flip-flop interaction

$$\overset{1 \quad 3}{\underset{2 \quad 4}{\times}} \;\rightarrow\; \overset{1 \quad 3}{\underset{2 \quad 4}{\rangle\langle}} \qquad (4)$$

which transforms knotted states to unknotted ones with some probability. In order to have stable knotted states, several possibilities exists:

<u>Possibility[1]</u> Suppress the interaction (4) by the repulsive interaction between hard cores of strings. Whether we have the hard core or not depends on the ingredients from which the strings are made up . To solve this problem we should deeply consider whether the above ingredients really exists as subquarks.

<u>Possibility[2]</u> Suppress the interation (4) through the effect of random scatterers located densely outside of the string.

<u>Possibility[3]</u> Consider a knot which requires an infinite number of repeated interactions of (4) to change it to be unknotted.

We will consider in the next section the situation relating to the possibility[2], where strings are propagating in the dense, randomly located, weakly-interating scatterers. On the other hand, knots or arcs having the property given in th Possibility[3] are called <u>wild</u>, and the examples of them have been given by E.Artin and R.H.Fox. A wild knot of Fox reads

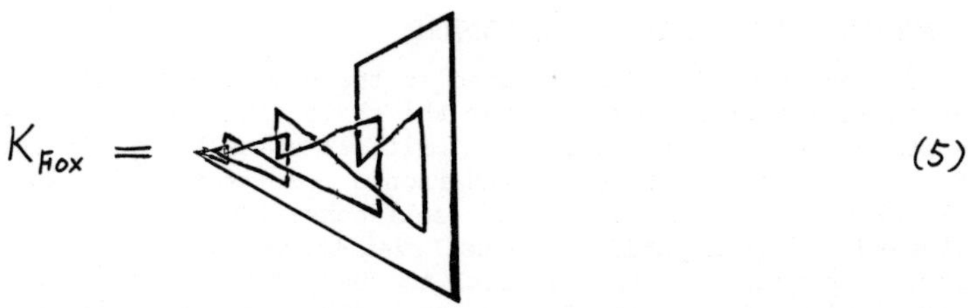

which we know from Fig.169 in J.Stillwell's book in Ref. 11). The knot group of this wild knot is

$$G_{Fox} = < b_0, b_1, b_3, \cdots ; b_1 b_0 b_1^{-1} = b_2 b_1 b_2^{-1} = b_3 b_2 b_3^{-1} = \cdots > \quad (6)$$

It is also interesting to know the connection between knots and singularities of the algebraic curves[12]. In the 2 dimensional complex plane $\mathbb{C}^2 = \{(z_1, z_2)\}$, an algebraic equation

$$F : \quad f(z_1, z_2) = 0 \quad (7)$$

gives a complex curve. In terms of real variables, Eq. (7) can be viewed as a world sheet of string, living in the 4 dimensional space-time. possible types of singularities of the surface F are classified as iterated knots of type

$$\left(\frac{m_1}{n_1}, \frac{m_2}{n_2}, \cdots, \frac{m_g}{n_g} \right), \quad (8)$$

where $n_1 > 1$, m_i and n_i are relatively prime, and $m_1/n_1 < m_2/n_1 n_2 < \cdots < m_g/n_1 n_2 \cdots n_g$. A simple example is the case for $f(z_1 z_2) = z_2^2 - z_1^3$. Considering a small 3 dimensional sphere $S_\varepsilon^3 \approx \{|z_1|=\varepsilon, |z_2| \le \varepsilon\} \cup \{|z_1| \le \varepsilon, |z_2| = \varepsilon\}$, the intersection $F \cap S_\varepsilon^3$ gives a trefoil torus knot, having the knot group in Eq.(2) . Here we can view S_ε^3 as our 3 dimensional space at time $t = \varepsilon$,where the radius of the sphere ε plays the role of time, and at $t=0$ the trefoil knot is created and it propagates, leaving the world sheet, $z_2^2 - z_1^3 = 0$. Resolution of the singularities may represent smooth formation of the torus knot at $t=0$, by changing the singularity from a point to a complex line through blowing up. If we come back again to real 4 dimensions, we can make an orientable surface called Seifert surface whose boundary coincides with the knot. This Seifert surface is also a world sheet producing the knot from the vacuum, but it is not homeomorphic to D^2 , having some singularities.

Therefore, we have a number of interesting concepts on the knots which are devloped in mathematics. We are hoping to have the knotted strings in the real world. They may be produced in the early universe, or they may play certain roles in the generation problem, since we already know that strings located at the singularities of the orbifolds[13] having twisted boundary conditions, correspond to some generations whose mixing matrix is diagonal.

3. STRINGS IN THE DISORDERED MEDIUM

If a point particle is located in the disordered medium, its wave-function is localized any time for the space time dimension $D=2$, or when its energy is below a certain critical value (mobility edge) E_c for $D \geq 3$. This is the well-known Anderson's localization[6]. To understand the localization of electrons, the self-consistent field theory of S.F.Edwards[7] seems familiar to us, even though it is an approximate theory. The framework is identical to that of his earlier work[8] on the statistical mechanics of polymers having interaction between constituent monomers, where the polymer length L corresponds to the Wick rotated time it of the electron. Depending whether it is a good or poor solvent, the polymers feel self-avoiding or self-attracting interaction when two different monomers inside a chain happen to meet with each other. A configuration of polymers can be viewed as a path of self-avoiding (or self-attracting) random walk, which is a non-Markovian process. The self-consistent field approximation, however, transforms the non-Markovian process into the Markovian one, which is the random walk starting from the origin under the effect of a self-consistent potential $U_{sc}(r)$. For the infinitely long polymer, asymptotic solution of $U_{sc}(r)$ is given by[8]

$$U_{sc}(r) \sim r^{-4/3} \quad . \tag{9}$$

Application of the potential to the motion of electrons in the dense, weakly-interacting, random scatterers, leads us to Anderson's localization for $E < E_c$.[7] For this purpose, one condition should be satisfied, namely, that we have large density of the scatterers $\rho (\gg 1)$ and small strength of the random potential $v (\ll 1)$, but the product ρv^2 should remain to be finite.

Now, we will apply the above mentioned method to the problem of string motion in the dense, weakly-interacting, random scatterers which also consist of strings. Such circumstances might be realized in the early universe where the following discussion is applicable. The problem may be viewed as that of self-interacting random surfaces, a generalization of self-interacting random walk discussed above.

If we take open string as an example, propagator of a string, starting from $\vec{X}_\perp(\sigma)^{(1)}$ at proper time $\tau^{(1)}$ and arriving at $\vec{X}_\perp(\sigma)^{(2)}$ at $\tau^{(2)}$, under the influence of an interaction $V[\vec{X}_\perp(\tau,\sigma) - \vec{R}_a(\sigma')]$ with randomly-located solvent strings $\vec{R}_a(\sigma)$ $[a = 1 \sim \infty]$, can be described as

$$\begin{aligned} & K_{p+}(\vec{X}_\perp(\sigma)^{(1)}, \vec{X}_\perp(\sigma)^{(2)}; \tau^{(1)}, \tau^{(2)}) \\ &= \lim_{N\to\infty} \prod_{a=1}^{N} \int \mathcal{D}\vec{R}_a(\sigma) \int_{-\infty}^{+\infty} dp_{a+}\, P[\vec{R}_a(\sigma), p_{a+}] \\ &\times \int_{\vec{X}_\perp(\sigma,\tau^{(1)}) = \vec{X}_\perp(\sigma)^{(1)}}^{\vec{X}_\perp(\sigma,\tau^{(2)}) = \vec{X}_\perp(\sigma)^{(2)}} \mathcal{D}\vec{X}_\perp(\sigma,\tau)\, \exp\int_{\tau^{(1)}}^{\tau^{(2)}} d\tau \left(\int_0^{2\pi p_+} d\sigma\, \frac{1}{2} [\dot{\vec{X}}_\perp^2 - \vec{X}_\perp'^2] \right. \\ & \left. - \int_0^{2\pi p_+} d\sigma \int_0^{2\pi p_{a+}} d\sigma'\, V[\vec{X}_\perp(\sigma,\tau) - \vec{R}_a(\sigma')] \right) . \end{aligned} \tag{10}$$

Here we have used the light-cone gauge formalism by S.Mandelstam and by M.Kaku and K.Kikkawa[14], where the fixing $X^+=\tau$ restricts the dynamical variables to transverse modes $X_\perp^i(\sigma,\tau)$ $(i=1\sim D-2)$ as well as the zero mode x^- of $X^-(\tau,\sigma)$ $[X^\pm=\frac{1}{\sqrt{2}}(X^0\pm X^{D-1})]$. Usually p_+, the conjugate momentum of x^- is used to label the string. Therefore, the notation $P[\vec{R}(\sigma),P_+]$ should be the probability distribution to find solvent strings in the transverse configuration $\vec{R}(\sigma)$ with P_+.

In Eq.(10), sum over the random configuration of solvent strings will be simplified, by introducing the density ρ of solvent strings, (The density ρ here is the number of strings with various shapes having the same center of mass inside a unit volume of (D-2) dimensions), and by using $\lim_{N\to\infty}(1+\frac{\rho}{N}V)^N=\exp(\rho V)$: The relevant part yields

$$\exp\frac{\rho}{\mathcal{N}}\int\mathcal{D}\vec{R}(\sigma)\int dP_+'\left(e^{-\int_{\tau^{(1)}}^{\tau^{(2)}}d\tau\int_0^{2\pi P_+}d\sigma\int_0^{2\pi P_+'}d\sigma' V[\vec{X}_\perp(\sigma,\tau)-\vec{R}_\perp(\sigma')]}-1\right), \quad (11)$$

with a normalization $\mathcal{N}=\int\mathcal{D}'\vec{R}(\sigma)\int dP_+'$, where $\mathcal{D}'$ denotes omitting of the zero mode integration. So far we are considering the solvent to be strings, but to avoid an extra mathematical difficulty we will assume later that the size of the solvent is viewed to consist of equivalent particles. Now, we will adopt Edwards' assumption, namely

$$\rho V^2\to\text{ finite for }\rho\to\infty\text{ and }V\to 0. \quad (12)$$

Then, the term of 1st order in V simply shifts the origin of the energy by $-2\pi P_+\rho\hat{V}_{\vec{K}=0}$ which is absorbed by the renormalization of the ground state energy, and the 2nd order term will be

$$\exp\left(+\frac{\rho}{2}\int_{\tau^{(1)}}^{\tau^{(2)}}d\tau\int_{\tau^{(1)}}^{\tau^{(2)}}d\tau'\int_0^{2\pi P_+}d\sigma\int_0^{2\pi P_+}d\sigma'\, W[\vec{X}_\perp(\sigma,\tau)-\vec{X}_\perp(\sigma',\tau')]\right), \quad (13)$$

where the momentum representation of W and V are related as $\hat{W}_{\vec{K}}=|\hat{V}_{\vec{K}}|^2$. Equation(13) represents that different points on the world sheet of the string interact with each other by W, which is attractive for the white noise case that $\hat{V}_{\vec{K}}=v$ (const). Edwards noted[7], in the case of self-interacting random walk, that if a maximum density ρ is permitted this will modify the potential to have an inner repulsive region and an outer attraction. If this works also in our problem, necessary <u>short distance repulsion</u> to reduce the flip-flop interaction in Eq.(4) will be obtained, leading to stabilize the knotted strings [Remind the possibility [2] raised after Eq.(4)]. Detailed analysis is, however, indispensable for this. Here, we concentrate on the outer attractive interaction which will fold world sheets with large area within a small region.

The interaction (13) is non-local, but the relating difficulty will be removed by introducing a random potential $U[X_\perp(\sigma,\tau)]$ as

$$e^{\frac{\rho}{2}W}=\int\mathcal{D}U[X_\perp]\,e^{-\int_{\tau^{(1)}}^{\tau^{(2)}}d\tau\int_0^{2\pi P_+}d\sigma\,U[X_\perp(\sigma,\tau)]}\,e^{-\frac{1}{2\rho}\Phi W^{-1}\Phi}. \quad (14)$$

Then, the original propagator now becomes

$$K_{P_+}(\vec{X}_\perp(\sigma)^{(1)},\vec{X}_\perp(\sigma)^{(2)};\ \tau^{(1)},\tau^{(2)})$$

$$= \int \mathcal{D} U[\vec{X}_\perp]\ K_{p+}\left(\vec{X}_\perp(\sigma)^{(1)}, \vec{X}_\perp(\sigma)^{(2)}; \tau^{(1)}, \tau^{(2)}; U[X_\perp]\right)$$

$$\times \exp\left(-\frac{1}{2\rho}\int_{\tau^{(1)}}^{\tau^{(2)}} d\tau \int_{\tau^{(1)}}^{\tau^{(2)}} d\tau' \int_0^{2\pi P_+} d\sigma \int_0^{2\pi P_+} d\sigma'\ U[X_\perp(\sigma,\tau)]\, W^{-1}[X_\perp(\sigma,\tau)-X_\perp(\sigma',\tau')]\, U[X_\perp(\sigma',\tau')]\right)$$

$$\times \left(\int \mathcal{D} U \exp\left(-\frac{1}{2\rho} U W^{-1} U\right)\right)^{-1}, \qquad (15)$$

which is a Gaussian average over the random potentials, of the propagator $K_{p+}(X_\perp^{(1)}, X_\perp^{(2)}; \tau^{(1)}, \tau^{(2)}; U[X_\perp])$ obtained under the effect of the potential $U[\vec{X}_\perp]$. Most probable random potential can be extracted, by applying the steepest descent to Eq.(15), and we will call it $U_{sc}[\vec{X}_\perp]$, since it is obtained self-consistently. We now have two equations, relating self-consistently, as follows:

$$\left[\frac{\partial}{\partial \tau^{(2)}} + \int_0^{2\pi P_+} d\sigma \left\{\pi\left(-\frac{\delta^2}{\delta \vec{X}_\perp(\sigma)^2} + \frac{1}{(2\pi)^2}\vec{X}_\perp'(\sigma)^2\right)^{(2)} + U_{sc}[X_\perp^{(2)}(\sigma,\tau)]\right\}\right]$$

$$\times K_{p+}\left(X_\perp^{(1)}(\sigma), X_\perp^{(2)}(\sigma); \tau^{(1)}, \tau^{(2)}; U_{sc}[X_\perp]\right)$$

$$= \delta(\tau^{(1)} - \tau^{(2)}) \prod_\sigma \delta^{(D-2)}\left(\vec{X}_\perp^{(1)}(\sigma) - \vec{X}_\perp^{(2)}(\sigma)\right), \qquad (16)$$

and

$$U_{sc}[X_\perp(\sigma,\tau)] = -\rho \int \mathcal{D} Y_\perp(\sigma'\tau') \int_{\tau^{(1)}}^{\tau^{(2)}} d\tau' \int_0^{2\pi P_+} d\sigma'\ W[\vec{X}_\perp(\sigma,\tau) - \vec{Y}_\perp(\sigma',\tau')]$$

$$\times \frac{K_{p+}(\vec{X}_\perp^{(1)}, \vec{Y}_\perp; \tau^{(1)}, \tau'; U_{sc}) \cdot K_{p+}(\vec{Y}_\perp, \vec{X}_\perp^{(2)}, \tau', \tau^{(2)}; U_{sc})}{K_{p+}(\vec{X}_\perp^{(1)}, \vec{X}_\perp^{(2)}; \tau^{(1)}, \tau^{(2)}; U_{sc})} \qquad (17)$$

Although the expression of (17) is complicated, its physical meaning is clear: The string $\vec{X}_\perp(\sigma,\tau)$ with P_+ feels the potential from the string $\vec{Y}_\perp(\sigma,\tau)$ through the interaction W, where the weight of configurations should be a probability to find the original string at $\vec{Y}_\perp(\sigma)$ at τ, on the way of its propagation from $\vec{X}_\perp^{(1)}$ at $\tau^{(1)}$ to $\vec{X}_\perp^{(2)}$ at $\tau^{(2)}$. In general, U_{sc} depends on $\vec{X}_\perp^{(1)}(\sigma)$, $\vec{X}_\perp^{(2)}(\sigma)$, $\tau^{(1)}$, and $\tau^{(2)}$.

In the following we will restrict ourselves to the simplest case of white noise potential $V(\vec{X}_\perp(\sigma,\tau) - \vec{R}_a)$ which gives

$$W[\vec{X}_\perp(\sigma,\tau) - \vec{Y}_\perp(\sigma',\tau')] = w \prod_{\sigma,\tau} \delta^{(D-2)}(\vec{X}_\perp(\sigma,\tau) - \vec{Y}_\perp(\sigma',\tau')), \qquad (18)$$

where $w = |\hat{V}_{\vec{k}}|^2$ is constant and positive, giving an attractive potential. Our following discussion of the self-consistency between Eqs.(16)

and (17) is restricted in the neighborhood of a minimum point of U_{sc}, where we will approximate U_{sc} as a sum of harmonic oscillator potentials:

$$U[\vec{X}_\perp(\sigma)] \approx \frac{1}{8\pi p_+^2}\Big[K + \sum_{n=0}^{\infty} \lambda_n (\vec{x}_n - \vec{x}_n^{(0)})^2\Big], \tag{19}$$

with parameters K, λ_n, and $\vec{x}_n^{(0)}$, which will be fixed self-consistently. In Eq.(19), x_n^i corresponds to the string oscillation in the i-th direction with 2n nodes, i.e.,

$$X^i(\sigma) = x_0^i + 2\sum_{n=1}^{\infty} x_n^i \cos(n\sigma/2p_+) . \tag{20}$$

$$(0 \le \sigma \le 2\pi p_+)$$

This is consistent with the boundary condition of open strings, $X'(\sigma) = 0$ for $\sigma = 0$ and $2\pi p_+$. The Hamiltonian H having the potential(19),

$$\begin{aligned} 2p_+ H = &-\Big\{\Big(\frac{\partial}{\partial \vec{x}_0}\Big)^2 + \frac{1}{2}\sum_{n=1}^{\infty}\Big(\frac{\partial}{\partial \vec{x}_n}\Big)^2\Big\} + \frac{1}{2}\sum_{n=1}^{\infty} n^2 (\vec{x}_n)^2 \\ &+ \frac{1}{2}\sum_{n=0}^{\infty} \lambda_n (\vec{x}_n - \vec{x}_n^{(0)})^2 + \frac{1}{2}K \\ \equiv &\; 2p_+ \sum_{n=0}^{\infty} H_n , \end{aligned} \tag{21}$$

is the sum over different oscillation modes. Therefore we can obtain the eigen-values $\{E_\lambda\}$ and eigen-functions $\{\psi_\lambda\}$ completely using Hermete polynomials.

The propagator $K_{p_+}(X^{(1)}, X^{(2)}; \tau^{(1)}, \tau^{(2)}; U)$ can be written as

$$\begin{aligned} &K_{p_+}(\{\vec{x}_n^{(1)}\}, \{\vec{x}_n^{(2)}\}; \tau^{(1)}, \tau^{(2)}; U[\{\vec{x}_n\}]) \\ &= \sum_\lambda \psi_\lambda(\{\vec{x}_n^{(2)}\})\, \psi_\lambda^*(\{\vec{x}_n^{(1)}\})\, e^{-(\tau^{(2)} - \tau^{(1)})E_\lambda} , \end{aligned} \tag{22}$$

where the labeling of the state is done by identifying the occupation number m_n^i of the n-th oscillator in the i-th direction as

$$\lambda = \{m_n^i\} \quad (n = 1, 2, \cdots ; \; i = 1 \sim D-2).$$

Since the eigen-value for $\lambda = \{m_n^i\}$

$$E_{\{m_n^i\}} = \frac{1}{4p_+}\left\{\begin{aligned} &\sqrt{2\lambda_0}\sum_{i=1}^{D-2}(2m_0^i+1) + \sum_{n=1}^{\infty}\sqrt{n^2+\lambda_n}\sum_{i=1}^{D-2}(2m_n^i+1) \\ &+ K + \sum_{n=1}^{\infty}\frac{n^2\lambda_n}{n^2+\lambda_n}(\vec{x}_n^{(0)})^2 \end{aligned}\right\}, \tag{23}$$

becomes larger as the occupation numbers increase, the dominent contribution to (22) in the asymptotic case of $\tau^{(2)}-\tau^{(1)} \gg 1$ comes from the ground state wave function Ψ_0 having $m_n^i = 0$. Then, the propagator in this asymptotic limit has a factorized from

$$K_{P+}\left(\{\vec{x}_n^{(1)}\}, \{\vec{x}_n^{(2)}\}; U[\{\vec{x}_n\}]\right)\Big|_{\tau^{(2)}-\tau^{(1)} \gg 1}$$
$$= \Psi_0(\{\vec{x}_n^{(2)}\})\, \Psi_0^*(\{\vec{x}_n^{(1)}\})\, e^{-E_0(\tau^{(2)}-\tau^{(1)})} , \tag{24}$$

where

$$\Psi_0(\{\vec{x}_n\}) = \left(\frac{\lambda_0}{2\pi^2}\right)^{(D-2)/8} e^{-\frac{1}{2}\sqrt{\lambda_0/2}\,(\vec{x}_0-\vec{x}_0^{(0)})^2}$$
$$\times \prod_{n=1}^{\infty}\left(\frac{n^2+\lambda_n}{\pi^2}\right)^{(D-2)/8} e^{-\frac{1}{2}\sqrt{n^2+\lambda_n}\left(\vec{x}_n-\frac{\lambda_n}{n^2+\lambda_n}\vec{x}_n^{(0)}\right)^2} , \tag{25}$$

and its energy is

$$4P_+E_0 = (D-2)\left\{\sqrt{2\lambda_0}+\sum_{n=1}^{\infty}\sqrt{n^2+\lambda_n}\right\} + K + \sum_{n=1}^{\infty}\frac{n^2\lambda_n}{n^2+\lambda_n}(\vec{x}_n^{(0)})^2 . \tag{26}$$

Now, we will determine the parameters K , $\{\lambda_n\}$, and $\{\vec{x}_n^{(0)}\}$ so that the Eq.(17) can hold consistently. Under the asymptotic situation that $\tau'-\tau^{(1)} \gg 1$ and $\tau^{(2)}-\tau' \gg 1$, with the white noise potential(18), Eq.(17) is reduced to

$$U(\{\vec{x}_n\}) = -\rho w \,|\, \Psi_0(\{\vec{x}_n\})|^2 , \tag{27}$$

which gives

$$K + \sum_{n=0}^{\infty}\lambda_n(\vec{x}_n-\vec{x}_n^{(0)})^2$$
$$\approx 8\pi P_+^2\rho w\left(\frac{\lambda_0}{2\pi^2}\prod_{n=1}^{\infty}\frac{n^2+\lambda_n}{\pi^2}\right)^{\frac{D-2}{4}}$$
$$\times\left\{-1+\sqrt{\lambda_0/2}\,(\vec{x}_0-\vec{x}_0^{(0)})^2+\sum_{n=1}^{\infty}\sqrt{n^2+\lambda_n}\left(\vec{x}_n-\frac{\lambda_n}{n^2+\lambda_n}\vec{x}_n^{(0)}\right)^2\right\}. \tag{28}$$

for the choice of (19).

(i) A solution of (28) is obtained when $\vec{x}_n^{(0)}=0$ for $n\geqslant 1$. Then,

$$\lambda_0 = \tfrac{1}{2}K^2, \quad \text{and} \quad \lambda_n = \tfrac{1}{2}\left[1 \mp \sqrt{1+4n^2/K}\right] \quad (n\geqslant 1), \tag{29}$$

and K is determined so as to satisfy

$$K = -8\pi P_+^2\rho w\left(\frac{\lambda_0}{2\pi^2}\right)^{\frac{D-2}{4}}\prod_{n=1}^{\infty}\left(\frac{n^2+\lambda_n}{\pi^2}\right)^{\frac{D-2}{4}} . \tag{30}$$

We have positive λ_n $(n\geqslant 0)$ for the negative value of K , so that we may have the potential which localizes the wave functional of the string around $\vec{x}_0^{(0)}$ and $\vec{x}_n^{(0)} = 0 \; (n \geq 1)$. The solution suggests that the string may be completely folded in the neighborhood of $\vec{x}^{(0)}$, but it can extend in the x^- direction.

(ii) Another solution of (28) is possible if $|\lambda_n| \gg n^2 \; (n \geq 1)$ is satisfied under an assumption that higher oscillation modes $n > n_*$ are negligible. Then, we have

$$2\lambda_0 = \lambda_n = K^2 \qquad (n \geqslant 1)$$

and

$$K = 2\pi \left(\frac{-1}{4\gamma_+^2 \rho w} \right)^{\frac{1}{(D-2)(n_*+1)/2 - 1}} . \qquad (31)$$

Here, we have encountered the fact that all the parameters including $\lambda_n \; (n \geqslant 0)$ might not be real parameters. In any way, it is safer not to conclude that we have really folded string states confined in a small region due to the interaction with the background disordered medium, since the discussion given above based on the assumption (19) seems rather crude.

More serious study would be necessary to find a self-consistent potential $U_{SC}[\vec{X}_\perp(\sigma)]$ which is attractive at long distance, but repulsive at short distance. We can, however, expect to have such potential, when we recognize an analogy between our problem and the study of electron localization based on the path integral formulation of polymers. In the near future, we are hoping to come back to this problem.[15)]

5. ACKNOWLEDGEMENTS

The author is grateful to the members of the theory group at KEK,for helpful discussions about Anderson's locarization, especially to H. Iwasaki for his good instructions.

6. REFERENCES

1) Green, M. and Schwartz, J.H., Phys. Lett. 149B, 117 (1984); Nucl. Phys. B255, 93 (1985).

2) Gross, D.J., Harvey, J.A., Martinec, E., and Rohm, R., Phys. Rev. Lett. 54 502 (1985); Nucl. Phys. B256, 253 (1985).

3) Gross, D.J., Harvey, J.A., Martinec, E., and Rohm, R., to be published in Nucl. Phys. B.

Yahikozawa, S., Phys. Lett. 166B, 135 (1986).

4) Sugamoto, A., and Suzuki, H., KEK preprint, KEK-TH 125 March (1986), revised on June (1986).

Suzuki, H., KEK preprint, KEK-TH 126, April (1986).

5) Green, M.B., in Proceedings of The Lepton and Photon Conference held at Kyoto (1985), ed. by Konuma, M., and Takahashi, K. .

6) Anderson, P.W., Phys. Rev. 109, 1492 (1958).

Abrahams, E., Anderson, P.W., Licciardello,D.C., and Ramakrishnan, T.V., Phys. Rev. Lett. 42, 673 (1979).

7) Edwards, S.F., J. Phys. C3, L30 (1970); J. noncryst. Solids 4, 417 (1970).

Freed, K.F., J. Phys. C4, L331 (1971).

Abram, R.A., and Edwards, S.F., J. Phys. C5 1183 and 1196 (1972).

Economou, E.N., Cohen, M.H., Freed, K.F., and Kirkpatrick, E.S. in Amorphous and Liquid Semiconductors ed. by Tauc, J., plenum press (1979).

8) Edwards, S.F., Proc. Phys. Soc. (London) 85, 613 (1965).

Freed, K.F., J. Chem. Phys. 55, 3910 (1971).

de Gennes, P.-G., Scaling Concepts in Polymer Physics, Cornell Univ. press (1979).

9) Seo, K., and Sugamoto, A., Phys. Rev. D24, 1630 (1981).

10) Yamaguchi, Y., "Development and Present Status of High Energy Physics in Europe and Others." (PHY-UT, YY-1) July (1968), and "Particle Physics Laboratory in Japan." (PHY-UT, YY-2) July (1968).

11) General references in this section are

Stillwell, J., Classical Topology and Combinatorial Group Theory, Springer-Verlag, (1980).

Encycropedic Dictionary of Mathematics, Vol 1, MIT press (1977), ed. by Iyanaga, S., and Kawada, Y. .

12) Hironaka, H., Sugaku Seminar 1, 6 (1985), (in Japanese).

13) Dixson, L., Harvey, J.A., Vafa, C., and Witten, E., Nucl. Phys. B261, 678 (1985), and Princeton preprint (1986).

Harvey, J.A., talk at Santa Barbara Workshop (1985).

Strominger, A., i.b.d. (1985).

14) Mandelstam, S., Nucl. Phys. B64, 205 (1973).

Kikkawa, K., and Kaku, M., Phys. Rev. D10, 1110, and 1823 (1974).

15) Iwasaki, H., and Sugamoto, A., (in preparation).

TOPOLOGICAL ANOMALIES IN FIELD AND STRING THEORIES

Kazuo Fujikawa
Research Institute for Theoretical Physics
Hiroshima University, Takehara, Hiroshima 725

ABSTRACT

Some of the anomalies in field theory are related to the topological properties. The best known example is the chiral anomaly in the Weinberg-Salam theory in the presence of instantons, which was analysed by 't Hooft. A closely related topological anomaly in the string theory is the Faddeev-Popov ghost number anomaly, which is a local version of the Riemann-Roch theorem. We discuss the analogy between these two anomalies and the structure of the topological vacua appearing in the string path integral.

I. Introduction

The chiral anomaly in field theory is related to the topological properties carried by the background Yang-Mills fields. The mathematical manifestation of this connection is known as the Atiyah-Singer index theorem.[1] In the language of particle physics, this connection is known as the fermion number non-conservation[2] in the presence of instantons.[3] A detailed analysis of this problem has been performed by 't Hooft in 1976. The interesting physical implication of this analysis is that the baryon decays into leptons in the presence of instantons. This baryon decay can take place even without the explicit baryon number non-conserving interactions, which appear, for example, in the conventional grand unification schemes.

On the other hand, a very geometrical formulation of string theories has been given by Polyakov[4], as an extention of the conventional Nambu-Goto action. If one applies the conventional Faddeev-Popov prescription to this path integral, one finds the BRS symmetry which gives rise to the generalized Virasoro conditions.[5] Another interesting property of this approach is that the Faddeev-Popov ghost number contains an anomaly,[5] which is proportional to the Euler number of the string world sheet. Since the ghost number anomaly has an appearance closely related to the chiral anomaly in the gauge theory, it was suggested[5] that the ghost number anomaly may play a role similar to the chiral anomaly in the θ-vacuum[6] when one sums the surfaces with non-trivial topologies in the string theory. It was later shown by Friedan[7] and by Alvarez[8] that the Faddeev-Popov ghost zero-modes are related to the Riemann-Roch theorem. From the mathematical view point, one can thus clearly understand the connection of the fermion number anomaly in the Weinberg-Salam theory (Atiyah-Singer index theorem) and the ghost number anomaly in the string path integral (Riemann-Roch theorem).

In the present essay, I would like to further discuss the analogy between these two topological anomalies in field and string theories and the implications of the topological anomaly on the vacuum structure in the string theory. We note that the implications of the ghost number anomaly have also been discussed recently by Friedan, Martinec and Shenker[9], and by Witten[10] in connection with the covariant formulation of string field theories.

II. Fermion Number Non-Conservation in Feild Theory

We study the simplified version of the Weinberg-Salam theory

$$\mathcal{L} = \bar{\psi}_L i\gamma^\mu(\partial_\mu + W_\mu)\psi_L + \bar{\psi}_R i\not{\partial}\psi_R - \frac{1}{4} F^a_{\mu\nu}F^{a\mu\nu} \tag{2.1}$$

with

$$\psi = \binom{u}{d}$$

$$W_\mu = -igW^a_\mu T^a, \quad [T^a, T^b] = if^{abc}T^c, \quad \mathrm{Tr}\, T^aT^b = \frac{1}{2}\delta^{ab} .$$

It can be shown that the fermion number current in (2.1) contains the anomaly[2)]

$$\partial_\mu [\bar{\psi}_L \gamma^\mu \psi_L + \bar{\psi}_R \gamma^\mu \psi_R] = \mathrm{Tr}\, \frac{i}{16\pi^2}\, {}^*F_{\mu\nu} F^{\mu\nu} \tag{2.2}$$

In the presence of the instanton, we have the relation

$$\int d^4x\, \frac{-i}{16\pi^2}\, {}^*FF = \nu = 1\ . \tag{2.3}$$

Eqs.(2.2) and (2.3) indicate that

$$Q(+\infty) - Q(-\infty) \equiv \int_{-\infty}^{\infty} dt\, \partial_0 \int d^3x j^0(x) = -\nu\ . \tag{2.4}$$

Namely, the fermion number is not conserved in the presence f instantons. This is the basic observation of 't Hooft.[2)]

To discuss this problem in more detail, one may define

$$W_\mu \equiv \hat{W}_\mu(\tau,\, x) + A_\mu(x) \tag{2.5}$$

with $\hat{W}_\mu(\tau, x)$ the instanton solution, and τ generically stands for the collective coordinates of the instanton; A_μ is the fluctuation around the instanton solution. The vacuum-to-vacuum amplitude is then given by

$$<-\nu|0> = \int [\mathcal{D}\hat{W}_\mu]_{(\nu)} \mathcal{D}A_\mu \mathcal{D}\bar{\psi}\mathcal{D}\psi e^{S+\int \mathcal{L}_g dx} \tag{2.6}$$

where $\mathcal{L}_g$ stands for the gauge fixing and compensating terms; the ghost variables are suppressed here, as they are not important for the discussions in this section.. The amplitude (2.6) as it stands vanishes, which can be understood by considering the change of integration variables in (2.6) as

$$\psi = e^{i\theta\gamma_5}\psi' , \qquad \bar{\psi} = \bar{\psi}' e^{i\theta\gamma_5}\ . \tag{2.7}$$

The chiral Jacobian factor then gives rise to[11)]

$$<-\nu|0> = \int [\mathcal{D}\hat{W}]_{(\nu)} \mathcal{D}A_\mu \mathcal{D}\psi' \mathcal{D}\bar{\psi}' e^{S+S_g - i\theta\nu}\ . \tag{2.8}$$

The parameter θ is arbitrary, and one can see that (2.6) vanishes.

This is a reffection of the fact that the fermion variables ψ and $\bar{\psi}$ contain zero modes specified by the Atiyah-Singer index theorem

$$\begin{aligned}\nu &= \sum_n \int \varphi_n(x)^\dagger \gamma_5 \varphi_n(x) dx \\ &= \sum_n \int \varphi_n(x)_R^\dagger \varphi_n(x)_R dx - \sum_n \int \varphi_n(x)_L^\dagger \varphi_n(x)_L dx\end{aligned} \tag{2.9}$$

where $\varphi_L(x)$ and $\varphi_R(x)$ stand for the zero modes in ψ_L and $\bar{\psi}_L$, respectively. The zero modes do not appear in the action S, but they appear in the path integral measure

$$\prod_x \mathcal{D}\bar{\psi}_L(x)\mathcal{D}\psi_L(x) = \prod_n da_n db_n \tag{2.10}$$

with[11])

$$\begin{aligned}\psi_L(x) &= \sum_{\lambda_n=0} a_n\left(\frac{1-\gamma_5}{2}\right)\varphi_n(x) + \sum_n{}' a_n\left(\frac{1-\gamma_5}{\sqrt{2}}\right)\varphi_n(x) \\ \bar{\psi}_L(x) &= \sum_{\lambda_n=0} b_n \varphi_n(x)^\dagger\left(\frac{1+\gamma_5}{2}\right) + \sum_n{}' b_n \varphi_n(x)^\dagger\left(\frac{1+\gamma_5}{\sqrt{2}}\right)\end{aligned} \tag{2.11}$$

where the prime indicates the summation over non-zero eigen-values in

$$\not{D}\varphi_n(x) \equiv \gamma^\mu(\partial_\mu + W_\mu)\varphi_n(x) = \lambda_n \varphi_n(x) \ . \tag{2.12}$$

The integral (2.6) thus vanishes for $\nu \neq 0$.

To obtain the non-vanishing result, one has to investigate the amplitude such as[2])

$$<-\nu|\bar{\psi}\gamma_5\psi \ \ldots\ldots \ |0> \neq 0 \tag{2.13}$$

so that the phase factor from the operators in (2.13) arising from the transformation (2.7) cancels the phase factor $e^{i\nu\theta}$ in (2.8). This gives the so-called chirality selection rule.

A technical detail in this scheme, which becomes instructive in the later discussions, is the transformation of the instanton measure as[2])

$$\prod \mathcal{D}\hat{W}_\mu(\tau, x) = J \prod_k d\delta\tau_k \tag{2.14}$$

with the Jacobian factor J and the collective coordinates $\delta\tau_k$. We also note that the so-called θ-vacuum prescription[6] corresponds to

$$\sum_{\nu}\int d\mu_{(\nu)} e^{S+S_g+i\theta\nu} . \tag{2.15}$$

III. Ghost Number Non-Conservation in String Theory

In the path integral of the Polyakov-type string theory, one encounters the ghost number anomaly. This is seen by starting with the Lagrangian

$$\mathcal{L} = -\frac{1}{2}\sqrt{g}\, g^{\alpha\beta}\partial_\alpha X^\mu \partial_\beta X^\mu \tag{3.1}$$

and the conformal gauge (which is well-defined at least locally)

$$g_{\alpha\beta} = \rho\delta_{\alpha\beta} . \tag{3.2}$$

The path integral in the conventional Faddeev-Popov prescription then becomes[5]

$$Z = \int \mathcal{D}\sqrt{\rho}\,\mathcal{D}\tilde{X}^\mu \mathcal{D}\xi\,\mathcal{D}\tilde{\eta}\, \exp\{\int [-\frac{1}{2}\partial_\alpha(\frac{\tilde{X}^\mu}{\sqrt{\rho}})\partial_\alpha(\frac{\tilde{X}^\mu}{\sqrt{\rho}}) + \xi\sqrt{\rho}\,\not{\partial}\frac{1}{\rho}\tilde{\eta}]d^2\sigma\} \tag{3.3}$$

where ξ and η stand for the Faddeev-Popov anti-ghost and ghost, respectively, and

$$\xi = \begin{pmatrix} \xi_1 \\ \xi_2 \end{pmatrix}, \qquad \tilde{\eta} \equiv \begin{pmatrix} \rho\eta^1 \\ \rho\eta^2 \end{pmatrix}$$

$$\not{\partial} \equiv \sigma^1\partial_1 + \sigma^3\partial_2 \tag{3.4}$$

with Pauli matrices σ^k.

By considering the ghost number transformation

$$\tilde{\eta} \to e^{\alpha(\sigma)}\tilde{\eta}$$

$$\xi \to \xi e^{-\alpha(\sigma)} \tag{3.5}$$

one finds the anomalous identity[5]

$$\partial_\mu[\xi\sqrt{\rho}\,\sigma^\mu \frac{1}{\rho}\tilde{\eta}] = \frac{3}{4\pi}\partial^2 \ln\rho = -\frac{3}{4\pi}\sqrt{g}\,R\,. \tag{3.6}$$

The relation (3.6) corresponds to the local version of the Riemann-Roch theorem, just as (2.2) corresponds to the local version of the Atiyah-Singer index theorem. The gauge (3.2) we used is well-defined for any local coordinate patch of the Riemannian surface, but it becomes singular for a general class of Riemannian surfaces when considered globally. One thus has to use a more general class of background gauges to see the connection of (3.6) with the Riemann-Roch theorem. The general choice of coordinates has been discussed in detail by Alvarez.[8)]

The essence of the analysis of Alvarez is summarized in our language by

$$\sum_\chi \int d\mu_{(\chi)} e^{S_{eff}+\theta\chi} \tag{3.7}$$

in analogy with (2.15). Here χ stands for the Euler number of the Riemannian surfaces, which is given by

$$\chi = 2 - 2g \tag{3.8}$$

for the closed orientable surface with g handles, for example.

For D=26, one can discard the detailed specification of path integral variables, and one may define the path integral measure

$$d\mu_{(\chi)} = \Pi[\mathcal{D}\,\hat{g}_{\alpha\beta}(\tau,\sigma)]_{(\chi)}\mathcal{D}h_{\alpha\beta}\mathcal{D}X^\mu\mathcal{D}\xi\,\mathcal{D}\eta\,\mathcal{D}B \tag{3.9}$$

with

$$g_{\alpha\beta}(\sigma) \equiv \tilde{g}_{\alpha\beta}(\tau,\sigma) + h_{\alpha\beta}(\sigma) \tag{3.10}$$

and $h_{\alpha\beta}$ is the fluctuation around the background metric $\hat{g}_{\alpha\beta}(\tau,\sigma)$,

$$h_{\alpha\beta} = h_{\beta\alpha}\,, \qquad g^{\alpha\beta}h_{\alpha\beta} = 0\,. \tag{3.11}$$

The effective action in (3.7) is then given by

$$S_{eff} = \int -\frac{1}{2}\sqrt{g}\,g^{\alpha\beta}\partial_\alpha X^\mu\partial_\beta X^\mu d^2\sigma + \int[iB_1h_{12} + iB_2\,\frac{1}{2}(h_{11} - h_{22})]d^2\sigma$$

$$+ \int \xi P(g_{\alpha\beta}) \eta d^2\sigma \tag{3.12}$$

where $P(g_{\alpha\beta})$ stands for the Faddeev-Popov operator. The ghost number anomaly (3.6) is then written as

$$\partial_\mu G^\mu = - \frac{3}{4\pi} \sqrt{g}\, R \tag{3.13}$$

by using the ghost number current defined from (3.12).

We here emphasize the difference between our prescription (3.7), (3.9) and (3.12) and that of Alvarez.[8] In the definition of Alvarez, one does not include the zero modes of ξ and η in the path integral, whereas we include all the possible zero modes in the path integral. This inclusion of all the zero modes of ξ and η is essential to keep the close analogy with the instanton calculation in Section 2.

As for the integration over the Teichmüller deformation[8] τ, one can rewrite (3.9) as

$$\Pi \mathcal{D} \hat{g}_{\alpha\beta}(\tau,\sigma) = J \prod_k d\delta\tau_k \tag{3.14}$$

with a suitable Jacobian factor. The Teichmüller parameters, which specify the deformation of the background Riemannian surfaces, correspond to the collective coordinates of the instanton solution (2.14) in Section 2.

IV. Topological Vacua in Open String Theory

We illustrate the implications of the ghost number anomaly for the simplest case in the string theory. For the tree-level open string amplitudes, one considers a disc or hemi-sphere, to the boundary of which n-external particles are attached. One can visualize this scattering amplitude as

$$n_i\text{-particles} \rightarrow \text{hemi-sphere} \rightarrow n_f\text{-particles} . \tag{4.1}$$

Namely, n_i external particles enter the vacuum and they are absorbed into the hemi-sphere, and then n_f external particles are emitted with $n = n_i + n_f$, the total number of external particles. In this process, the string (hemi-sphere) appears and then disappears only inside the vacuum; the hemi-sphere is then regarded as the "instanton" in this scattering process. The presence of the "instanton" inside the vacuum influences the topological properties of the scattering amplitude. In fact, the ghost number anomaly (3.6) suggests

$$Q_g(\infty) - Q_g(-\infty) = -\frac{3}{4\pi}\int\sqrt{g}\, R d^2\sigma = -3 \tag{4.2}$$

for the hemi-sphere which gives $\chi = 1$.

The vacuum-to-vacuum amplitude then vanishes

$$<-3|0> = \int [d\mu]_{\chi=1} e^{S_{eff}+\theta\chi} = 0 \ . \tag{4.3}$$

In the present case, (4.3) can also be written as

$$<-\frac{3}{2}|-\frac{3}{2}> = 0 \tag{4.4}$$

since the ghost "number" is the scale transformation, not the phase transformation, just to keep the real ghost variables ξ and η to be real. This is a major difference from the instanton theory, where the fermion number is the phase transformation. To make the amplitude (4.4) non-vanishing, one may insert 3 ghost fields just as we inserted fermion operators into (2.13). Namely

$$<-\frac{3}{2}|\eta(\sigma_1)\eta(\sigma_2)\eta(\sigma_3)|-\frac{3}{2}> \neq 0 \ . \tag{4.5}$$

Geometrically, (4.5) corresponds to the fact that the hemi-sphere contains 3 conformal killing vectors[8)] and thus the Faddeev-Popov ghost fields contain 3 Euclidean zero modes, saturating the ghost number anomaly (4.2).

The conformal killing vectors are generated by the SL(2,R) (Moebius) transformation in the upper-half complex plane when the unit hemi-sphere is mapped into the complex plane. In this case, the 3 conformal killing vectors are spanned by

$$1, \quad z, \quad z^2 \tag{4.6}$$

in terms of the complex coordinate z. (4.5) then gives rise to

$$<-\frac{3}{2}|\eta(z_1)\eta(z_2)\eta(z_3)|-\frac{3}{2}>$$

$$= <-\frac{3}{2}|\eta_1^{(0)}\eta_2^{(0)}\eta_3^{(0)}|-\frac{3}{2}> \det\begin{pmatrix} 1 & z_1 & z_1^2 \\ 1 & z_2 & z_2^2 \\ 1 & z_3 & z_3^2 \end{pmatrix} \tag{4.7}$$

$$= <-\frac{3}{2}|\eta_1^{(0)}\eta_2^{(0)}\eta_3^{(0)}|-\frac{3}{2}> \times (z_1 - z_2)(z_2 - z_3)(z_3 - z_1)$$

since the path integral measure contains the factor

$$\int d\eta_1^{(0)} d\eta_2^{(0)} d\eta_3^{(0)} \tag{4.8}$$

with ghost zero-modes $\eta_1^{(0)} \sim \eta_3^{(0)}$. Namely, we need at least 3 zero-modes in the integrand to get the non-vanishing result. The integration over the Grassmann numbers anti-symmetrizes the order of ghost fields, and thus giving rise to the determinant factor in (4.7). The factor in (4.7) agrees with the Jacobian factor for the transformation from the factor $dz_1dz_2dz_3$ to the coefficients ε_1, ε_2 and ε_3 in the infinitesimal SL(2,R) transformation

$$\delta z = \varepsilon_1 + \varepsilon_2 z + \varepsilon_3 z^2 . \tag{4.9}$$

A systematic analysis of the ghost number anomaly and associated ghost zero modes thus gives rise to the same result as in the formulation of Alvarez.[8] The present approach, which maintains a close analogy with the instanton calculation, has certain advantages when one compares the path integral formalism with the operator formalism. In fact, Friedan, Martinec and Shenker[9], and also Witten[10] have recently shown that the considerations of the ghost number anomaly are quite important for the covariant formulation of string scattering amplitudes and string field theories.[12]

When one considers the more general string configurations with handles, one encounters the anti-ghost zero modes which are in one-to-one correspondence with the Teichmüller parameters.[8] The calculation then closely resembles the instanton calculation, and one has to include the anti-ghost zero modes also into the amplitude such as in (4.5). The relation between the number of (anti-) ghost zero modes and the ghost member anomaly in the case of closed strings is well-described in Ref.(8).

V. Conclusion

It is quite amusing to find a close analogue of the instanton calculation in the path integral formulation of string theories. In fact, the string path integral provides a very concrete and understandable example of the general "instanton" calculation.[13] It will be very interesting to find further examples of the general "instanton" calculations associated with topological anomalies.

I thank Professor Yoshio Yamaguchi for his challenging and critical comments on the various aspects of my past study in particle physics, and I wish him many happy returns.

References

1) M. F. Atiyah and I. M. Singer, Ann. Math. 87, 484 (1968).
2) G. 't Hooft, Phys. Rev. Lett. 37, 8 (1976); Phys. Rev. D14, 3432 (1976).
3) A. A. Belavin et al., Phys. Lett. 59B, 85 (1975).
4) A. M. Polyakov, Phys. Lett. 103B, 207 (1981); 103B, 211 (1981).
5) K. Fujikawa, Phys. Rev. D25, 2584 (1982).
6) R. Jackiw and C. Rebbi, Phys. Rev. Lett. 37, 172 (1976).
C. Callan, R. Dashen and D. Gross, Phys. Lett. 63B, 334 (1976).
7) D. Friedan, in Recent Advances in Field Theory and Statistical Mechanics, J. B. Zuber and R. Stora, eds. (Elsevier, 1984).
8) O. Alvarez, Nucl. Phys. B216, 125 (1983).
9) D. Friedan, E. Martinec and S. H. Shenker, Phys. Lett. 160B, 55 (1985) and Chicago preprint 85-89 (1985).
10) E. Witten, Princeton preprints, October, 1985 and March, 1986.
11) K. Fujikawa, Phys. Rev. Lett. 42, 1195 (1979); Phys. Rev. D21, 2848 (1980).

12) As for the review of the string field theory, see T. Yoneya, Talk given at Seventh Workshop on Grand Unification, Toyama, April 16 ∿ 18, 1986, and references therein.

13) J. Polchinski, Comm. Math. Phys. 140, 37 (1986).

IV. NUCLEAR PHYSICS

MICROSCOPIC THEORY OF NUCLEAR COLLECTIVE DYNAMICS

Toshio Marumori and Fumihiko Sakata*)

Institute of Physics, University of Tsukuba, Ibaraki 305, JAPAN

*)Institute for Nuclear Study, University of Tokyo,
Tanashi, Tokyo 188, JAPAN

ABSTRACT

A systematic microscopic theory of nuclear collective dynamics, which has been developed under INS-TSUKUBA joint research project supported by Theoretical Physics Division of INS, is outlined by placing emphasis on its underlying physical pictures.

§1. INTRODUCTION

The aim of the microscopic theory of nuclear collective dynamics is to make clear non-linear dynamics involving the collective and single-particle modes of motion on the basis of the nuclear many-body problem. Since the nucleus is an isolated finite many-body quantal system in which the self-consistent mean field is realized, characteristic difficulties in exploring the dynamics come from the following fact: The single-particle orbits have to adjust their features self-consistently in accordance with time-evolution of the mean field associated with the collective motion under consideration. Such a dynamical change of the single-particle modes in turn induces remarkable alteration of the microscopic structure of collective operators characterizing the collective motion, because the collective operator is expressed as coherent sum of the single-particle modes of motion.

The first step to explore such highly non-linear dynamics, requiring the self-consistency strongly, was to employ the time-dependent Hartree-Fock (TDHF) theory.[1),2)] In §2 we formulate our theory within the framework of the TDHF theory.[3)~6)] Since the TDHF equation is known[4),7),8)] to be expressed as a set of canonical equations of motion in classical mechanics in the TDHF symplectic manifold (phase space), the use of the TDHF theory enables us to investigate basic ingredients of the non-linear dynamics in terms of TDHF trajectories. Thus, collective, dissipative and chaotic (stochastic) behaviors in the nuclear collective dynamics are discussed in connection with the geometry of an "optimum" collective submanifold (surface) extracted by our theory out of the TDHF manifold.

By explicitly employing clear-cut correspondence between the TDHF symplectic manifold and the boson space used in the nuclear boson-expansion theory,[9] in §3 a full quantum theory of nuclear collective dynamics is developed as a direct extension of the "classical" theory discussed in §2. The theory properly defines an "optimum" collective subspace and "global" collective operators providing the subspace. The global collective operators have to have a definite sense not only in the vicinity of the "vacuum" with a stable mean field but also in a higher excited region far from it, i.e., a "large-amplitude region".

§2. NUCLEAR COLLECTIVE DYNAMICS IN THE TDHF THEORY

[2.1] TDHF Equation in Canonical-Variable Representation

The basic equation of the TDHF theory is

$$\delta\langle f|\left(i\frac{\partial}{\partial t}-\hat{H}\right)|f\rangle = 0 \qquad (2.1)^{*)}$$

where the time-dependent Slater determinant $|f\rangle$ is given by

$$|f\rangle = e^{i\hat{F}}|\phi_0\rangle,\quad \hat{F} = \sum_{\mu i}\{f_{\mu i}(t)a^\dagger_\mu b^\dagger_i + f^*_{\mu i}(t)b_i a_\mu\}. \qquad (2.2)^{**)}$$

Here $|\phi_0\rangle$ denotes a Hartree-Fock stationary state, and $a^\dagger_\mu$ (μ=1,2,···, m) and $b^\dagger_i$ (i=1,2,···,n) mean the particle- and hole-creation operators with respect to $|\phi_0\rangle$. Instead of 2mn (≡ 2K) variables $\{f_{\mu i}(t), f^*_{\mu i}(t)\}$, we introduce a new set of variables $\{C_{\mu i}(t), C^*_{\mu i}(t)\}$ through a variable transformation $f_{\mu i} = f_{\mu i}(C, C^*)$. The symplectic structure of the TDHF theory always enables us to choose the variable $\{C_{\mu i}, C^*_{\mu i}\}$ to be of the canonical-variable representation,[4),7),8)] in which the TDHF equation (2.1) can be expressed as the canonical equations of motion in classical mechanics;

---

*) Throughout this paper, we adopt the convention of using $\hbar$=1.

**) We use the convention of denoting occupied single-particle states of $|\phi_0\rangle$ by indices i, j, ···, and unoccupied states by μ, ν, ···.

$$i\dot{C}_{\mu i} = \partial H/\partial C^*_{\mu i}, \quad i\dot{C}^*_{\mu i} = -\partial H/\partial C_{\mu i}, \quad H = \langle f|\hat{H}|f\rangle ,$$
$$q_{\mu i} = \frac{1}{\sqrt{2}}(C^*_{\mu i}+C_{\mu i}), \quad p_{\mu i} = \frac{i}{\sqrt{2}}(C^*_{\mu i}-C_{\mu i}). \tag{2.3}$$

The necessary condition for the canonical-variable representation is that the local infinitesimal generators, $\hat{O}^\dagger_{\mu i} \equiv \{\partial e^{i\hat{F}}/\partial C_{\mu i}\}\cdot e^{-i\hat{F}}$, (which are one-body operators) have to satisfy the "weak" boson-like commutation relations

$$\langle f|[\hat{O}_{\mu i}, \hat{O}^\dagger_{\nu j}]|f\rangle = \delta_{\mu\nu}\delta_{ij}, \quad \langle f|[O_{\mu i}, O_{\nu j}]|f\rangle = 0. \tag{2.4}$$

Since the TDHF equation (2.1) can be written as $\delta\langle f|\hat{H}-i\Sigma\{\dot{C}_{\mu i}\hat{O}^\dagger_{\mu i} - \dot{C}^*_{\mu i}\hat{O}_{\mu i}\}|f\rangle = 0$, we can easily see that Eq.(2.4) is the necessary condition for obtaining Eq.(2.3), by taking $|\delta f\rangle \propto :\hat{O}_{\mu i}:|f\rangle$ and $:\hat{O}^\dagger_{\mu i}:|f\rangle$. Since Eq.(2.3) is invariant under the canonical transformations of the variables $\{C_{\mu i}, C^*_{\mu i}\}$, there is a freedom of choice of the set of canonical variables $\{C_{\mu i}, C^*_{\mu i}\}$ satisfying Eq. (2.4). One of the simplest is to choose a set so as to satisfy the following canonical-variable condition;[3),4)]

$$\langle f|\hat{O}^\dagger_{\mu i}|f\rangle = \frac{1}{2} C^*_{\mu i}, \quad \langle f|\hat{O}_{\mu i}|f\rangle = \frac{1}{2} C_{\mu i}. \tag{2.5}$$

As shown in §3, this choice of the set of canonical variables $\{C_{\mu i}, C^*_{\mu i}\}$ enables us to make unique correspondence to the set of bosons $\{B_{\mu i}, B^\dagger_{\mu i}\}$ employed in the nuclear boson-expansion theory.[9)]

[2.2] "Optimum" Collective Submanifold

The solutions of Eq.(2.3) give trajectories (called TDHF trajectories) in a 2K-dimensional TDHF symplectic manifold (phase space) denoted by $M^{2K}:\{C_{\mu i}, C^*_{\mu i}\}$. If a group of many TDHF trajectories gathering together around a certain small-dimensional submanifold (i.e., surface) Σ^{2L} with L<<K at an instant can be always confined in a small domain of the neighborhood of the submanifold for a long enough time, we may naturally say that the system under consideration displays "large-amplitude collective motion" along an "optimized" trajectory on the surface Σ^{2L}, which is a representative of the group of TDHF trajectories. Hereafter we call the optimized trajectory a

collective trajectory and call the surface Σ^{2L} an "optimum" collective submanifold. Consequently, the collective trajectory should be described by a set of L-pairs of collective variables $\{\eta_r, \eta_r^*; r=1, 2,\cdots,L\}$. The use of the collective variables $\{\eta_r, \eta_r^*\}$ implies an introduction of a 2L-dimensional manifold $M^{2L}:\{\eta_r, \eta_r^*\}$, which is diffeomorphic to the optimum collective submanifold Σ^{2L} embedded in the TDHF manifold M^{2K}. Thus, the central theme of the theory of nuclear collective dynamics consists of the following problems:

(i) We have to find a prescription how to extract the 2L-dimensional optimum collective submanifold Σ^{2L} out of the TDHF manifold M^{2K}, i.e., how to define the diffeomorphic mapping

$$M^{2L} \to \Sigma^{2L} \text{ embedded in } M^{2K}. \tag{2.6}$$

In addition, we have to obtain the collective Hamiltonian $\mathcal{H}_{coll}(\eta,\eta^*)$ defined in M^{2L}. A solution of the small dimensional coupled equations of motion

$$i\dot{\eta}_r = \partial\mathcal{H}_{coll}/\partial\eta_r^*, \quad i\dot{\eta}_r^* = -\partial\mathcal{H}_{coll}/\partial\eta_r \quad (r=1,\cdots,L) \tag{2.7}$$

gives a trajectory in M^{2L}, which is mapped onto Σ^{2L} with the aid of the mapping in Eq.(2.6). The resultant mapped trajectory should coincide with the collective trajectory in the TDHF manifold M^{2K}. (Fig. 1).

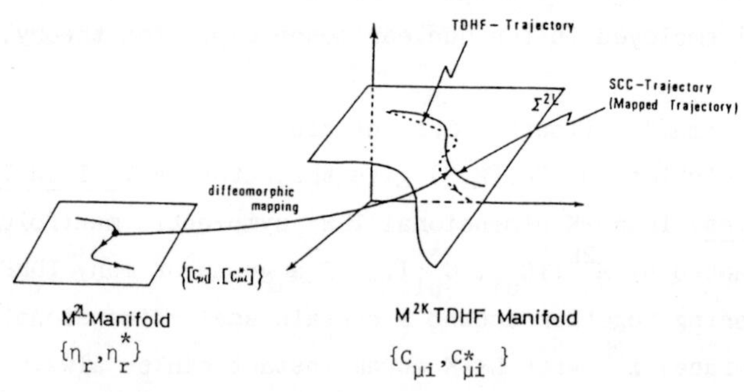

[Fig. 1]

(ii) We have to determine dynamical conditions, under which the optimum collective submanifold Σ^{2L} can be regarded as an approximate integral surface to guarantee "stability" of the collective trajectories.

[2.3] Self-Consistent Collective-Coordinate Method

The first problem (i) for the nuclear collective dynamics has been performed under the name of Self-Consistent Collective-Coordinate (SCC) Method.[3] The basic principle of the SCC method is to define the optimum collective submanifold (surface) Σ^{2L} in such a way that the total energy H on the surface is stationary with respect to variations perpendicular to the surface. This principle is called the maximal-decoupling condition[3] formulated in a form named "the invariance principle of the time-dependent Schrödinger equation."[2]

Let us consider a general variable transformation from the canonical variables $\{C_{\mu i}, C^*_{\mu i}\}$ to a set of variables $\{\eta_r, \eta^*_r; \xi_\alpha, \xi^*_\alpha: \alpha=1,\cdots,K-L\}$ including the collective variables $\{\eta_r, \eta^*_r\}$;

$$C_{\mu i} = C_{\mu i}(\eta,\eta^*;\ \xi,\xi^*), \qquad C^*_{\mu i} = C^*_{\mu i}(\eta,\eta^*;\ \xi,\xi^*), \tag{2.8}$$

where new variables $\{\xi_\alpha, \xi^*_\alpha; \alpha=1,\cdots,K-L\}$ are called intrinsic degrees of freedom. If the transformation were an exact canonical transformation satisfying

$$i\dot{\eta}_r = \frac{\partial H}{\partial \eta^*_r}, \quad i\dot{\eta}^*_r = -\frac{\partial H}{\partial \eta_r}, \quad i\dot{\xi}_\alpha = \frac{\partial H}{\partial \xi^*_\alpha}, \quad i\dot{\xi}^*_\alpha = -\frac{\partial H}{\partial \xi_\alpha}, \tag{2.9}$$

the local infinitesimal generators defined by

$$\begin{aligned} \hat{O}^\dagger_r &= \{\frac{\partial}{\partial \eta_r} e^{i\hat{F}}\}\cdot e^{-i\hat{F}} = \sum_{\mu i}\{\frac{\partial C_{\mu i}}{\partial \eta_r}\hat{O}^\dagger_{\mu i} - \frac{\partial C^*_{\mu i}}{\partial \eta_r}\hat{O}_{\mu i}\}, \\ \hat{O}^\dagger_\alpha &= \{\frac{\partial}{\partial \xi_\alpha} e^{i\hat{F}}\}\cdot e^{-i\hat{F}} = \sum_{\mu i}\{\frac{\partial C_{\mu i}}{\partial \xi_\alpha}\hat{O}^\dagger_{\mu i} - \frac{\partial C^*_{\mu i}}{\partial \xi_\alpha}\hat{O}_{\mu i}\} \end{aligned} \tag{2.10}$$

would have to obey the "weak" boson-like commutation relations similar to Eq.(2.4),

$$\langle f|[\hat{O}_r, \hat{O}^\dagger_s]|f\rangle = \delta_{rs}, \qquad \langle f|[\hat{O}_r, \hat{O}_s]|f\rangle = 0, \tag{2.11a}$$

$$\langle f|[\hat{O}_r, \hat{O}^\dagger_\alpha]|f\rangle = 0, \qquad \langle f|[\hat{O}_r, \hat{O}_\alpha]|f\rangle = 0, \tag{2.11b}$$

$$\langle f|[\hat{O}_\alpha, \hat{O}^\dagger_\beta]|f\rangle = \delta_{\alpha\beta}, \qquad \langle f|[\hat{O}_\alpha, \hat{O}_\beta]|f\rangle = 0. \tag{2.11c}$$

The SCC method does not require such an exact canonical transformation, but demand[4] that the intrinsic variables $\{\xi_\alpha, \xi^*_\alpha\}$ describe "small-amplitude motion" in a small domain of the neighborhood of the collective surface Σ^{2L}. We may then introduce the following Taylor expansion

$$C_{\mu i} = [C_{\mu i}] + \sum_\alpha\{[\frac{\partial C_{\mu i}}{\partial \xi_\alpha}]\xi_\alpha + [\frac{\partial C_{\mu i}}{\partial \xi^*_\alpha}]\xi^*_\alpha\} + \cdots ,$$
$$H = [H] + \sum_\alpha\{[\frac{\partial H}{\partial \xi_\alpha}]\xi_\alpha + [\frac{\partial H}{\partial \xi^*_\alpha}]\xi^*_\alpha\} + \cdots . \tag{2.12}$$

where the symbol [h] for any function $h(\eta,\eta^*; \xi;\xi^*)$ denotes a function on the expansion surface; $[h] \equiv h(\eta,\eta^*; \xi = \xi^* = 0)$. The local infinitesimal generators $\{\hat{O}_r, \hat{O}^\dagger_r; \hat{O}_\alpha, \hat{O}^\dagger_\alpha\}$ are also expanded with respect to (ξ, ξ^*), and the zeroth-order terms of Eq.(2.11) are given by

$$\langle g|[\hat{X}_r, \hat{X}^\dagger_s]|g\rangle = \delta_{rs} , \qquad \langle g|[\hat{X}_r, \hat{X}_s]|g\rangle = 0 , \tag{2.13a}$$

$$\langle g|[\hat{X}_r, \hat{X}^\dagger_\alpha]|g\rangle = 0 , \qquad \langle g|[\hat{X}_r, \hat{X}_\alpha]|g\rangle = 0 , \tag{2.13b}$$

$$\langle g|[\hat{X}_\alpha, \hat{X}^\dagger_\beta]|g\rangle = \delta_{\alpha\beta} , \qquad \langle g|[\hat{X}_\alpha, \hat{X}_\beta]|g\rangle = 0 , \tag{2.13c}$$

where $\{\hat{X}_r, \hat{X}^\dagger_r; \hat{X}_\alpha, \hat{X}^\dagger_\alpha\}$ are the local infinitesimal generators on the expansion surface, i.e., $\hat{X}^\dagger_r \equiv [\hat{O}^\dagger_r]$ and $\hat{X}^\dagger_\alpha \equiv [\hat{O}^\dagger_\alpha]$, and $|g\rangle$ is the zeroth-order term of the TDHF state defined through

$$|f\rangle = e^{i\hat{F}}|\phi_0\rangle = \{1 + \sum_\alpha(\hat{X}^\dagger_\alpha\xi_\alpha - \hat{X}_\alpha\xi^*_\alpha) + \cdots\}\cdot e^{i[\hat{F}]}|\phi_0\rangle ,$$
$$|g\rangle = e^{i[\hat{F}]}|\phi_0\rangle \equiv e^{i\hat{G}(\eta,\eta^*)}|\phi_0\rangle . \tag{2.14}$$

The basic equations of the SCC method consist of the zeroth-order term of the TDHF equation (2.1), called the <u>invariance principle of the time-dependent Schrödinger equation</u>,

$$\delta\langle g|(i\frac{\partial}{\partial t} - \hat{H})|g\rangle = \delta\langle g|i\sum_r\{\dot{\eta}_r\hat{X}^\dagger_r - \dot{\eta}^*_r\hat{X}_r\} - \hat{H}|g\rangle = 0 , \tag{2.15}$$

and the <u>canonical-variable condition</u>

$$\langle g|\hat{X}^\dagger_r|g\rangle = \frac{1}{2}\eta^*_r , \qquad \langle g|\hat{X}_r|g\rangle = \frac{1}{2}\eta_r, \tag{2.16}$$

which guarantees the "weak" boson-like commutation relations (2.13a). The collective Hamiltonian $\mathcal{H}_{coll}(\eta,\eta^*)$ in Eq.(2.7) is then defined by

$$\mathcal{H}_{coll}(\eta,\eta^*) = [H] \quad \langle g|\hat{H}|g\rangle \ . \qquad (2.17)$$

In the canonical-variable representation of the TDHF equation (2.3), the invariance principle (2.15) is simply expressed as the zeroth-order term of Eq.(2.3),

$$i\frac{d}{dt}[C_{\mu i}] = [\frac{\partial H}{\partial C_{\mu i}}] \ , \qquad i\frac{d}{dt}[C^*_{\mu i}] = - [\frac{\partial H}{\partial C_{\mu i}}] \ . \qquad (2.18)$$

By taking $\delta|g\rangle \propto :\hat{X}^\dagger_r:|g\rangle$ and $:\hat{X}_r:|g\rangle$ and by taking $\delta|g\rangle \propto :\hat{X}^\dagger_\alpha:|g\rangle$ and $:\hat{X}_\alpha:|g\rangle$ in Eq.(2.15), respectively, and then by using Eqs. (2.13a) and (2.13b), we can easily see that Eq.(2.18) can be decomposed into the <u>canonical equations of collective motion</u>[4)]

$$i\dot{\eta}_r = \partial[H]/\partial\eta^*_r \ , \quad i\dot{\eta}^*_r = -\partial[H]/\partial\eta_r \ , \qquad (r=1,2,\cdots,L) \quad (2.19)$$

and the <u>maximal-decoupling condition</u> for the optimum collective submanifold Σ^{2L},

$$[\partial H/\partial\xi_\alpha] = -\langle g|[\hat{X}^\dagger_\alpha, \hat{H}]|g\rangle = 0 \ ,$$
$$[\partial H/\partial\xi^*_\alpha] = \langle g|[\hat{X}_\alpha, \hat{H}]|g\rangle = 0 \ , \qquad (\alpha=1,2,\cdots,K-L) \ . \qquad (2.20)$$

As has been shown in Refs. 3) and 10), Eqs. (2.15) (i.e., (2.18)) and (2.16) enable us to determine the functional forms of $\{[C_{\mu i}], [C^*_{\mu i}]\}$ as well as the collective Hamiltonian [H] with respect to (η,η^*), provided that a specific "boundary condition" appropriate for the collective motion under consideration is set up.[10),11)] Namely, we can obtain the mapping (2.6);

$$M^{2L}:\{\eta_r, \eta^*_r\} \rightarrow \Sigma^{2L}:\{[C_{\mu i}], [C^*_{\mu i}]\}. \qquad (2.21)$$

The trajectory determined by Eq.(2.19) is thus mapped on the collective submanifold Σ^{2L}, and it is called the <u>SCC trajectory</u> hereafter. As is easily seen from the relation (2.12), the maximal-decoupling condition (2.20) demonstrates that the Hamiltonian H is stationary at each point on the collective submanifold Σ^{2L} with respect to the variations toward <u>intrinsic directions</u> characterized by $\{\hat{X}_\alpha, \hat{X}^\dagger_\alpha\}$, which have to be orthogonal to the variations toward <u>collective</u>

directions characterized by $\{\hat{X}_r, \hat{X}_r^\dagger\}$, i.e., the tangential directions of the collective submanifold Σ^{2L}.

[2.4] "Local" Canonical-Variable Approximation for Intrinsic Variables

We are now at a stage to discuss the second problem (ii) for the nuclear collective dynamics discussed at the end of [2.2]. In order to identify the SCC trajectory with the optimized collective trajectory, which is a representative of the group of TDHF trajectories always confined in a small domain of the neighborhood of the collective surface Σ^{2L}, we have to find specific conditions characterizing the "collectivity" of the group of TDHF trajectories. For this purpose, we here require the following condition for $\{\xi_\alpha, \xi^*_\alpha; \alpha=1,2,\cdots,K\text{-}L\}$ describing the "small-amplitude" intrinsic motion in the small domain of the neighborhood of Σ^{2L}: The intrinsic variables $\{\xi_\alpha, \xi^*_\alpha\}$ should satisfy the canonical equations of motion within the first-order of the (ξ, ξ^*)-Tayler expansion.[5),6)] This requirement means that first-order term of the TDHF equation (2.3), i.e.,

$$i\frac{d}{dt}\sum_\alpha\{\xi_\alpha[\frac{\partial C_{\mu i}}{\partial \xi_\alpha}] + \xi^*_\alpha[\frac{\partial C_{\mu i}}{\partial \xi^*_\alpha}]\} = \sum_\alpha\{\xi_\alpha[\frac{\partial^2 H}{\partial \xi_\alpha \partial C^*_{\mu i}}] + \xi^*_\alpha[\frac{\partial^2 H}{\partial \xi^*_\alpha \partial C^*_{\mu i}}]\} , \quad (2.22)$$

should be reduced to the canonical equations of motion

$$i\dot{\xi}_\alpha = \frac{\partial H}{\partial \xi^*_\alpha} \Rightarrow \sum_\beta\{\xi_\beta[\frac{\partial^2 H}{\partial \xi^*_\alpha \partial \xi_\beta}] + \xi^*_\beta[\frac{\partial^2 H}{\partial \xi^*_\alpha \partial \xi^*_\beta}]\} , \quad (2.23)$$

under the maximal-decoupling condition (2.20). In order to uniquely define the intrinsic variables, we further suppose that the intrinsic variables $\{\xi_\alpha, \xi^*_\alpha\}$ are of the normal modes satisfying

$$[\partial^2 H/\partial \xi_\alpha \partial \xi_\beta] = [\partial^2 H/\partial \xi^*_\alpha \partial \xi^*_\beta] = 0, \quad [\partial^2 H/\partial \xi_\alpha \partial \xi^*_\beta] = \delta_{\alpha\beta}[\partial^2 H/\partial \xi_\alpha \partial \xi^*_\alpha], \quad (2.24a)$$

$$\xi_\alpha(t) = \xi_\alpha(0)e^{-i\omega_\alpha t}, \quad \xi^*_\alpha(t) = \xi^*_\alpha(0)e^{i\omega_\alpha t}. \quad (2.24b)$$

The above two requirements (2.23) and (2.24) constitute a <u>"local" canonical-variable approximation</u>[5),6)] for the intrinsic variables $\{\xi_\alpha, \xi^*_\alpha\}$.

After straightforward but cumbersome calculations, it can be shown[5),6)] that the local canonical-variable approximation is satisfied under Eqs. (2.13b), (2.13c) and the following condition

$$|\langle g|[\hat{X}_\alpha,\ i\frac{\partial}{\partial t}\hat{X}^\dagger_\beta]|g\rangle/\omega_\alpha| \ll 1, \quad |\langle g|[\hat{X}_\alpha, i\frac{\partial}{\partial t}\hat{X}_\beta]|g\rangle/\omega_\alpha| \ll 1,$$
$$\frac{\partial}{\partial t}\hat{X}_\beta \equiv \sum_r(\dot{\eta}_r\frac{\partial}{\partial\eta_r}\hat{X}_\beta + \dot{\eta}^*_r\frac{\partial}{\partial\eta^*_r}\hat{X}_\beta), \quad (\omega^2_\alpha > 0,) \tag{2.25}$$

and the eigen frequency ω_α is determined, together with the intrinsic directions $\{[\partial C_{\mu i}/\partial\xi_\alpha],[\partial C^*_{\mu i}/\partial\xi_\alpha]\}$, by an RPA-type eigenvalue equation

$$\omega_\alpha\begin{bmatrix}[\partial C_{\mu i}/\partial\xi_\alpha]\\ [\partial C^*_{\mu i}/\partial\xi_\alpha]\end{bmatrix} = \sum_{\nu j}\begin{bmatrix}[\partial^2 H/\partial C^*_{\mu i}\partial C_{\nu j}], & -[\partial^2 H/\partial C^*_{\mu i}\partial C^*_{\nu j}]\\ [\partial^2 H/\partial C_{\mu i}\partial C_{\nu j}], & -[\partial^2 H/\partial C_{\mu i}\partial C^*_{\nu j}]\end{bmatrix}\begin{bmatrix}[\partial C_{\nu j}/\partial\xi_\alpha]\\ [\partial C^*_{\nu j}/\partial\xi_\alpha]\end{bmatrix}, \tag{2.26}$$

under the constraint condition (2.13b).[5),6)] By definition, the eigenvalues ω_α are defined to be locally dependent on each point of the collective submanifold Σ^{2L}, i.e., $\omega_\alpha = \omega_\alpha(\eta,\eta^*)$. This implies that the <u>concept of the intrinsic variables</u> $\{\xi_\alpha, \xi^*_\alpha\}$ under the "local" canonical-variable approximation can be considered as a <u>local concept</u> valid only in the neighborhood of the respective point on Σ^{2L}. Thus, the concept of "local" canonical intrinsic variables $\{\xi_\alpha, \xi^*_\alpha\}$, compatible with the "global" canonical collective variables $\{\eta_r, \eta^*_r\}$, are definable as long as the condition (2.25) is satisfied. In this case, the group of TDHF trajectories is always confined in the neighborhood of Σ^{2L}, each trajectory being on a torous like the Kolmogorov-Arnold-Moser (KAM) torous.[12)] The group of TDHF trajectories has then <u>well organized collectivity</u>, so that the SCC trajectory on Σ^{2L} can be uniquely identified with the optimized collective trajectory representing the goup.[11),13)]

The condition (2.25) means that the terms in it with non-local dependence of the intrinsic modes, such as $\{\partial\hat{X}_\alpha/\partial\eta_r,\ \partial\hat{X}^\dagger_\alpha/\partial\eta^*_r$, etc.$\}$, should be fairly small compared with $|\omega_\alpha|$. In other words, time-dependent variations in the internal structure of the intrinsic excitation modes $\{\hat{X}_\alpha, \hat{X}_\alpha^\dagger\}$ should be fairly small in comparison with their intrinsic excitation energies $|\omega_\alpha|$. Here the time-dependent variations of the internal structure of the intrinsic excitation modes are induced by the time-evolution of the collective variables $\{\eta_r, \eta_r^*\}$. This non-locality is a manifestation of a kind of coupling between the collective and intrinsic degrees of freedom: The large non-local effect does not allow us to derive the RPA-type equation (2.26) for the "local" intrinsic motion. Thus, the condition (2.25) can be called a separability condition[5),6)] between the collective and intrinsic motion.

[2.5] Collective, Dissipative and Chaotic Behaviors of TDHF Trajectories

In the previous subsections, we have treated the collective and intrinsic motion independently: The former is described within the SCC method and the latter is treated within the RPA-type procedure.

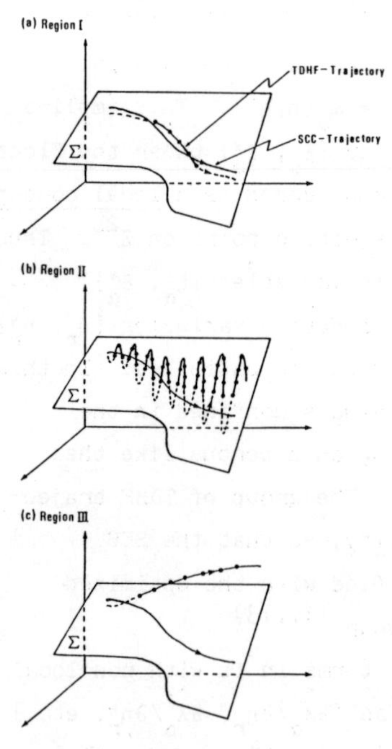

[Fig. 2]

This separate treatment has become possible under the separability condition (2.25).

If the excitation energies $|\omega_\alpha(\eta,\eta^*)|$ of the intrinsic motion take relatively large values, the restoring forces toward the intrinsic directions perpendicular to the collective submanifold Σ^{2L} become large. Consequently, the group of TDHF trajectories in the small domain of the neighborhood of Σ^{2L} is always confined in the small domain due to the large restoring forces. (Fig. 2(a).) In this case, a large amount of TDHF trajectories are expected to accumulate around Σ^{2L} and the collectivity of Σ^{2L}

increases to a fairly large extent. We may thus expect that the small domain in the neighborhood of Σ^{2L} consists of only approximate invariant tori like the KAM tori.

If $\omega_\alpha^2(\eta,\eta^*)$ are relatively small, the TDHF trajectories starting from the small domain at an initial instant can travel toward the intrinsic directions with a fairly large amount of deviations from Σ^{2L}. (Fig. 2(b)). In this case, the separability condition (2.25) is not well satisfied, so that the "local" canonical approximation for the intrinsic variables may be no more valid. Because there are non-negligible energy transfer between the collective and intrinsic degrees of freedom, the extraction of the collective submanifold by the SCC method does not have a definit sense in comparison with the case of large $\omega_\alpha^2(\eta,\eta^*)$. However, it has a sense that "time-averaged" property or "phase-space averaged" property of the group of TDHF trajectories is represented on the submanifold Σ^{2L}.[11),13),14)] In this sense, the collectivity of the system is still expected to survive even there are significant dissipative effects.

If $\omega_\alpha(\eta,\eta^*)$ become imaginary, there do not exist any restoring forces toward the intrinsic directions, and the TDHF trajectories starting from the small domain in the neighborhood of Σ^{2L} at an initial instant prefer to escape the small domain, having lost their initial focussing for a long enough time. (Fig. 2(c).) In this case, the SCC trajectory is running on an unstable ridge-line and we can not expect any other trajectories accumulated to the submanifold Σ^{2L}. This implies that the collective motion under consideration has disappeared and the system shows a chaotic (stochastic) behavior.[13),14)]

By summarizing the above discussion, the "optimum" collective submanifold Σ^{2L} obtained by the SCC method may be characterized by the following three regions,[13)]

$$\begin{aligned}
&\text{Region I}: && \omega_\alpha^2(\eta,\eta^*) > 0 \quad \text{and} \quad 1 > I_\alpha^\beta(\eta,\eta^*) \geq 0,\\
&\text{Region II}: && \omega_\alpha^2(\eta,\eta^*) > 0 \quad \text{and} \quad I_\alpha^\beta(\eta,\eta^*) \geq 1,\\
&\text{Region III}: && \omega_\alpha^2(\eta,\eta^*) < 0,
\end{aligned} \tag{2.27}$$

$$I^{\beta}_{\alpha}(\eta,\eta^*) \equiv |\langle g|[\hat{X}_{\alpha}, i\frac{\partial}{\partial t}\hat{X}^{\dagger}_{\beta}]|g\rangle/\omega_{\alpha}| .$$

Hereafter we call Region I, Region II and Region III on Σ^{2L} <u>collective, dissipative and chaotic regions</u>, respectively.

[2.6] Illustrative Example with A Simple Model Hamiltonian[13),14)]

In order to vizualize the collective, dissipative and chaotic behaviors of the TDHF trajectories, we adopt a simple three-level model, which is a modification of the SU(3) model used by Li, Klein and Dreizler.[15)] The Hamiltonian is given by

$$\hat{H} = \epsilon_0\hat{K}_{00}+\epsilon_1\hat{K}_{11}+\epsilon_2\hat{K}_{22}+\frac{1}{2}V_1\{\hat{K}_{10}\hat{K}_{10}+\text{h.c.}\} + \frac{1}{2}V_2\{\hat{K}_{20}\hat{K}_{20}+\text{h.c.}\}. \quad (2.28)$$

There are three levels with energies $\epsilon_0 < \epsilon_1 < \epsilon_2$ and each level has N-hold degeneracy. The fermion pair operators $\hat{K}_{\alpha\beta}$ are defined by

$$\hat{K}_{\alpha\beta} \equiv \sum_{m=1} c^{\dagger}_{\alpha m}c_{\beta m}, \quad (\alpha,\beta = 0,1,2)$$

$$[\hat{K}_{\alpha\beta}, \hat{K}_{\gamma\delta}] = \delta_{\beta\gamma}\hat{K}_{\alpha\delta} - \delta_{\alpha\delta}\hat{K}_{\gamma\beta} . \quad (2.29)$$

We consider a system with N particles and the lowest-energy state $|\phi_0\rangle$ <u>without interaction</u> (i.e., $V_1 = V_2 = 0$) is given by $|\phi_0\rangle = \Pi^N_{m=1}c^{\dagger}_{0m}|0\rangle$, where $|0\rangle$ denotes the vacuum of the fermion operators $\{c_{\alpha m}, c^{\dagger}_{\alpha m}\}$.

The TDHF Slater determinant in this case is

$$|f\rangle = e^{i\hat{F}}|\phi_0\rangle, \quad \hat{F} = \{f_1(t)\hat{K}_{10}+f_2(t)\hat{K}_{20}\} + \text{h.c.}, \quad (2.30)$$

and the canonical-variable representation $\{C_j(t), C^*_j(t);\ j=1,2\}$ is obtained by the following variable transformation

$$f_j = -i\frac{C_j}{\sqrt{C_1C^*_1+C_2C^*_2}}\sin^{-1}\sqrt{\frac{C_1C^*_1+C_2C^*_2}{N}}, \quad (2.31)$$

which is obtained from the canonical-variable condition

$$\langle f|\hat{O}^{\dagger}_j|f\rangle = C^*_j/2, \quad \hat{O}^{\dagger}_j = \{\partial e^{i\hat{F}}/\partial C_j\}\cdot e^{-i\hat{F}}. \quad (2.32)$$

In the canonical-variable representation, the TDHF equation is simply reduced to the "classical" canonical equations of motion

$$i\dot{C}_j = \partial H/\partial C^*_j, \quad i\dot{C}^*_j = -\partial H/\partial C_j \quad (2.33)$$

where

$$H \equiv \langle f|\hat{H}|f\rangle - \langle \phi_0|\hat{H}|\phi_0\rangle = (\epsilon_1 - \epsilon_0)C_1^*C_1 + (\epsilon_2 - \epsilon_0)C_2^*C_2$$

$$+ \frac{N-1}{N}\{N-(C_1^*C_1+C_2^*C_2)\}\{\frac{V_1}{2}(C_1^*C_1^*+C_1C_1)+ \frac{V_2}{2}(C_2^*C_2^*+C_2C_2)\}. \quad (2.34)$$

Consequently, the TDHF equation (2.33) determines TDHF trajectories in a four-dimensional TDHF manifold (phase space) M^4 given by

$$\{p_1,q_1,p_2,q_2\}; \quad p_j \equiv \frac{i}{\sqrt{2}}(C_j^*-C_j), \quad q_j \equiv \frac{i}{\sqrt{2}}(C_j^*+C_j). \quad (2.35)$$

In the simple model under consideration, the basic equations (2.15) and (2.16) of the SCC method, with a single pair of collective variables (η,η^*), can be analytically solved. Namely, we can analytically solve the mapping

$$M^2:(\eta,\eta^*) \to \Sigma^2:([C_j],[C_j^*];j=1,2), \quad (2.36)$$

with a boundary condition appropriate for the collective motion under consideration.[13] For example, when we choose the boundary condition for Σ^2 in such a way that the SCC trajectory on it can be described by only p_1 and q_1 degrees of freedom in its small amplitude limit, we obtain the following simple results;[13]

$$[C_1] = \eta, \quad [C_1^*] = \eta^*, \quad [C_2] = [C_2^*] = 0, \quad \text{i.e.,}$$

$$[p_1] = p, \quad [q_2] = q, \quad [p_2] = [q_2] = 0,$$

$$p \equiv \frac{i}{\sqrt{2}}(\eta^*-\eta), \quad q \equiv \frac{1}{\sqrt{2}}(\eta^*+\eta), \quad (2.37)$$

which demonstrates that the global canonical variables (p,q) describing the collective submanifold Σ^2 are simply related with the variables (p_1,q_1). The equations of motion of the collective variables are then given by

$$i\dot{\eta} = \partial[H]/\partial\eta^*, \quad i\dot{\eta}^* = -\partial[H]/\partial\eta,$$

$$[H] = (\epsilon_1-\epsilon_0)\eta^*\eta + \frac{V_1}{2}\frac{N-1}{N}(\eta\eta+\eta^*\eta^*)(N-\eta^*\eta). \quad (2.38)$$

The obtained trajectory is embedded into the four-dimensional $\{p_j, q_j; j=1,2\}$ phase space through the mapping (2.36), and by definition, the resultant mapped trajectory is the SCC trajectory.

The intrinsic excitation energy $\omega(\eta,\eta^*)$ in Eq.(2.26) and the quantity $I(\eta,\eta^*) \equiv |\langle g|[\hat{X}_{intr}, i\frac{\partial}{\partial t}\hat{X}^\dagger_{intr}]|g\rangle/\omega|$ in Eq.(2.27) are also obtained analytically, so that we can characterize the collective submanifold Σ^2 by the collective, dissipative and chaotic regions, according to Eq.(2.27).

In order to vizualize the behaviors of the TDHF trajectories in these regions, we solved the TDHF equation (2.33) with the initial condition; $p_1=p_0$, $q_1=0$, $p_2=q_2=0.1 \ll p_0$ at t=0, where p_0 is chosen in such a way that the system has a given total energy E. The point ($p_1=p_0$, $q_1=0$, $p_2=q_2=0.1 \ll p_0$) at t=0 is of course in the small domain of the neighborhood of Σ^2. We illustrate the trajectory in a three-dimensional coordinate space (X,Y,Z), by using the cartesian product of $S_1 \times S_2$, defined as

$$X = (R_1+R_2\cos\theta_2)\cos\theta_1, \quad Y = (R_1+R_2\cos\theta_2)\sin\theta_1,$$

$$Z = R_2\sin\theta_2 \; ; \quad R_1 = \sqrt{p_1^2+q_1^2}, \quad R_2 = \sqrt{p_2^2+q_2^2}, \tag{2.39}$$

$$\cos\theta_1 = q_1/R_1, \; \cos\theta_2 = q_2/R_2, \; \sin\theta_1 = p_1/R_1, \; \sin\theta_2 = p_2/R_2.$$

In Fig. 3, the results of the typical trajectory in each region are shown.[13),14)] In these calculations we have adopted the following values of parameters in the model; $\epsilon_0=-1$, $\epsilon_1=0$, $\epsilon_2=1$, $V_1=-1/5$, $V_2=-1/3$ and N=10. The case with E=8 corresponds to the TDHF trajectory in the collective region, i.e., Region I. We also show the corresponding SCC trajectory. It can be easily seen that the SCC trajectory represents the TDHF trajectory quite well. The case with E=5 corresponds to the TDHF trajectory in the dissipative region, i.e., Region II, and the TDHF trajectory in the chaotic region, i.e., Region III is shown with the case of E=1.

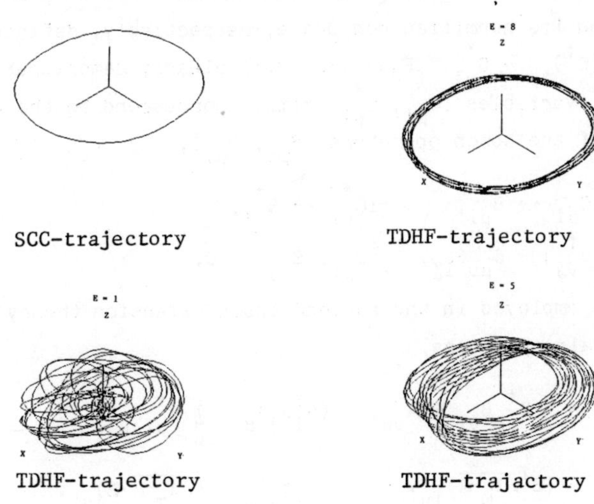

SCC-trajectory TDHF-trajectory

TDHF-trajectory TDHF-trajactory

[Fig. 3]

§3. QUANTUM THEORY OF NUCLEAR COLLECTIVE DYNAMICS

[3.1] TDHF Theory and Boson-Expansion Theory

The theory of nuclear collective dynamics formulated in §2 within the framework of the TDHF theory can be extended to a full quantum theory, when once we recognize clear-cut correspondence between the canonical-variable representation of the TDHF theory and the nuclear boson-expansion theory.

By direct calculations in terms of the canonical variable representation of the TDHF theory, we obtain[4)] for the fermion-pair operators $\{a^\dagger_\mu a_\nu,\ b^\dagger_i b_j,\ a^\dagger_\mu b^\dagger_i,\ b_i a_\mu\}$

$$\langle f|a^\dagger_\mu a_\nu|f\rangle = \sum_i C^*_{\mu i}C_{\nu i} \equiv A_{\nu\mu}, \quad \langle f|b^\dagger_i b_j|f\rangle = \sum_\mu C^*_{\mu i}C_{\mu j},$$

$$\langle f|a^\dagger_\mu b^\dagger_i|f\rangle = -i\{C^\dagger\sqrt{I_m - A}\}_{i\mu}, \quad \langle f|b_i a_\mu|f\rangle = i\sqrt{I_m - A}\cdot C\}_{\mu i}, \qquad (3.1)$$

where I_m denotes an m by m unit matrix and C and $C^\dagger$ means an m by n matrix and its hermitian conjugate, respectively, defined by $(C)_{\mu i} = C_{\mu i}$ and $(C^\dagger)_{i\mu} = C^*_{\mu i}$. Equation (3.1) clearly demonstrates that the canonical variables $\{C_{\mu i}, C^*_{\mu i}\}$ simply correspond to the c-number version of the boson operators $\{B_{\mu i}, B^\dagger_{\mu i}\}$,

$$iC_{\mu i} \leftrightarrow B_{\mu i}, \qquad -iC^*_{\mu i} \leftrightarrow B^\dagger_{\mu i},$$
$$[B_{\mu i}, B^\dagger_{\nu j}] = \delta_{\mu\nu}\delta_{ij}, \qquad [B_{\mu i}, B_{\nu j}] = 0, \tag{3.2}$$

which are employed in the nuclear boson-expansion theory[9] of the fermion-pair operators;

$$(a^\dagger_\mu a_\nu)_B = \sum_i B^\dagger_{\mu i} B_{\nu i} \equiv \mathbb{A}_{\nu\mu}, \qquad (b^\dagger_i b_j)_B = \sum_\mu B^\dagger_{\mu i} B_{\mu j},$$
$$(a^\dagger_\mu b^\dagger_i)_B = (\mathbb{B}^\dagger \sqrt{\mathbb{I}_m - \mathbb{A}})_{i\mu}, \qquad (b_i a_\mu)_B = (\sqrt{\mathbb{I}_m - \mathbb{A}} \cdot \mathbb{B})_{\mu i}. \tag{3.3}$$

This correspondence becomes more transparent from the following fact: The boson expansion is determined so as to keep the Lie algebra of the fermion-pair operators,

$$[b_i a_\mu, a^\dagger_\nu b^\dagger_j] = \delta_{\mu\nu}\delta_{ij} - \delta_{\mu\nu} b^\dagger_j b_i - \delta_{ij} a^\dagger_\nu a_\mu,$$
$$[a^\dagger_\mu a_\nu, a^\dagger_\sigma b^\dagger_r] = \delta_{\nu\sigma} a^\dagger_\mu b^\dagger_r, \quad [b^\dagger_i b_j, a^\dagger_\mu b^\dagger_r] = \delta_{jr} a^\dagger_\mu b^\dagger_i, \quad \text{etc.}, \tag{3.4}$$

while we have a realization of the Lie algebra in terms of the canonical-variable representation $\{C_{\mu i}, C^*_{\mu i}\}$,

$$\{\langle b_i a_\mu\rangle, \langle a^\dagger_\nu b^\dagger_j\rangle\}_{P.B} = \delta_{\mu\nu}\delta_{ij} - \delta_{\mu\nu}\langle b^\dagger_j b_i\rangle - \delta_{ij}\langle a^\dagger_\nu a_\mu\rangle,$$
$$\{\langle a^\dagger_\mu a_\nu\rangle, \langle a^\dagger_\sigma b^\dagger_r\rangle\}_{P.B} = \delta_{\nu\sigma}\langle a^\dagger_\mu b^\dagger_r\rangle, \quad \text{etc.}, \tag{3.5}$$

where $\langle a^\dagger_\mu b^\dagger_i\rangle \equiv \langle f|a^\dagger_\mu b^\dagger_i|f\rangle$, etc., and the symbol $\{A,B\}_{P.B}$ denotes the Poisson bracket $\{A,B\}_{P.B} \equiv \Sigma_{\mu i}(\partial A/\partial C_{\mu i} \cdot \partial B/\partial C^*_{\mu i} - \partial B/\partial C_{\mu i} \cdot \partial A/\partial C^*_{\mu i})$.

Depending on the collective motion under consideration, we may of course adopt an appropriate set of the boson operators $\{B_j, B^\dagger_j: j = 1,2,\cdots,K\}$, according to a linear transformation such as $B^\dagger_j = \Sigma_{\mu i}\{\psi_j(\mu i)B^\dagger_{\mu i} - \phi_j(\mu i)B_{\mu i}\}$. Thus, a general many-fermion system with even particle number can be described in a boson space characterized by the set of the boson operators $\{B_j, B^\dagger_j\}$. Correspondingly, the

Hamiltonian $\hat{H}$ of the fermion system can be mapped onto the boson space;

$$\hat{H} \rightarrow H(B^{\dagger},B), \tag{3.6}$$

whose c-number version corresponds to $H \equiv \langle f|\hat{H}|f\rangle$ in Eq.(2.3) of the theory. The boson space is spanned by an orthonormal set of states $\{|n_1,n_2,\cdots,n_K)\}$ with

$$|n_1,n_2,\cdots,n_K) = \prod_{j=1}^{K} \frac{1}{\sqrt{n_j}}(B_j^{\dagger})^{n_j}|0), \quad B_j|0) = 0, \tag{3.7}$$

where n_j denotes the number of bosons with index j.

[3.2] "Optimum" Collective Subspace[16)]

To make clear the correspondence with the TDHF theory, we hereafter denote the boson space consisting of the states (3.7) by $S^{2K}:\{|n_1,n_2,\cdots,n_K)\}$, where K is equivalent to the total number of boson indices and so 2K corresponds to the dimension of the TDHF manifold M^{2K}. Now, let us suppose the many-body system under consideration exhibits a class of well-defined collective states described by L-collective degrees of freedom with L<<K. If there is an ideal case where these collective states constitute an invariant subspace of the Hamiltonian, it may be possible to introduce a set of basis <u>collective</u> boson operators $\{\beta_r,\beta_r^{\dagger};\ r=1,2,\cdots,L\}$ constituting the invariant subspace. The set of the <u>ideal</u> collective bosons corresponds to the set of the <u>ideal</u> collective variables $\{\eta_r,\ \eta_r^*;\ r=1,2,\cdots,L\}$ in Eq.(2.9) in the TDHF theory. We also introduce basis <u>intrinsic</u> boson operators $\{\alpha_k,\alpha_k^{\dagger};\ k=L+1,L+2,\cdots,K\}$ for the other degrees of freedom. The set of the <u>ideal</u> intrinsic bosons corresponds to the set of the <u>ideal</u> intrinsic variables $\{\xi_\alpha,\xi_\alpha^*;\ \alpha= 1,2,\cdots,K-L\}$ in Eq.(2.9) in the TDHF theory. Corresponding to the fact that the <u>ideal</u> set of the collective and intrinsic variables $\{\eta_r,\ \eta_r^*;\ \xi_\alpha,\xi_\alpha^*\}$ is defined by an <u>exact</u> canonical transformation from the set $\{C_{\mu i},\ C_{\mu i}^*\}$, the ideal bosons $\{\beta_r,\beta_r^{\dagger};\ \alpha_k,\alpha_k^{\dagger}\}$ can be defined from the original bosons $\{B_j,B_j^{\dagger}\}$ in terms of a general <u>non-linear</u> unitary transformation;

$$\beta_r = e^{-F(B^\dagger,B)} B_{j=r} e^{F(B^\dagger,B)}, \quad \alpha_k = e^{-F(B^\dagger,B)} B_{j=k} e^{F(B^\dagger,B)},$$

$$B_{j=r} = e^{F(\beta^\dagger,\beta;\alpha^\dagger,\alpha)} \beta_r e^{-F(\beta^\dagger,\beta;\alpha^\dagger,\alpha)}, \quad B_{j=k} = e^{F(\beta^\dagger,\beta;\alpha^\dagger,\alpha)} \alpha_k e^{-F(\beta^\dagger,\beta;\alpha^\dagger,\alpha)},$$

$$F^\dagger = -F, \quad e^{F(B^\dagger,B)} = e^{F(\beta^\dagger,\beta;\alpha^\dagger,\alpha)} \tag{3.8}$$

Hereafter we call the representation with $\{\beta_r, \beta_r^\dagger;\ \alpha_k, \alpha_k^\dagger\}$ a <u>dynamical representation</u>.

By definition, the basis states $\{|m_1, m_2, \cdots, m_L\rangle\}$ of the collective subspace are given in terms of the collective basis operators $\{\beta_r, \beta_r^\dagger;\ r=1,2,\cdots,L\}$;

$$|m_1, m_2, \cdots, m_L\rangle = \prod_{r=1}^{L} \frac{1}{\sqrt{m_r}} (\beta_r^\dagger)^{m_r} |0\rangle, \tag{3.9}$$

where the ground state $|0\rangle$ in the dynamical representation is defined by $\beta_r|0\rangle = \alpha_k|0\rangle = 0$, i.e., $|0\rangle = \exp\{-F(B^\dagger,B)\}|0)$. With the use of Eq.(3.9), the projection operators P and Q onto the collective subspace and its orthogonal complement are defined, respectively, as

$$P = \sum_{m_1,\cdots,m_L} |m_1, m_2, \cdots, m_L\rangle\langle m_1, m_2, \cdots, m_L|, \quad Q = I-P, \tag{3.10}$$

where I denotes the identity operator in the full space. In the ideal case under consideration, the collective subspace is the invariant subspace of the Hamiltonian $H(B^\dagger,B)$, i.e., the complete decoupling condition, $PH(B^\dagger,B)Q = 0$, must be satisfied.

The above discussion is valid only for the ideal case. In actual cases, it is impossible to ascertain whether or not such an invariant subspace of the Hamiltonian exists without solving the eigenvalue equation of $H(B^\dagger,B)$ in the full space. We therefore introduce a concept of <u>"optimum" collective subspace</u> $S^{2L}:\{|m_1, m_2, \cdots, m_L\rangle\}$ as an <u>approximate</u> invariant subspace of the Hamiltonian, by extending the SCC method into a quantum theory.

[3.3] Quantum Version of SCC Method

In the same way as the SCC method, here we demand that the basis intrinsic operators $\{\alpha_k,\alpha_k^\dagger\}$ describe "small-amplitude motion" in the neighborhood of the optimum collective subspace $S^{2L}:\{|m_1,m_2,\cdots,m_L\rangle\}$. Corresponding to Eq.(2.12), we then introduce the following expansion with respect to the basis intrinsic operators $\{\alpha_k,\alpha_k^\dagger;\ k=L+1,\cdots,K\}$ <u>around S^{2L}</u>;

$$\begin{aligned} H(B^\dagger,B) &= \mathcal{H}^{(0)}(\beta^\dagger,\beta) + \sum_{n=1} \mathcal{H}^{(n)}, \\ \mathcal{H}^{(1)} &= \sum_k \{\alpha_k^\dagger P[\alpha_k, H(B^\dagger,B)]P + \text{h.c.}\}, \\ \mathcal{H}^{(2)} &= \sum_{k\neq\ell} \alpha_k^\dagger P[\alpha_k,[H(B^\dagger,B),\alpha_\ell^\dagger]]P\alpha_\ell \\ &\quad + \sum_{k,\ell} N_{k\ell}^{-1}\{\alpha_k^\dagger\alpha_\ell^\dagger P[\alpha_\ell\alpha_k, H(B^\dagger,B)]P + \text{h.c.}\},\quad \text{etc.,} \end{aligned} \tag{3.11}$$

where $N_{k\ell} \equiv \langle 0|\alpha_\ell\alpha_k\alpha_k^\dagger\alpha_\ell^\dagger|0\rangle$. The same expansion around S^{2L} is made for an arbitrary operator. For example, we have

$$B_j = B_j^{(0)}(\beta^\dagger,\beta) + \sum_{n=1} B_j^{(n)}. \tag{3.12}$$

Hereafter we call the expansion an <u>$(\alpha,\alpha^\dagger)$-expansion around S^{2L}</u>

Now, the basic idea in extracting the optimum collective submanifold Σ^{2L} by the SCC method[3] in [2.3] is to perform a <u>projection</u> of the TDHF equation (2.1) in the full TDHF manifold M^{2K} onto a neighborhood of Σ^{2L}, where the TDHF equation is reduced to Eq. (2.15) called "the invariance principle of the Schrödinger equation."[2] The optimum collective submanifold Σ^{2L} is then determined within the neighborhood in such a way that the Hamiltonian on Σ^{2L} is <u>stationary</u> with respect to the variations $\delta_\perp|g\rangle \propto :\hat{X}_\alpha^\dagger:|g\rangle$ and $:\hat{X}_\alpha:|g\rangle$ perpendicular to the surface Σ^{2L}. The quantum correspondent of the neighborhood of Σ^{2L} is defined by a subspace specified by a projection operator $\mathring{I} \equiv P + Q^{(1)}$, where $Q^{(1)}$ is the first term of the projection operator Q in Eq.(3.10) with respect to the $(\alpha,\alpha^\dagger)$-expansion around S^{2L};

$$Q = \sum_{n=1} Q^{(n)}, \qquad Q^{(1)} = \sum_k \alpha_k^\dagger P\alpha_k,$$

$$Q^{(2)} = \sum_{k,\ell} N_{k\ell}^{-1}\alpha_k^\dagger\alpha_\ell^\dagger P\alpha_\ell\alpha_k, \quad \text{etc.,} \tag{3.13}$$

We call hereafter this subspace specified by $\overset{\circ}{I} \equiv P + Q^{(1)}$ an <u>SCC subspace</u>.

In the canonical-variable representation of the TDHF theory, the basic equation (2.15) in the SCC method is simply expressed by Eq. (2.18); i.e.,

$$i\frac{d}{dt}[C_{\mu i}] \Leftrightarrow [\frac{\partial H}{\partial C^*_{\mu i}}] \quad \text{and} \quad \text{c.c.,} \tag{3.14}$$

from which we have obtain the maximal-decoupling condition (2.20), i.e., $[\partial H/\partial\xi_\alpha] = [\partial H/\partial\xi^*_\alpha] = 0$, to determine the optimum collective submanifold Σ^{2L}. As a natural extension of the SCC method, we may expect that a decisive condition to extract the optimum collective subspace S^{2L} can be derived from a quantum correspondent of Eq.(3.14). In the Schrödinger picture under consideration, the time-displacement of any operator A is defined by $idA/dt \equiv [A,H(B^\dagger,B)]$. Thus, the quantum correspondent of Eq.(3.14) is simply expressed as

$$i\frac{d}{dt}B_j^{(0)}(\beta^\dagger,\beta) \equiv [B_j^{(0)},H(B^\dagger,B)] \Leftrightarrow \{[B_j,H(B^\dagger,B)]\}^{(0)}, \tag{3.15}$$

where $B_j^{(0)}(\beta^\dagger,\beta)$ is defined in Eq.(3.12) and the $\{[B_j,H(B^\dagger,B)]\}^{(0)}$ is the zeroth-order term of the $(\alpha,\alpha^\dagger)$-expansion around S^{2L},

$$[B_j,H(B^\dagger,B_j)] = \{[B_j^\dagger,H]\}^{(0)} + \sum_{n=1}\{[B_j,H]\}^{(n)}. \tag{3.16}$$

It is shown that, <u>with the restriction within the SCC subspace</u>, Eq. (3.15) is simply reduced to the maximal-decoupling condition $\mathcal{H}^{(1)} = 0$, i.e.,

$$P[\alpha_k, e^{-F(\beta^\dagger,\beta;\alpha^\dagger,\alpha)}H(\beta^\dagger,\beta;\alpha^\dagger,\alpha)e^{F(\beta^\dagger,\beta;\alpha^\dagger,\alpha)}]P = 0. \tag{3.17}$$

This is just a quantum correspondent of Eq.(2.20), so that the subspace $S^{2L}:\{|m_1,m_2,\cdots,m_L\rangle\}$ (specifying P) satisfies the "stationary" condition.

It can be demonstrated[16),17)] that, <u>with the restriction within the SCC subspace</u>, the maximal-decoupling condition (3.17) enables us to determine the functional form of the generator $F(\beta^\dagger,\beta;\alpha^\dagger,\alpha)$, provided that a specific "boundary condition" appropriate for the collective motion under consideration is set up. With the aid of the obtained function $F(\beta^\dagger,\beta;\alpha^\dagger,\alpha)$, we then can determine the collective Hamiltonian $\mathcal{H}_{coll} \equiv \mathcal{H}^{(0)}(\beta^\dagger,\beta)$ in Eq.(3.11) as well as the functional forms $\{B_j^{(0)}(\beta^\dagger,\beta),\ B_j^{\dagger(0)}(\beta^\dagger,\beta)\}$ in Eq.(3.12) with respect to the collective boson operators $\{\beta_r,\beta_r^\dagger;\ r=1,2,\cdots,L\}$. Namely, we can extract the optimum collective subspace $S^{2L}:\{|m_1,m_2,\cdots m_L\rangle\}$ out of the full space S^{2K}.

[3.4] "Local" Approximation for Intrinsic Modes

The second problem of the theory is to determine dynamical conditions, under which the optimum collective subspace S^{2L} can be regarded as an approximate invariant subspace of the Hamiltonian $H(B^\dagger, B)$. Namely, we have to investigate to what extent the eigenstates $\{|\sigma\rangle\}$ satisfying

$$\mathcal{H}_{coll}(\beta^\dagger,\beta)|\sigma\rangle = E_\sigma|\sigma\rangle, \qquad |\sigma\rangle \in S^{2L}, \tag{3.18}$$

satisfactorily reproduce the large-amplitude collective motion governed by the original Hamiltonian $H(B^\dagger,B)$. For this purpose, we here require the following condition for $\{\alpha_k,\alpha_k^\dagger\}$, which is completely analogous to that used in defining the "local" canonical variables $\{\xi_\alpha,\xi_\alpha^*\}$ in [2.4]: The "small-amplitude motion" of the intrinsic operators $\{\alpha_\kappa,\alpha_k^\dagger;\ k=L+1,\cdots,K\}$ around S^{2L} should be determined within the first-order of the $(\alpha,\alpha^\dagger)$-expansion around S^{2L}. This requirement means that

$$i\frac{d}{dt}a_k \equiv [a_k,H(B^\dagger,B)] \Rightarrow [a_k,\mathcal{H}^{(2)}] \tag{3.19}$$

under the maximal decoupling condition (3.17), i.e., $\mathcal{H}^{(1)} = 0$. The operator $\mathcal{H}^{(2)}$ in Eq.(3.19) is defined in Eq.(3.11) and, with the use of $P = \sum_\sigma|\sigma\rangle\langle\sigma|$, it is decomposed into

$$\begin{gathered}\mathcal{H}^{(2)} = \mathcal{H}_d^{(2)} + \mathcal{H}_{n-d}^{(2)}, \qquad \mathcal{H}_d^{(2)} = \sum_\sigma h(\sigma),\\ h(\sigma) = \sum_{k,\ell}\ [\{f_{k\ell}(\sigma)\alpha_k^\dagger\alpha_\ell^\dagger + \text{h.c.}\} + g_{k\ell}(\sigma)\alpha_k^\dagger\alpha_\ell],\\ f_{k\ell}(\sigma)=N_{k\ell}^{-1}\langle\sigma|[\alpha_\ell\alpha_k,H(B^\dagger,B)]\sigma\rangle,\ g_{k\ell}(\sigma)=\langle\sigma|[\alpha_k,[H(B^\dagger,B),\alpha_\ell^\dagger]]|\sigma\rangle,\end{gathered} \tag{3.20}$$

where, by definition, $\mathcal{H}_d^{(2)}$ is the diagonal part of $\mathcal{H}^{(2)}$ with respect to the collective eigenstates $|\sigma\rangle$ and $\mathcal{H}_{n-d}^{(2)}$ denotes the non-diagonal parts of $\mathcal{H}^{(2)}$.

Here we introduce "local" intrinsic normal modes described by

$$X_\gamma^\dagger(\sigma) = \sum_k \{\psi_k^\gamma(\sigma)\alpha_k^\dagger - \phi_k^\gamma(\sigma)\alpha_k\}, \tag{3.21}$$

which satisfies the RPA-type eigenvalue equation

$$\omega_\gamma(\sigma)X_\gamma^\dagger(\sigma) = [h(\sigma),\ x_\gamma^\dagger(\sigma)], \quad \omega_\gamma(\sigma) > 0. \tag{3.22}$$

Needless to say, Eq.(3.22) simply corresponds to Eq.(2.26) which defines the "local" canonical-variable approximation for the intrinsic variables $\{\xi_\alpha,\xi_\alpha^*\}$. Thus, the quantum correspondent of the separability condition (2.25) is now given by

$$|\langle\sigma|[\alpha_\ell\alpha_k, H(B^\dagger,B)]|\sigma'\rangle/\omega_\gamma(\sigma)| \ll 1,$$
$$|\langle\sigma|[\alpha_k,[H(B^\dagger,B),\alpha_\ell^\dagger]]|\sigma'\rangle/\omega_\gamma(\sigma)| \ll 1 \quad \text{for} \quad \sigma \neq \sigma', \tag{3.23}$$

which ensures that the eigenstates $\{|\sigma\rangle\}$ determined by Eq.(3.18) are not seriously disturbed by the non-diagonal effects due to $\mathcal{H}_{n-d}^{(2)}$.

§4. CONCLUDING REMARKS

By placing emphasis on its underlying physical pictures, we have outlined the framework of the systematic microscopic theory of nuclear collective dynamics, which has been developed under INS-TSUKUBA joint research project. Since the theory has shown its ability in clarifying the microscopic mechanism in nuclear collective dynamics, it will give an interesting subject to apply the theory to the realistic problems,[18] such as large-amplitude collective motion of soft nuclei, high-spin states, heavy-ion reactions, fissions, etc.

It is authors' great pleasure to dedicate the present paper to Prof. Y. Yamaguchi, in honor of his 60th birthday, who has enabled the authors to make this INS-TSUKUBA joint project with the encouragement of research as Director of INS, together with Prof. M. Muraoka, Head of Theoretical Physics Division of INS.

REFERENCES

1) Baranger, M. and Vénéroni, M., Ann. of Phys. A114, 123 (1978).
Villars, F., Nucl. Phys. A285, 269 (1977).
Moya de Guerra and Villars, F., Nucl. Phys. A285, 297 (1977).
Goeke, K. and Reinhard, P.G., Ann. of Phys. 112, 328 (1978).
Rowe, D.J. and Besserman, R., Can. J. Phys. 54, 1941 (1976).
Bonche, P., Koonin, S. and Negele, J.W., Phys. Rev. C13, 1226 (1976).
2) Marumori, T., Prog. Theor. Phys. 57, 112 (1977).
3) Marumori, T., Maskawa, T., Sakata, F. and Kuriyama, A., Prog. Theor. Phys. 64, 1294 (1980).
Marumori, T., Sakata, F., Maskawa, T., Une, T. and Hashimoto, Y., "Nuclear Collective Dynamics" (Lectures of the 1982 International Summer School of Nuclear Physics, Brasov, Romania) ed. by Bucurescu, D.,Ceausescu, V. and Zamfir, N.V., (World Scientific), 1 (1983).
4) Sakata, F., Marumori, T., Hashimoto, Y. and Une, T., Prog. Theor. Phys. 70, 424 (1983).
5) Sakata, F., Marumori, T., Muramatsu, K. and Hashimoto, Y., Prog. Theor. Phys. 74, 51 (1985).
6) Sakata, F., Marumori, T., Hashimoto, Y., Muramatsu, K. and Ogura, M., INS-Rep.-570, March 1986; to be published in Prog. Theor. Phys.
7) Reinhard, P.G. and Goeke, K., Nucl. Phys. A312, 121 (1978).
8) Kuriyama, A. and Yamamura, M., Prog. Theor. Phys. 66, 2130 (1981).
9) Beliaev, S.T. and Zelevinsky, V.G., Nucl. Phys. 39, 582 (1962).
Marumori, T., Yamamura, M. and Tokunaga, A., Prog. Theor. Phys. 31, 1009 (1964). Marshalek, E.R., Nucl. Phys. A347, 253 (1980).
Ring, P. and Schuck, P., "The Nuclear Many-Body Problem" (Springer-Verlag), (1980). Marumori, T., Takada, K. and Sakata, F., Prog. Theor. Phys. Suppl. 71, 1 (1981).
10) da Providencia, J. and Urbano, J.N., Proceedings of 1982 INS International Symposium on "Dynamics of Nuclear Collective Motion", (INS, Japan), 361.
11) Sakata, F., Hashimoto, Y., Marumori, T. and Une, T., Prog. Theor. Phys. 70, (1983) 163. Hashimoto, Y., Sakata, F. and Marumori, T., Prog. Theor. Phys. 73, 386 (1985).
12) Arnold, V.I. and Avez, A., "Ergodic Problems in Classical Mechanics" (New York, Benjamin) (1968).
13 Muramatsu, K., Sakata, F., Yamamoto, Y. and Marumori, T., submitted to Prog. Theor. Phys. Sakata, F., Marumori, T., Hashimoto, Y. and Muramatsu, K., Invited talk at Topical Meeting on "Phase Space Approach to Nuclear Dynamics" (Trieste, Italy, 1985).
14) Yamamoto, Y., Master Thesis, University of Tsukuba, January (1986).
15) Li, S.Y., Klein, A. and Dreizler, R.M., J. of Math. Phys. 11, 975 (1970).
16) Sakata, F., Marumori, T. and Ogura, M., INS-Rep.-571, March 1986; to be published in Prog. Theor. Phys.
17) Ogura, M., Doctor Thesis, University of Tsukuba, January (1986).

18) Matsuo, M. and Matsuyanagi, K., Prog. Theor. Phys. 74, 1227 (1985).
19) Shimizu, Y.R. and Matsuyanagi, K., Prog. Theor. Phys. 74, 1346 (1985).

THE STUDY OF EFFECTS OF THREE-NUCLEON FORCE

Tatuya Sasakawa
Department of Physics, Tohoku University
Sendai 980, Japan

ABSTRACT

The study of effects of three-nucleon force has achieved a great leap in recent a few years. Here, we review the underlying concept, the methods of calculations which were presented and employed by us, a succinct description of what have been done so far, and what should be done in a future.

1. INTRODUCTION

A question as to where we can see the effect of quark or gluon is of a fundamental interest in nuclear physics. Naturally, the deuteron, the simplest of nuclei, renders a test ground. However, the deuteron is certainly too simple a system to obtain a definitive conclusion about this question. For example, the deuteron has no relation to dynamics such as represented by a three-body force, iso-vector current etc. As a result, this system might not be sufficient to make an overall checking of the quark-gluon effects in nucleus. Therefore, we might have to make recourse to the next-simplest nucleus, the triton.

However, the triton is already too complicated a system, which is, at least, at a present moment almost impossible tc be handled by the quark-gluon picture. Indeed, if we try to say something about the quark-gluon effect in the triton, we will have inevitably to introduce some fantastic imagination or some drastic approximations which are difficult to justify, and as a result, it will be almost impossible to draw any convincing conclusion. Therefore, we abandon this ambitious approach and are taking a conservative but safer way to study the triton as throughlly as possible in terms of realistic two- and three-nucleon potentials. After studying several phenomena concerning triton, I have a feeling that the quark-gluon description in not a "must", at least low energy phenomena are concerned. Even at high energy, a calculation[1] of the EMC effect, which was once thought to be a realization of the quark-gluon picture, suggests that we might need not to have recourse to this picture. In Fig.1, the EMC effect calculated from

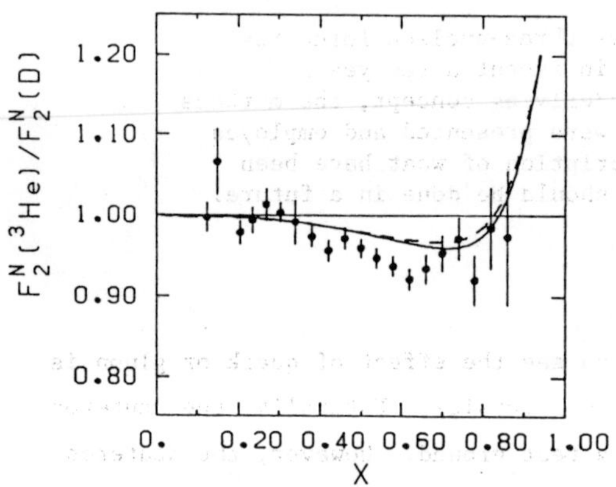

Fig.1 The EMC effect calculated by using our 18-channel triton wave function for the Reid soft core (RSC) potential and the two-pion exchange Tucson-Melbourne (TM) three-nucleon potential.

our triton wave function without a quark is shown together with experimental values for ^{4}He. We hope that an experiment for ^{3}He is done urgently to see if our way of approach actually works well to the EMC effect, too.

Even we turn aside from the above interesting question, the triton should give important information about four fundamental assumptions in "classical" nuclear physics: (1) Two-nucleon potentialsa (2NP) acting between a pair of nucleons are elements of basic dynamics in nuclei, and we need not to take account of three-nucleon potentials. (2) Since the kinetic energy of a nucleon inside nucleus is much smaller than the mass of a nucleon, we may neglect the relativistic effects. (3) A nucleon can be treated as an elementary particle and there is no nucleon excited state, or need not to be taken into account, if any. (4)

In the bound state of nucleus, pions are virtual. Therefore, we usually don't take account of the pion degrees of freedom explicitly, but treat them as a potential. We should study as to what extent these four assumptions are valid. For this purpose, a-few-nucleon systems should render the most desirable testing prove, because we can obtain (numerically) exact wave functions for a given two- and three-body potentials. However, putting aside the discussions for assumptions (2) - (4), we confine ourselves to the problem of three-body force in the present article.

2. THREE-BODY FORCE

Some three-body forces[2,3] proposed in the past are now considered to be inadequate. Fujita-Miyazawa (FM)[4] first proposed a "contemporary" three-body force in which a nucleon get excited to a virtual delta after absorbing a virtual p-wave pion. Later, the Tucson-Melbourne group has proposed a two-pion exchange three-nucleon potential (2PE3NP) now called the Tucson-Melbourne (TM) potential[5].

In formulating the TM potential, (hereafter referred to simply as TM), we don't have recourse to any dynamical model, but make use of PCAC and current algebra and at the same time the phenomenological input so that TM reproduces the pion-nucleon scattering amplitude on mass-shell as well as reasonable off-mass-shell behaviors[6]. A recent derivation of the three-body potential by Friar-Coon[7] is worth while to read. Starting from PV-coupling and Seagull Lagrangian, they performed the Weinberg transformation so that it introduces the PS-coupling too. Then the Foldy-Wouthysen transformation is applied. The way of derivation is different from TM, yet the result is almost equal to TM except a few non-local terms. In any way, this potential is "model independent". The ambiguity as to the mixing ratio of the PS- versus PV-coupling exists in the three-body potential as in the two-nucleon potential.

The TM potential looks like a sum of the (p-wave) FM potential and the potential from the s-wave pion. However, since any dynamical model is not used, the blob in TM in Fig.1 includes not only delta but also any effect as shown in Fig.2.[8] As a result, the TM potential is of a very different nature from a two-body potential. By this

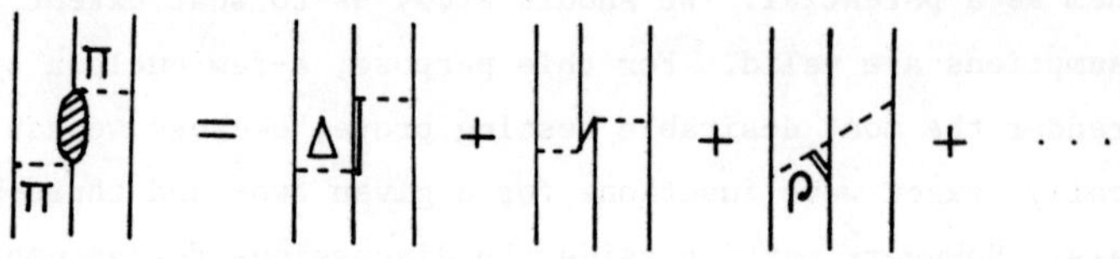

Fig.2 The Tucson-Melbourne potential

reason, the triton calculation using a realistic two-nucleon potential and TM is called a hybrid.[9] After calculated various triton physical properties, I feel that no one will be able to go beyond the hybrid; no one will be able to perform a calculation on the way of the "first principle". If one tried this way, thing will be too complicated. We will have to calculate many terms as shown on the right hand side of Fig.2, with introducing assumptions and approximations, and yet it will be almost impossible to exhaust all conceivable effects.

I got this feeling from the calculation by Hajduk-Sauer[10], in which the three-body force is not a bubble as in TM but a "real" delta, desribed by the Rarita-Schwinger propagator. This term produces the additional binding energy of 0.3 MeV, where more than 0.5 MeV is needed. To get more binding, they might have to add not only the second and third diagrams on the right hand of Fig.2, but other diagrams such as illustrated in Fig.3. Before doing it, we have to answer questions such as how much is the mass of delta in a virtual state, how is the off-shell behavior of $\pi\Delta\pi$-process, etc. On the other hand, by utilizing the TM potential, we can save,

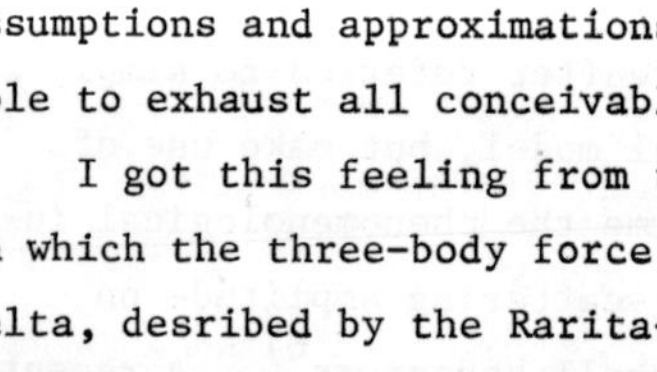

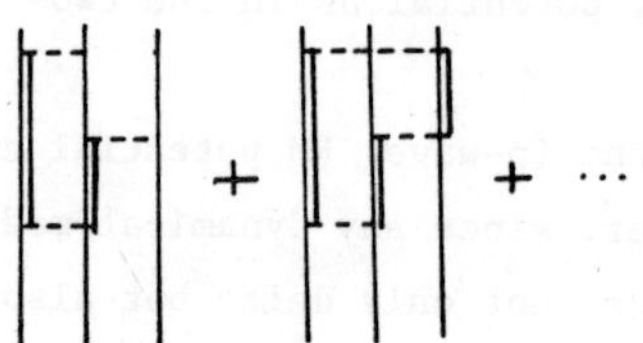

Fig.3 A higher order diagram of the three-body force

to some extent, a formidable task of summing up various diagrams. Here, the strengh of the semi-phenomenology using TM lies. If we tried to calculate the triton by the entirely quark-gluon approach, things will be more complicated than the Hajduk-Sauer's approach. By this reason, I would say that the entirely quark-gluon approach stops to work at the triton, and here nuclear physics takes off particle physics.

3. FADDEEV EQUATION

We have to solve the Faddeev equation to get the solution of a three-body problem. Apart from the mathematical rigour of the Faddeev equation, a remarkable characteristic of this equation is that its formulation is a true realization of quantum mechanics having the wave-particle duality: Waves are linearly superposed and they interfer. Therefore, the multiple scattering process of particles, too, should be described in this way.

Suppose that two particles 2 and 3 are fixed at each position, and another particle 1 comes in, as shown in Fig.4a. Subsequent

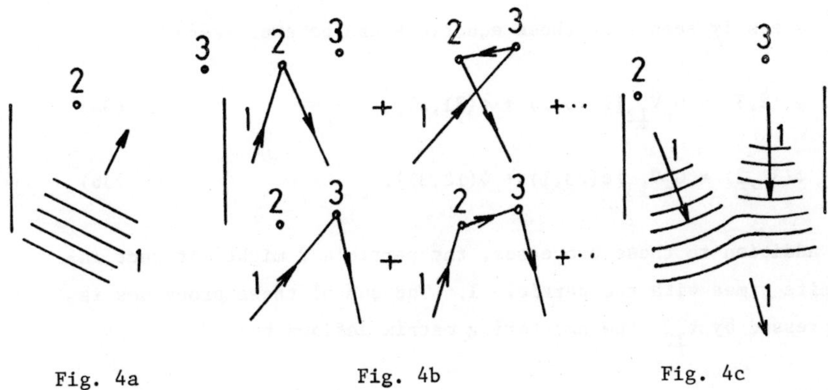

Fig. 4a Fig. 4b Fig. 4c

processes are illustrated in Fig.4b. On the upper row of Fig.4b, we show the processes in which the particle 1 comes out after making the final collision with the particle 2, while on the lower row, we show the processes in which the particle 1 comes out after making the final collision with the particle 3. Since the particle 1 is at the same time wave, these two processes interfer as shown in Fig.4c.

Let us try to express this process in a form of equations. We denote the initial state as shown in Fig.4a by $\phi(23,1)$, where the notation 23 stands for the target, while 1 denotes that the particle 1 is incoming. We designate the propagation of the particle 1 by G_o and the potential between 1 and 2 by V_{12}. All processes on the upper row of Fig.4b end after the interaction V_{12}. If we express the sum of these processes as $\Phi(12,3)$, it takes the form

$$\Phi(12,3) = G_o V_{12} \phi(23,1) + G_o V_{12} G_o V_{13} G_o V_{12} \phi(23,1) + \dots \qquad (1)$$

Similarly, the sum of processes on the lower row of Fig.4b is expressed as

$$\Phi(13,2) = G_o V_{13} \phi(23,1) + G_o V_{13} G_o V_{12} G_o V_{13} \phi(23,1) + \dots \qquad (2)$$

It is easily seen that these equations can be expressed

$$\Phi(12,3) = G_o V_{12} [\phi(23,1) + \Phi(31,2)], \qquad (3a)$$

$$\Phi(31,2) = G_o V_{13} [\phi(23,1) + \Phi(12,3)]. \qquad (3b)$$

In addition to these processes, the particle 2 might interact infinite times with the particle 1. The sum of these processes is expressed by t_{12}, the scattering matrix defined by

$$t_{12} = V_{12} + V_{12} G_o V_{12} + V_{12} G_o V_{12} G_o V_{12} + \dots$$

$$= V_{12} + V_{12} G_o t_{12} \; . \qquad (4)$$

Then, Eqs.(1) and (2) to be more general, we replace all V in these equations by t. As a result, Eq.(3) reads

$$\phi(12,3) = G_o t_{12}[\phi(23,1) + \phi(31,2)],$$
$$\phi(31,2) = G_o t_{13}[\phi(23,1) + \phi(12,3)]. \qquad (5)$$

In Eq.(5), we did not include $\phi(23,1)$ itself. In Fig.4a, the particle 1 passes through without making any interaction with other particles. Therefore, the final state interaction of this "process is between 2 and 3. Thus, we can write

$$\Phi(23,1) = \phi(23,1) \quad . \qquad (6)$$

Since the particle 1 is wave, all wave functions (5) and (6) interfere. Therefore, the total wave function of the system should be a linear superposition of these wave functions

$$\Psi = \Phi(12,3) + \Phi(31,2) + \Phi(23,1) \quad . \qquad (7)$$

So far, we considered the case that the particles 2 and 3 are fixed. If these particles can move, we should extend Eq.(5) and (6) as

$$\Phi(12,3) = \qquad G_o t_{12}(\Phi(31,2) + \Phi(23,1)) \qquad (8a)$$

$$\Phi(31,2) = \qquad G_o t_{13}(\Phi(23,1) + \Phi(12,3)) \qquad (8b)$$

$$\Phi(23,1) = \phi(23,1) + G_o t_{23}(\Phi(12,3) + \Phi(31,2)) \qquad (8c)$$

This set of equations is called the Faddeev equation. It is a simple exercise to show that Eq.(8) satisfies the original three-body Schrödinger equation

$$[E - H_o - (V_{12} + V_{23} + V_{31})]\Psi = 0, \qquad (9)$$

where H_o is the sum of the kinetic energy operators subtracted that of the center of mass motion of the total system.

For identical particles, Eq.(8) assumes a symmetric form

$$\Phi(12,3) = \phi(12,3) + G_o t_{12}(\Phi(31,2) + \Phi(23,1)) \qquad (10)$$

and its cyclic permutations of 1,2 and 3. Since the functional form of the Faddeev components $\Phi(ij,k)$ are the same, we need to solve one equation of the form (10).

In the quantum mechanical three-body problem, there are nine degrees of freedom. This number is reduced to six, if we subtract three degrees of freedom pertaining to the center of mass motion. Further, if we bring the system to the body fixed frame, the rotation of the total system reduces three more degrees of freedom, leaving three degrees of freedom, which are usually given by a set of Jacobi coordinates (x_i, y_i, θ_i), for i = 1,2 or 3, where

$$\vec{x}_1 = \vec{r}_2 - \vec{r}_3 \quad , \vec{y}_1 = \vec{r}_1 - (\vec{r}_2 + \vec{r}_3)/2, \cos\theta_1 = \vec{x}_1\vec{y}_1/|\vec{x}_1||\vec{y}_1|.$$

Therefore, solving the Faddeev equation is to solve Eq.(8) with three variables. Actually, if we make the partial wave decomposition, angles θ_i (i = 1,2 or 3) appear only in the quadrature. Therefore, solving the Faddeev equation means to solve an integral (or differential) equation of two variables. One of technical difficulties in solving the Faddeev equation is how to make the coordinate transformation from one set to another in the course of calculations. In fact, this is very difficult if the system involves a Coulomb interaction.

So far, I have discussed an intuitive way of deriving the Faddeev equation. True difficulty is how to convince physicists, who are studying nuclear reactions in a conventional way, that in treating the breakup processes we must work with the Faddeev equation. Many articles were published attempting to explain the mathematical rigour of the Faddeev equation. However, since physics is in some sense a practical business, those physicists do not pay much attention to the argument of mathematical rigour. Recently, Sawada[11] wrote an article in which the necessity of using the Faddeev equation is demonstrated in a convincing way based on the argument of the asymptotic behavior of the equation. I am hoping this article makes end the belief in the conventinal treatment of the breakup process.

If a three-body force is acting in the system, we extend the Faddeev equation to accommodate it. Since the three-body force acts as in Fig.5, where the solid line shows the line of the strongest force, the use of the Jacobi coordinates is not suitable

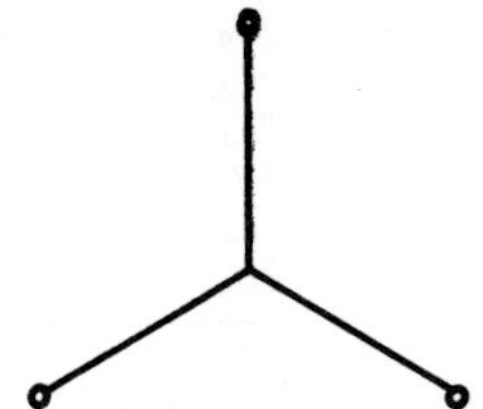

Fig.5 Three-body force

to handle the three-body force, although we are forced to use it. As a result, we have to perform calculations with a large number of states until we get a satisfactory convergence.

4. 34-CHANNEL CALCULATIONS

At this point, to aid a reader who is not familiar with three-nucleon problems, let us explain the meaning of "channel", used in the three-body calculations. With the condition that the total spin of the triton is 1/2, the interacting pair of nucleons as well as the spectator nucleon can take various angular momenta. Table 1 shows the first thirty four possible angular momenta.[12] In solving the Faddeev equation, if we take the first 5 (18, 26, 34) states, we say that we performed the 5 (18,26,34)-channel calculation. If we peform the 5-channel calculation for the Reid soft core (RSC)[13] potential, we express it as RSC5. If we perform the 34-channel calculation for RSC and TM, we write it as (RSC + TM)34. Since the TM depends on the cut-off mass Λ, we express it also as (RSC + TM(Λ))34.

Channel number	Interacting pair $^{2S+1}L_J$	Spectator l_j	Channel number	Interacting pair $^{2S+1}L_J$	Spectator l_j
1	$^1S_0$	$s_{1/2}$	18	$^3F_2$	$f_{5/2}$
2	$^3S_1$	$s_{1/2}$	19	$^1F_3$	$f_{5/2}$
3	$^3D_1$	$s_{1/2}$	20	$^1F_3$	$f_{7/2}$
4	$^3S_1$	$d_{3/2}$	21	$^3D_3$	$d_{5/2}$
5	$^3D_1$	$d_{3/2}$	22	$^3G_3$	$d_{5/2}$
6	$^3P_0$	$p_{1/2}$	23	$^3D_3$	$g_{7/2}$
7	$^1P_1$	$p_{1/2}$	24	$^3G_3$	$g_{7/2}$
8	$^1P_1$	$p_{3/2}$	25	$^3F_3$	$f_{5/2}$
9	$^3P_1$	$p_{1/2}$	26	$^3F_3$	$f_{7/2}$
10	$^3P_1$	$p_{3/2}$	27	$^3G_4$	$g_{7/2}$
11	$^3D_2$	$d_{3/2}$	28	$^3G_4$	$g_{9/2}$
12	$^3D_2$	$d_{5/2}$	29	$^1G_4$	$g_{7/2}$
13	$^1D_2$	$d_{3/2}$	30	$^1G_4$	$g_{9/2}$
14	$^1D_2$	$d_{5/2}$	31	$^3F_4$	$f_{7/2}$
15	$^3P_2$	$p_{3/2}$	32	$^3H_4$	$f_{7/2}$
16	$^3F_2$	$p_{3/2}$	33	$^3F_4$	$h_{9/2}$
17	$^3P_2$	$f_{5/2}$	34	$^3H_4$	$h_{9/2}$

Table 1. 34 channels in the triton

In Fig.6, the convergence of the 5-, 18-, 26- and 34-channel calculations is illustrated for the triton binding energy (B_3)' and the D-state probability P_D. Here, the cut-off mass of the πNN form factor in the TM three-body force is taken to be Λ = 700 MeV. We see that the 5-channel calculations are decisively insufficient and we should perform 34 channel calculations to obtain reliable results, as first found by the Los Alamos-Iowa group.[14] For convergence of other quantities, see ref.15.

5. METHOD OF CALCULATION

As known from exercises for integral equation, a separable kernel reduces one dimensional integral equation to an algebraic equation. Based on this idea, Yamaguchi introduced a separable potential to handle two-nucleon problems.[16] This Yamaguchi potential was utilized for handling a three-body problem by a countless number of authors. Later, several methods were proposed for making separable expansions of a local potential or, in general, the separable expansion of the kernel.

I made every effort to solve the Faddeev equation in another way. The basic idea was two-fold.

(1) We should invent a new perturbation theory in which

(a) the lowest order term should represent the main result,

(b) the lowest order term should be calculated easily, e.g., taking a separable form, and

(c) the remainder should be calculated as a perturbation.

If a reader simply reads these statements, he might have an impression that the existing perturbation theory also satisfies these requirements. It is true. In my theory, however, the following condition (2) should be imposed in addition to (1).

In the usual perturbation theory, the Hamiltonian is divided into the unperturbed part and the perturbation. In the integral expression, the usual Green's function is used. What I had in my mind was different at this point. I was thinking to invent a different Green's function and establish a new perturbation method with this Green's function. The idea came from the following thought.

(2) In any system, we know the asymptotic behavior analytically. If we formulate the theroy so that we may fully make use of it, the calculation should be performed only inside the collision complex, not extending to the asymptotic region. This is, of course, desirable from the point of view of numerical calculations. In mathematical term, our kernel constructed in this way is a non-singular kernel, because it does not involve any delta function, due to the vanishing asymptotic behavior.

My belief is that we should adhere to the perturbational treatment and should not make recourse to an expansion by an ortho-normal set of functions, e.g., the Sturm-Liouville functions. If we try to calculate a set of (Sturm-Liouville) functions, it becomes simply impossible due to increasing number of nodes. As a result, we are forced to truncate the expansion without knowing whether we are coming to a convergent value or not, if we make use of only a finite number of functions, and do not exhaust infinite number of functions. On the other hand, we can perform a perturbational calculation until we get a convergent value, if we invent a way of making any series converge. I should say that in the computer age, one should not have recourse to the method of expansion by a complete orthonormal set of functions.

Honestly, I must say that these ideas were not formed in one night, but took a long time. Taking this occasion, let me write a short history of development of ideas of myself.

First, I tried to extend the Jost function method[17)] to a three-nucleon problem. In fact, the Green's function of this method vanishes at large distances and hence it is a non-singular kernel. The convergence is guaranteed. However, there is a drawback in this method; the wave function that is singular at the origin is used and it makes the treatment of coupled channels very difficult[18)]. This makes the application of this method to a three-body problem imposssible. Therefore, we abandoned this method.

Next, I tried to use a complete orthonormal set of Sturm-Liouville functions.[19)] However, as stated before, the computation of Sturm-Liouville functions becomes simply impossible with increasing number of nodes, and we should abandon the method in which the kernel is expanded by a set of Sturm-Liouville functions.

Therefore, I have come to a method, in which a two-body t-matrix is expressed as a sum of first Sturm-Liouville function which is represented by one separable term, and the remainder which can be treated as the perturbation.[20,21,22)] We call this method "a method of partially separable t-matrix". A three-body problem consists of many such t-matrices. By this method, the separable term represents the attractive effect of a potential, while the effect of the core is not included in this term, but in the perturbation. As a result, in the case of RSC, where the repulsive potential is large, the perturbation series diverges oscillatory. This is illustrated by the dotted line in Fig. 6, where the quantity "DIFF" is plotted against the order of the perturbation. If DIFF becomes zero at a given energy, this value is the binding energy of the three-body system. [In our theory, we canobtain the binding energy if we solve N homogeneous equations, or if we obtain the solution of the N×N determinantal equation $||a_{ij}|| = 0$. For this purpose, we take the first component of the vector as 1, and obtain the solution x_j of a set of coupled N - 1 simultaneous equations $a_{i1} + \Sigma_j a_{ij} x_j = 0$ $(2 \leqq i \leqq N,\ 2 \leqq j \leqq N)$. Then, we use the solution x_j to calculate $a_{11} + \Sigma_j a_{1j} x_j$ $(2 \leqq j \leqq N)$, whose value is called DIFF. All matrix elements a_{ij} are calculated by the perturbation expansion.]

We have overcome this difficulty by introducing the method of acceleration.[23] As shown by a solid line of Fig.6, we get a very quick convergence. This method was employed in performing the 18-channel calculation.[24]

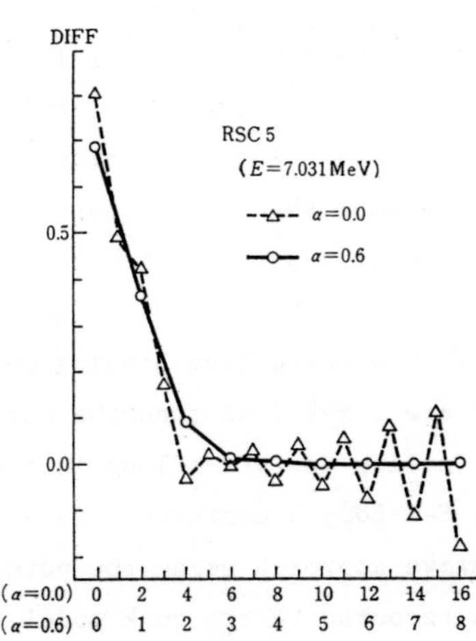

Fig.6 Perturbational calculation with and without the method of acceleration.

This success did not last, however, when we solve the Faddeev equation not only with a two-body force, but also a three-body force, where the repulsive effect is larger than the case only with the two-body force. To avoid this difficulty, we proposed the method of continued fractions,[25] and extended it to the calculation of the three-nucleon bound state.[12] Here, we do not demonstrate this method. Now, we feel that after a long effort as stated above, we have reached the final method by which we can calculate a three-body problem very accurately and very fast. Indeed, by employing this method, we can solve the 34 (26, 18)-channel Faddeev equation with 2NP + TM in 600 (300, 180) seconds by a non-super computer available at Tohoku University (ACOS 1000). This should be compared with other methods for 18-channel calculations: Böhmelburg[26] takes two hours to perform the usual perturbation iteration in momentum space by CYBER. Payne et al.[27] takes twenty minutes to perform his coordinate space calculation by CRAY II.

6. A BREAKTHROUGH

6.1 Five channel calculations in 1983

In 1983, four groups performed five channel calculation of the Faddeev equation for RSC by completely different methods. These groups obtained the triton binding energy of from 7.02 MeV - 7.04 MeV. In view of big computer programs, the difference in these values may not be taken seriously. Rather, this agreement shows that apart from values themselves we are now solving the three-body problem without a mistake.

These groups also calculated the first order perturbation correction due to the TM three-body force. The calculated binding energies are shown in TAble 2. These values are 1 MeV less than the experimental value. As a result, a pessimistic feeling was prevailing in the effect of the three-nucleon force. At the Few-Body X Conference in Karlsruhe, 1983, people felt that the conservative approach using the potential model is dying and we might have to recourse to the quak model to get a correct nuclear force at short distances.

	Purdue[28]	Bochum[29]	Los Alamos[30]	Sendai-Osaka[24]
RSC5	6.98	7.04	7.02	7.03
$-E_3$	- 0.12	0.16	0.41	0.47
Sum	6.86	7.20	7.43	7.50

Table 2. Binding energy of triton calculated from RSC and the first order perturbation energy for TM.

6.2 Eighteen channel calculations in 1984

After the Karlsruhe conference, we performed the 18-channel calculations. Table 3 shows the triton binding energy for various realistic potentials. This Table shows that although calculated binding energy increases with increasing number of channels, the value obtained from calculations with a two-body potential alone is about 1 MeV less than

Number of channels	RSC[13)]	URG[31)]	UG[32)]	TRS[33)]	PARIS[34)]
3	6.389	6.945	6.844	6.829	6.850
5	7.031	7.480	7.395	7.464	7.479
18	7.24	7.50	7.44	7.49	7.56

Table 3. Binding energy of triton calculated for various two-nucleon potentials (without TM).

the experimental value of 8.48 MeV. However, if we calculate the first order perturbation energy for TM, we obtain the values which are very close to the experimental value as shown in Table 4.

Number of channels	RSC	URG	PARIS
5	7.50	7.55	7.73
18	8.13	8.00	8.23

Table 4. Binding energy of triton calculated for various two-nucleon potentials. The first order perturbation value for TM is added to the two-nucleon binding energy.[24)]

the πNN form factor appeared in TM is assumed to be the monopole form $[(\Lambda^2 - \mu^2)/(\Lambda^2 + q^2)]$, with the cut off mass of Λ = 800 MeV. This eighteen channel calculation with TM as the first order perturbation first cast a hope of getting the triton binding energy.[24)]

In ref. 24, we discussed that the contributions from small matrix elements are very important, and the five channel calculations do not take enough number of matrix elements. We also found that the calculated binding energy strongly depends on the πNN form factor and its cut-off mass Λ.

7. SOLUTION OF THE FADDEEV EQUATION INCLUDING TM

Stimulated by ref.24, the Los Alamos-Iowa group has solved the 34-channel problem in 1985 and concluded that we get a convergent value at the 34-channel calculation.[35] Almost simultaneously, we have solved the Faddeev equation with TM for 18 channels,[36] 26-channels[37] and 34-channels at the end of 1985 and found that if we take Λ to be 700 MeV, we obtain the experimental triton binding energy.[12]

We performed our calculations by the method of continued fractions. This method is more or less flexible, and we can consider some various versions of it.[25] In any case, the basic structure is the following.

Let b_n be a given matrix element, which is a function of an assumed binding energy. Assuming that a_{N+1} be zero, we calculate a_n from the equation

$$a_n = \frac{b_n^2}{b_n - a_{n+1}} \quad , \; n = N, N-1, \ldots, 2, 1 \qquad (11)$$

succesively, and obtain a_1. If a_1 thus obtained and a given matrix element b_o satisfy the equation

$$a_1 - b_o = 0, \qquad (12)$$

the assumed binding energy is what we want to obtain. In the same way, we can calculate the scattering amplitude. In fact, the left hand side of Eq.(12) is the denominator of the scattering amplitude, whose zero is the pole; a bound state.

The triton binding energy B_3 from 34-channel calculations with and without TM are given in Table 5.[12,15] This result shows that the effect of the three-body force is about 10% and if we choose Λ to be 700 MeV, we get the triton binding energy as desired. However, we should keep in mind the need to include three-nucleon components such as the ρ-exchange potential. In view of this remark, the values in Table 5 as well

Potential	2NP34	(2NP + TM)34 Λ=700	Λ=800
AV[38]	7.68	8.42	9.29
PARIS	7.64	8.32	9.18
TRS	7.55	8.47	9.71

Table 5. The triton binding energy obtained from 34-channel calculations with and without TM.

as in the following tables should be taken as temporary results.

8. EFFECTS OF THREE-BODY FORCE

Now, having the wave function in which the effect of the three-body force is fully taken, we can study the triton properties. We found that the low energy properties of the three-nucleon system are simply correlated to the calculated binding energy.

[A] Asymptotic normalization constants[37,39] $[C_o, C_2]$

Without a three-nucleon potential, the ratio C_2/C_o is at most 0.04. Inclusion of the TM force in the Faddeev equation makes C_2/C_0 increase about 10%. There is a linear relationship between C_2/C_o and B_3 as shown in Fig.7. Using the experimantal value for B_3, we estimate the value of C_2/C_o to be

$$C_2/C_o = 0.0432 \pm 0.0015.$$

A recent ETH experimental value of C_2/C_o agrees with our value.

[B] Charge radius[15] $[r_{ch}(^3H)]$

Inclusion of the TM force makes B_3 larger, and as a result $r_{ch}(^3H)$ smaller. This is shown in Table 6.

Potential	2NP34	(2NP + TM)34 Λ=700	Λ=800
AV	1.80	1.79	1.68
PARIS	1.79	1.74	1.68
TRS	1.80	1.73	1.64

Table 6. Calculated charge radius.

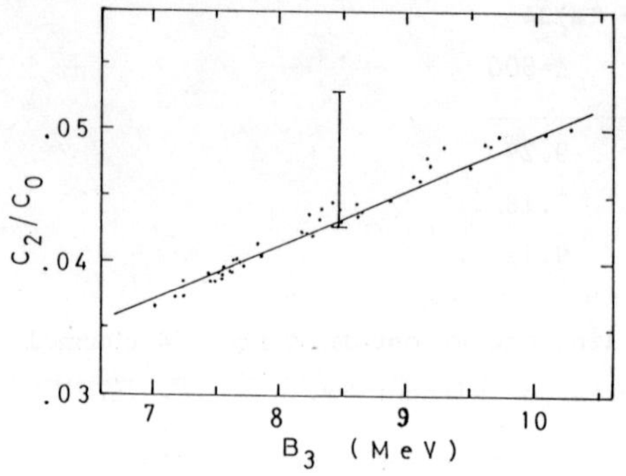

Fig.7
Ratio C_2/C_0 of asymptotic normalization constants vs B_3 for various potential models.

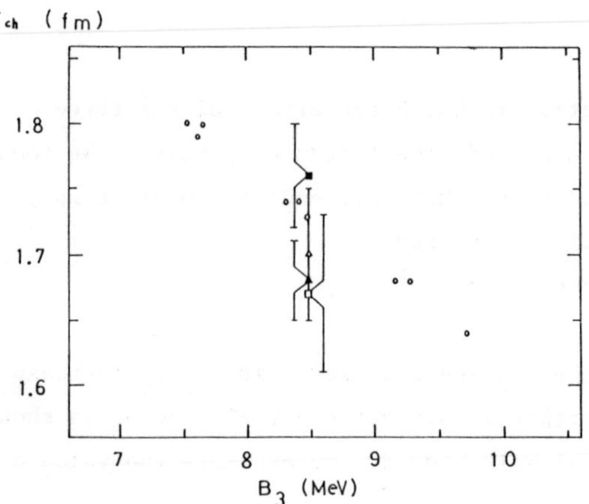

Fig.8
Charge radius vs B_3 for various potential models.

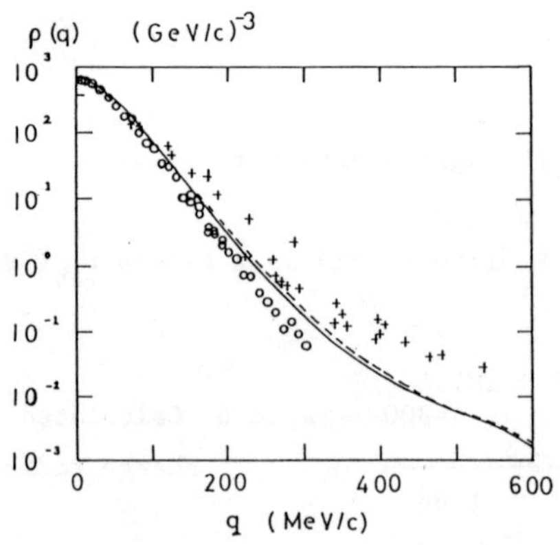

Fig. 9
Momentum distribution of the spectator nucleon relative to the deuteron in the triton obtained from (e,e') (marked o) and (p,p') (marked by +) experiments compared with AV34 (solid line) and (AV + TM)34 (dashed line).

For small q, the calculated chrge form factor $F_{ch}(q)$ is in good agreement with experimental results, when the calculation yields the correct binding energy. The charge radius r_{ch} is obtained by the formula $F_{ch}(q) = 1 - r_{ch}q^2/6 + O(q^4)$. Plotted in Fig.8 are the charge radius vs B_3 obtained from various potential models. We see that if the triton binding energy is correctly given by the calculation, the experimental value of r_{ch} is reproduced.

[C] Momentum distribution[15] [$\rho(q)$]

At low momentum, we can reproduce the experimental momentum distribution of the spectator nucleon relative to the deuteron in the triton, again provided that the triton binding energy is correctly given by the calculation. In Fig. 9, the momentum distribution obtained from the (e,e') scattering (marked by o) and (p,p') scattering (marked by +) experiments are compared with the calculated results for AV34 (solid line) and (AV + TM)34 (dashed line). At large momentum transfer, (e,e') and (p,p') results do not agree with each other. This might be due to the breakdown of the PWIA assumption in reducing the momentum distribution from experimental data.

At medium momentum, the effect of a three-body force is within the experimental error. At high momentum, the effect of the three-body force is important.

[D] Charge and magnetic form factors[40,41] [F_c and F_M]

About the charge form factor of ^{3}He and ^{3}H, we were annoyed, for a long time, having too low second maximum and the first minimum at too large q^2. This was improved greately by the three-body force.[40]

Calculated ^{3}He form factors are in good agreement with experimental values, but ^{3}H form factors are not quite. At this moment, we can not get rid of this discrepancy. A more consistent calculation (between the exchange current and the potential) might be necessary, and/or we might have to have a correct nucleon form factor.

In concluding this section, we do need a three-nucleon potential in addition to a two-nucleon potential to obtain experimental properties at low energy. The $\pi\pi$-exchange three-nucleon potential produces a 10%-efect for any physical quantity. Since these quantities depend only on B_3

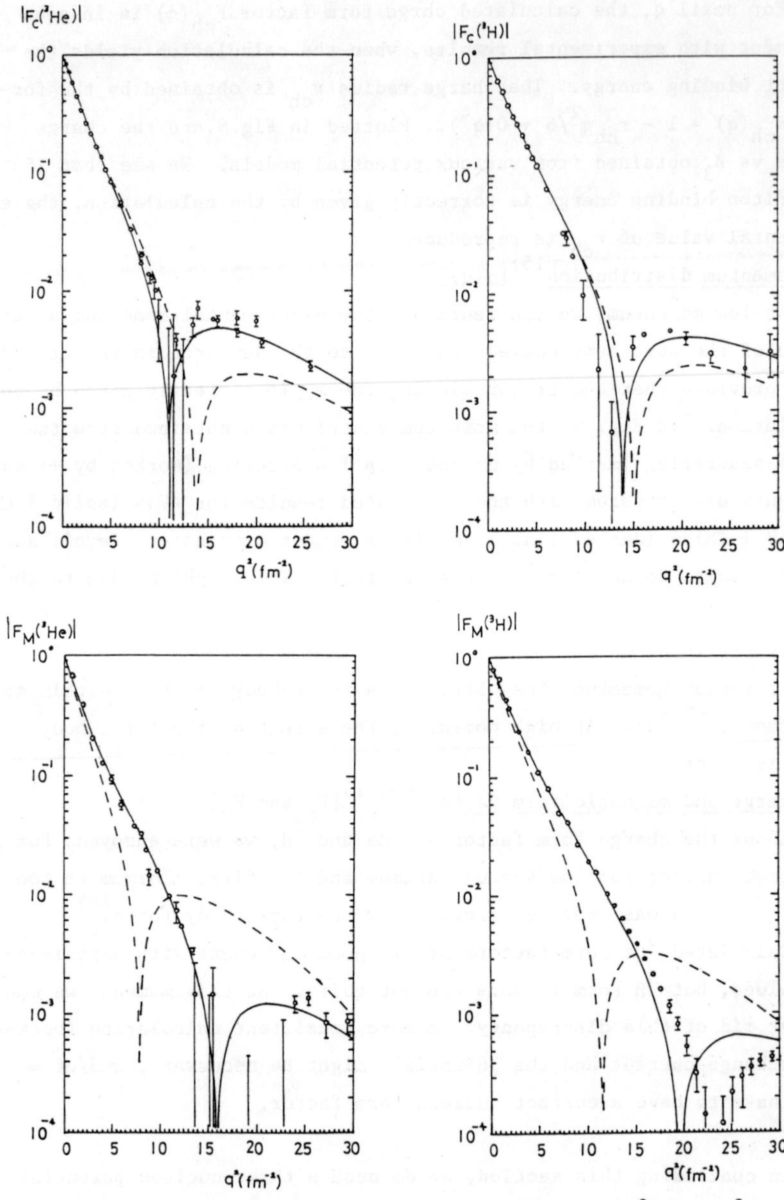

Fig.10 Charge and magnetic form factors of ^{3}He and ^{3}H.

in a simple manner, we can not say much about details of the triton dynamics by measuring these quantities. Any potential model that yields the correct triton binding energy will yield the experimental value of these quantities correctly.

9. WHAT ARE TO BE DONE

So far, so good. However, this is not the end of the story. We have several important subjects lying before us.

As we have seen, the calculations were done for various realistic potential models. However, we did not exhaust studying all conceivable two-body and three-body potential models. We have to study a very sophisticated Bonn potential.[24] (I think that the Bonn potential is the "last potential" in its sophistication.) It might be possible that the Bonn potential changes the conclusion as to a three-body force that it should account for the missing 10% of various quantities. This is because the D-state probability of the deuteron for the Bonn potential is very small (4.25%, but later corrected as 4.65%), and as a result, this potential might yield the triton binding energy to the extent to almost annul the necessity of the three-body force.

About the three-body force, our calculation on the inclusive electron scattering and the y-scaling[43,44] seems suggesting the necessity of the ρ- and ω-exchange three-body forces.

With these potentials, we have to perform the calculation not only of ^3H but also ^3He. Once we have made the 3-channel calculations for ^3He,[20] but now for 34 channels. The calculation should yield important information about the charge symmetry breaking force, which should spell out the Nolen-Schiffer anomaly. Also, once we obtain the wave function of ^3He, we can calculate the β-decay, by which dynamics of the weak interaction in nuclei should be studied. Also, we have to come back to the problem of charge and magnetic form factors to see how the quark structure of a nucleon manifests itself as the nucleon form factor.

All of these problems, which are fundamental in nuclear physics, will yet require further years to come.

References

1) Uchiyama, T. and Saito, K., Private communication.

2) Primakoff, H. and Holstein, T., Phys. Rev. 55, 1218 (1939).

3) Drell, S.D. and Huang, K., Phys. Rev. 91, 1527 (1953).

4) Fujita, J.I. and Miyazawa, H., Prog. Theor. Phys. 17, 360 (1957).

5) Coon, S.A., Barrett, B.R., Scadron, M.D., Blatt, D.W.E. and McKeller, B.H.J., in "Few-Body Dynamics", ed. by A.N. Mitra, et al. (North Holland, Amsterdam, 1976) p.739. Coon, S.A., Scadron, M.D., McNamee, P.C., Barrett, B.R., Blatt, D.W.E. and McKeller, B.H.J., Nucl. Phys. A317, 242 (1979). Coon, S.A. and Glöckle, W., Phys. Rev. C23, 1790 (1981). Glöckle, W., Phys. Rev. C31, 1045 (1985).

6) McKeller, B.H.J. and Glöckle, W., in "Few Body Problems in Physics", ed. by B. Zeitnitz (North Holland, Amsterdam, 1984) p. 435.

7) Coon, S.A. and Friar, J.L., Preprint.

8) McKeller, B.H.J., Talk at the International Symposium on the Three-Body Force in Three-Nucleon System, at the Geroge Washington University, April, 1986.

9) Sick, I., Talk at the International Symposiumon the Three-Body Force in Three-Nucleon System, at the George Washington University, April, 1986.

10) Hajduk, Ch. and Sauer, P.U., Nucl. Phys. A369, 321 (1981).

11) Sawada, T. and Thsushima, K., Preprint.

12) Sasakawa, T. and Ishikawa, S., Few-Body Systems 1, 3 (1986).

13) Reid, R.V., Ann. Phys. (N.Y.) 50, 411 (1968).

14) Chen, C.R., Payne, G.L., Friar, J.L., Gibson, B.F., Phys. Rev. Lett. 55, 374 (1985); Phys. Rev. C33, 1740 (1986).

15) Ishikawa, S. and Sasakawa, T., Few-Body Systems 1, no.3 (1986) to appear.

16) Yamaguchi, Y. and Yamaguchi, Y., Phys. Rev. 95, 1628 (1954); 95, 1635

(1954).

17) Sasakawa, T., Prog. Theor. Phys. Supple. 27, 1 (1963).

18) Sasakawa, T. and Sawada, T., Phys. Rev. C 11, 87 (1975).

19) Sasakawa, T., Nucl. Phys. A160, 321 (1971).

20) Sasakawa, T., Okuno, H. and Sawada, T., Phys. Rev. C 23, 905 (1981).

21) Sasakawa, T., Okuno, H., Ishikawa, S. and Sawada, T., Phys. Rev. C 26, 42 (1982).

22) Sasakawa, T., Horacek, J. and Sawada, T., Phys. Rev. C 27, 18 (1983) ; C 28, 439 (1983).

23) Horacek, J. and Sasakawa, T., Czech. J. Phys. B34, 1 (1984).

24) Ishikawa S., Sasakawa T., Sawada, T., and Ueda, T., Phys. Rev. Lett. 53, 1877 (1984).

25) Horacek, J. and Sasakawa, T., Phys. Rev. A 28, 2151 (1983); A 30, 2274 (1984); Phys. Rev. C 32, 70 (1985).

26) A. Bömelburg, Preprint.

27) Payne, G., Private communication.

28) Muslim, Kim, Y.E. and Ueda, T., Phys. Lett. 115B, 273 (1982); Nucl. Phys. A393, 399 (1983).

29) Bömelburg, A., Phys. Rev. C 28, 403 (1983); Bömelburg, A. and Glöckle, W., Phys. Rev. C 28, 2149 (1983).

30) Wiringa, R.B., Friar, J.L., Gibson, B.F., Payne, G.L. and Chen, C.R. , Phys. Lett. 143B, 273 (1984).

31) Ueda, T., Riew, F.E. and Green, A.E.S., Phys. Rev. C 17, 1763(1978).

32) Ueda, T. and Green, A.E.S., Phys. Rev. 174, 1304 (1968).

33) de Tourreil, R., Rouben, B., and Sprung, D.W.L., Nucl. Phys. A242, 445 (1975).

34) Lacombe, M., Loiseau, B., Richard, J.M., Vinh Mau, R., Cote, J., Pires, P. and de Tourreil, R., Phys. Rev. C 21, 861 (1980).

35) See, ref. 14.

36) Sasakawa, T., Proceedings of the Xth European Symposium on the Dynamics of Few-Body Systems, ed. by P. Doleschall et al. (Hungarian Academy of Science, 1986) p.71.

37) Sasakawa, T. and Ishikawa, S.,"Deuteron Involving Reactions and Polarization Phenomena" ed. by Y. Aoki et al. (World Scientific, 1986) p.341.

38) Wiringa, R.B., Smith, R.A., Ainsworth, T.A., Phys. Rev. C 29, 1207 (1984).

39) Ishikawa, S. and Sasakawa, T., Phys. Rev. Lett. 56, 317 (1986).

40) Sasakawa, T., Fukunaga, A. and Ishikawa, S., Czech. J. Phys. B36, 312 (1986).

41) Fukunaga, A., Private communication.

42) Machleidt, R., Holinde, K. and Elster, Ch., Preprint (to appear in Physics Report).

43) Sasakawa, T., A talk given at the International Symposium on the Three-Body Force in the Three-Nucleeon System, at the George Washington University, 1986.

44) Soutome, M., Master Thesis, Tohoku University, 1986.

RELEVANCE OF THE GAMOW-TELLER STRENGTH DISTRIBUTION IN ^{71}Ge TO THE SOLAR NEUTRINO DETECTOR

T. Oda
Department of Physics, Tokyo Institute of Technology
Oh-okayama, Meguro-ku, Tokyo 152

The full-scale solar neutrino experiment using ^{71}Ga as a detector has been proposed and discussed[1-5] in order to settle the discrepancy between the observation in the ^{37}Cl experiment[6] and the prediction of the standard theory of stellar evolution[7]. The Ga detector, first suggested by Kuzmin[8], has a sufficiently low threshold energy of 0.236 MeV and is considered to be sensitive to low-energy neutrinos from the basic reaction $p + p \rightarrow {}^{2}H + e^{+} + \nu_e$ of the pp chain. The neutrino flux from this basic reaction contributes more than 90% of the solar neutrino flux on the earth and can be calculated practically independent of the solar model[5]. The ^{37}Cl detector, on the other hand, is insensitive to this low-energy pp neutrino due to its high threshold energy of 0.814 MeV, and sensitive to high-energy neutrinos from ^{8}B decays. The flux of ^{8}B neutrinos is highly temperature-dependent, and thus dependent on the solar model.

Nuclear structure studies of the Ga solar neutrino detector have been intensively performed recently from both sides of experiment[9,10] and theoretical calculation[11,12,13]. The neutrino capture rate for the Ga detector is determined by the Gamow-Teller (GT) strength (B(GT)) distribution in ^{71}Ge up to a neutron emission threshold of 7.4 MeV.

The ground state ($3/2^-$ in ^{71}Ga) to ground state ($1/2^-$ in ^{71}Ge) GT strength can be obtained reliably from the ft value for ^{71}Ge (β^+) ^{71}Ga to be 0.09. One can see the role of Gamow-Teller strength distribution in ^{71}Ge for each solar neutrino source from the excitation energy dependence of $\phi<\sigma_{av}(E_{exc})>$, where ϕ is the neutrino flux and $<\sigma_{av}(E_{exc})>$ is the weighted average phase space factor for the neutrino absorption cross section for the fixed final state. From this excitation energy dependence of $\phi<\sigma_{av}(E_{exc})>$, one can see the following properties. First, the pp neutrino capture rate is determined only by the Gamow-Teller strength to the ground state. Therefore, the pp neutrino capture rate is reliably estimated by using the GT strength from the β decay ft value of ^{71}Ge. Secondly, the ^{7}Be neutrino capture rate is determined by the Gamow-Teller strength to the lowest three levels in ^{71}Ge. Orihara et al.[9] performed the (p,n) reactions with 35 MeV protons and deduced the GT matrix elements to the first three states in ^{71}Ge, assuming the proportionality of the (p,n) forward-angle cross sections to B(GT). Their deduced B(GT) to the first $5/2^-$ state is comparable to that to the ground state and gives the ^{7}Be neutrino capture rate comparable contribution as B(GT)

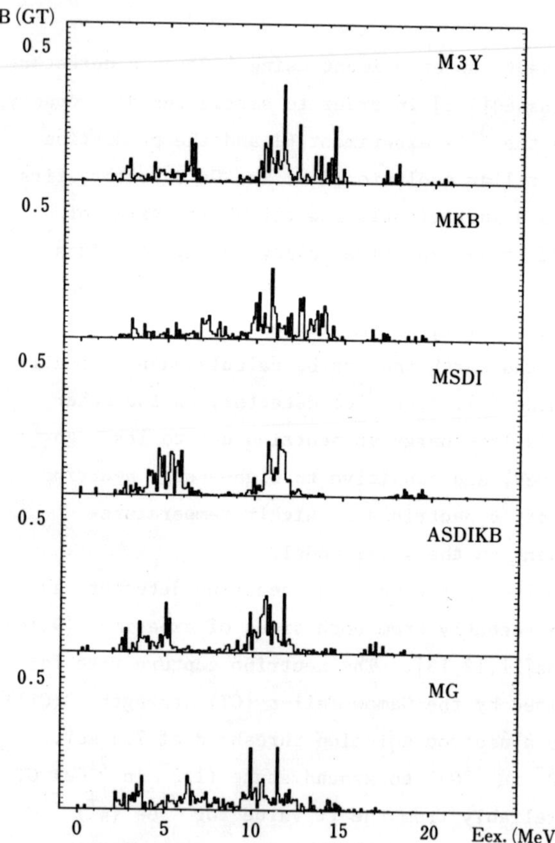

Fig. 1 Calculated GT strength distributions in ^{71}Ge.

to the ground state. Baltz et al.[12] discussed the (p,n) reaction mechanism on the proportionality above mentioned for a low energy projectile. Thirdly, the ^{8}B neutrino capture rate is very small if only those Gamow-Teller strength to the lowest three levels are taken into account. However, if there is large Gamow-Teller strength in the 2 ∿ 5 MeV region, the ^{8}B neutrino capture rate could be increased to the comparable order as the ^{7}Be neutrino capture rate. Klapdor et al. [11] first pointed out by BCS calculation that there is a bump of GT strength at 2 ∿ 6 MeV in ^{71}Ge and it increases considerably the ^{8}B neutrino capture rate. Krofcheck et al.[10] studied the ^{71}Ga (p,n) ^{71}Ge reaction at proton bombarding energies of 120 and 200 MeV, and deduced the GT strength distribution in ^{71}Ge up to ∿14 MeV. Their adopted upper limit of B(GT) to the first $5/2^-$ state is 0.009, which is one order smaller than that to the ground ($1/2^-$). Mathews et al. [13] performed shell model studies within truncated model spaces and with Kallio-Kolltveit interaction[14], and obtained similar conclusion as that of Grotz et al.[11] and Krofcheck et al.[10] that the GT stregth distribution for the excited states increases the ^{8}B neutrino capture rate by a factor of 10 compared with the estimate of Bahcall[2] which takes into account only the lowest three levels.

Table 1 Calculated solar neutrino capture rates (in SNU) in five sets of interactions

Neutrino source	M3Y	MKB	MSDI	ASDIMKB	MG
pp	65.5	65.5	65.5	65.5	65.5
pep	3.1	4.2	3.4	4.0	3.6
^{7}Be	32.6	39.4	34.4	38.4	35.5
^{8}B	18.8	12.8	25.6	21.0	21.7
^{13}N	3.1	3.6	3.2	3.5	3.3
^{15}O	4.7	5.9	5.0	5.7	5.2
Total ($\Sigma\sigma\phi$)	127.7	131.4	137.1	138.2	134.7

We have performed detailed shell model calculations of the Gamow-Teller strength distribution in ^{71}Ge within $(f_{5/2},p_{1/2},p_{3/2})^{15} + f_{7/2}^{-1}(f_{5/2},p_{1/2},p_{3/2})^{16}$ configuration space and with five sets of interaction in order to study the effective interaction dependence of this ^{8}B neutrino capture rate by the Ga detector. The effective interactions used are M3Y[15], Kuo-Brown (MKB)[16], modified surface delta interaction (MSDI) and adjusted surface delta interaction (ASDIMKB) of Koops-Glaudemans[17], and Mooy-Glaudemans (MG)[18].

The single particle energies and the monopole parts of diagonal matrix elements of the effective interactions are modified to reproduce the lowest three levels in ^{71}Ga and the Gamow-Teller giant resonance at 11 ∿ 12 MeV in ^{71}Ge. The calculated GT strength distributions in ^{71}Ge are shown in fig. 1. These are the calculated results in the truncated space, in which the summed value of seniority in each orbit for each configuration is limited in isospin formalism as equal to or smaller than 3. However, the total B(GT) calculated in this truncated space exhausts about 90% of that in the full space of $(f_{5/2},p_{1/2},p_{3/2})^{15} + f_{7/2}^{-1}(f_{5/2},p_{1/2},p_{3/2})^{16}$.

We have made the following modification on this strength distributions. For the ground state, we have normalized the calculated strength to be 0.09 which is reliably obtained from the β decay ft value of ^{71}Ge. We have multiplied this normalization factor to the calculated strength for the first $5/2^-$ and $3/2^-$ states. For the other excited states, we have multiplied the quenching factor 0.5.

The solar neutrino capture rates thus obtained are shown in table 1. The percentage of the pp neutrino capture rate in the total capture rate is almost constant (47 ∿ 51%). Our calculated B(GT) values to the first $5/2^-$ state are smaller than 0.004, and the ^{7}Be neutrino capture rate does not have the strong effective interaction dependence as shown in table 1. On the other hand, the ^{8}B neutrino capture rate depends on the GT strength distribution up to 7.4 MeV. Our calculated capture rate has a strong effective interaction dependence, and varies from 13 SNU to 26 SNU. According to this calculations, the solar model dependent "background" from ^{8}B neutrinos in a Ga solar neutrino experiment ranges - in the standard solar model - between 19.5 and 39% of the pp-rate - , i.e. a factor of 10 larger than the Bahcall expectation. It is to be noted that Gamow-Teller strength distribution should be carefully calculated with a reliable effective interaction in a reliable model space for the prediction of solar neutrino capture rate. It might be the same situation in the case of ^{205}Tl experiment.

The GALLEX European collaboration[19] has been funded, and started the construction of the laboratory for the Ga experiment under the Gran Sasso massif in January, 1986. They are going to collect 30t of Ga in two years, from the beginning of 1988 to the end of 1989. They will start its execution phase of the Ga experiment from the beginning of 1990, and finish it at the end of 1994. Their aims are firstly to get the first experimental evidence (if the capture rates will have exceeded $\sim$70 SNU) that hydrogen burning is the energy source of the sun, and/or secondly (if the capture rates will have went down lower than $\sim$70 SNU) to see the possibility of the neutrino oscillations[19].

References:

1. Bahcall, J. N. et al., Phys. Rev. Lett. 40, 1351 (1978).
2. Bahcall, J. N., Rev. Mod. Phys. 50, 881 (1978).
3. Bahcall, J. N. et al., Rev. Mod. Phys. 54, 767 (1982).
4. Hampel, W., in Proc. Int. Conf. Neutrino Physics and Astrophysics, Erice, Italy, 1980, ed. E. Fiorini (Plenum, New York, 1982), p.61.
5. Klapdor, H. V., Prog. Part. Nucl. Phys. 10, 131 (1983).
6. Davis. Jr., R., in Proc. Conf. Status and Future Solar Neutrino Research, ed. G. Friedlander (BNL 50879), Vol. 1, p.1.
7. Bahcall, J. N. et al., Phys. Rev. Lett. 45, 945 (1980).
8. Kuzmin, V. A., Zh. Eksp. Teor. Fiz. 49, 1532 (1965) [Sov. Phys. JETP 22, 1051 (1966)].
9. Orihara, H. et al., Phys. Rev. Lett. 51, 1328 (1983).
10. Krofcheck, D. et al., Phys. Rev. Lett. 55, 1051 (1985).
11. Grotz, K. et al., in Capture Gamma-Ray Spectroscopy and Related Topics - 1984, ed. S. Raman, AIP Conf. Proc. No. 125, p.793,

Phys. Rev. C33, 1263 (1986).

12. Baltz, A. J. et al., Phys. Rev. Lett. 53, 2078 (1984).
13. Mathews, G. J. et al., Phys. Rev. C32, 796 (1985).
14. Kallio, A. and Kolltveit, K., Nucl. Phys. 53, 87 (1964).
15. Love, W. G., in The (p,n) Reaction and the Nucleon-Nucleon Force, ed. C. D. Goodman et al., p.23.
16. Kuo, T. T. S. and Brown, G. E., Nucl. Phys. A114, 241 (1968).
17. Koops, J. E. and Glaudemans, P. W. M., Z. Physik A280, 181 (1977).
18. Mooy, R. B. M. and Glaudemans, P. W. M., Z. Physik A312, 59 (1983).
19. Kirsten, T., Proc. Int. Symposium on Beta Decay and Neutrino, Osaka 1986.

    Hampel, W., Proc. Int. Symposium on Weak and Electromagnetic Interactions in Nuclei, Heidelberg, 1986.

V. GRAVITY AND SPACE-TIME

GRAVITY: SOURCE OF ENERGY AND ENTROPY IN THE UNIVERSE

Humitaka SATO
Department of Physics, Kyoto University
Kyoto 606
JAPAN

Star light is extracted from matter by nuclear fusion but the matter gets finally more energy from gravitational energy. Isothermal and uniform state of the early universe is not a state of thermal death. Gravity can generate a higher entropy state. Our universe is in a state of "far from equilibrium".

1. SOURCE OF NUCLEAR ENERGY

Important application of nuclear physics was astrophysics. Now, everybody knows that star light is supplied by nuclear energy and elements are ashes of nuclear burning. But, this is only a half of the whole story concerning stellar evolution. In the later half, role of gravity is more important. As an example, let us consider a star which evolves to a neutron star. Through its evolution, the initial hydrogen rich matter is transformed into neutrons, that is, the matter gains some energy rather than it supplies energy. Who has paid this extra energy? That is gravitational energy.

As well known, star light is supplied by nuclear binding energy. We write this schematically as

$$\text{Process A:} \quad \int L_{star}\, dt = BN \quad ,$$

where stellar luminosity L_{star}, binding energy per nucleon B and nucleon number N. On the other hand, at the end of evolution, the star explodes as a supernova and leaves a neutron star and an expanding nebula. For the transformation from "ash" nuclei into neutrons, a vast energy is required. All the energies are supplied by a binding energy of the neutron star. We write this schematically as

$$\text{Process B:}\quad \frac{GM^2}{R} = (B + \Delta mc^2)\,N + W_{SN}\quad ,$$

where $\Delta m = m_n - m_p$ and supernova explosion energy W_{SN}.

We are talking about a supernova which leaves a neutron star. There is another type of supernova which does not leave a neutron star. The previous type of supernova is however more important, because it generates neutron stars, r-process elements, cosmic rays and stars in next generation.

The process A lasts over a long period but the process B happens instantaneously. Therefore, we are tempted to think the process A as a major process in Galaxy and the process B as an exceptional one. In fact, most stars do not evolve into supernova. However, if the process B is overlooked, we will miss an important rink of Galactic evolution.

For an example, let consider the synthesis of elements heavier than iron. Since the binding energy per nucleon decreases with mass number for those elements, they can not be ashes at fusion burning. They are considered to be synthesized through a rapid neutron capture process ("r-process"). The neutronization of matter occurs through a capture of energetic electrons via inverse β-process. Then, who has paid a cost for acceleration of these electrons? That is essentially gravitational energy of a collapsed star, $\Omega \simeq -GM^2/R$. For an equilibrium state of a star, the internal energy is $U = |\Omega|/2$ and the energy for the neutronization can be provided by the electron Fermi energy in U.

In a sense, "nuclear power" generated by a fission of uranium is a packed form of gravitational energy, like that "combustion energy" of oil is a packed form of an ancient solar energy through the process A.

2. ENTROPY OF THE UNIVERSE

Stellar evolution and element synthesis are major astronomical processes in our universe. Through these processes, the universe is aging in one-way. They must have started in a finite time ago, i.e., "start" is necessary. Otherwise, the universe must have aged completely. We have known now that "star" was provided by Big Bang.

We have observed a glory of ancient fire-ball by the microwave background radiation. Through the anisotropy observation, it has been checked also that the early universe was almost uniform and isothermal. In the thermodynamics, we learn that the uniform and isothermal state is the maximum entropy state or "thermal death" state. But, if the universe started from the thermal death, the evolution would have been impossible.

The above paradoxical result is resolved by including the following two overlooked ingredients:

I: Expansion of the system

II: Effect of gravity

The cosmic expansion is in a sense too fast to maintain a thermal equilibrium for most of the processes. It is not "adiabatic" and "over-cooled state" is realized. This departure from the complete equilibrium provides it with variety of our world: Compactification of higher dimensional space, GUT phase transition, baryon synthesis, hadronization, decoupling of neutrinos, element synthesis, decoupling by hydrogen recombination and so forth. All these phenomena belong to this category.

Gravity has an unique character in this respect. We can not define clearly a "decoupling of gravitational interaction" because of unshielded long range nature. Due to this feature, it is found that the isothermal and uniform density state is not a state of "thermal death". One of such evidences is "gravo-thermo catastroph". If the density contrast is large enough, the entropy of the system increases indefinitely for a further contraction. This situation is also understood as follows. If the energy generated by a contraction is large enough, thermal energy will be in the radiation and the entropy is written as

$$S = bT^3V = b(E - Mc^2 + \alpha \frac{GM^2}{R})^{3/4}V^{1/4} ,$$

where the total energy of the system is

$$E = Mc^2 + aT^4V - \alpha \frac{GM^2}{R} ,$$

α being a numerical factor. Then, S can increase by a decrease of R under constant E and V.

Quantum theoretical consideration of black hole has revealed that it has an entropy and a temperature given by

$$S_{BH} = 4\pi (\frac{M}{m_{pl}})^2$$

and

$$T_{BH} = \frac{m_{pl}{}^2c^2}{8\pi M} ,$$

where m_{pl} is the Planch mass. Relation between the gravothermo catastroph and the black hole entropy has not been clarified. Penrose once conjectured an idea of gravitational entropy given by the Weyl tensor but it was not successful.

Now, let consider a system which consists of one black hole and black body radiation with the same temperature. The entropy is given now as

$$S = 4\pi(\frac{M}{m_{pl}})^2 + b(E-Mc^2)^{3/4}V^{1/4}$$

with $T_r = T_{BH}$. The maximum entropy state under fixed E and V is realized when $M/E \neq 0$ if $y \equiv 3\pi(abV/G^4E^5)^{1/4} < 1$. We call such state as "black hole (BH) phase" and otherwise as "black body phase". If we write $E = aT^4V$, the criteria of the BH phase is $y \sim (vt^5)^{-1} < 1$ or $t > v^{-1/5}$, where $v = V/l_{pl}{}^3$ and $t = T/T_{pl}$. If v is taken as a horizon volume $vt^6 \sim 1$ for the radiation universe. Therefore, the universe has passed through the BH phase since the Planck epock. So, it has not been the entropy maximum phase where most of energy would have condensed into the black hole.

How large is the difference between the maximum entropy and the present entropy? Taking the horizon volume of $(10\ell y)^3$, the entropy of matter is given roughly by the number of the microwave background photons as

$$S_{BB} \sim 10^{90} .$$

But, if an appreciable part of the nucleons would have collapsed into black holes through stellar evolution, the entropy of the stellar black holes is estimated as

$$S_{\text{stellar BH}} \sim \frac{m_N S_r \bar{B}}{M_\odot} \left(\frac{M_\odot}{m_{pl}} \right)^2 \sim \left(\frac{m_{pl}}{m_N} \right) \bar{B}\, S_r \sim 10^{100} ,$$

using the relation $M_\odot \sim m_{pl}^3/m_N^2$, $\bar{B}$ and m_N being Baryon-Photon ratio and nucleon mass respectively. The maximum entropy is attainable when they merge into a single black hole. That is given as

$$S_{max} \sim \left(\frac{m_N \bar{B} S_r}{m_{pl}} \right)^2 \sim 10^{124} .$$

In a sense, we are enjoying this enormous entropy gap. However, our local system on the globe has been maintained through more subtle balance of entropy. Therefore, we must be careful enough to keep it.

ACKNOWLEDGEMENT

Even now, I clearly remember a scene when I first listened the lecture of Professor Yamaguchi in early 1960's. At the time, he gave a set of lectures at Department of Physics, Kyoto University to the graduate students including me. The reason why I remember it is because he did not talk on "physics" mostly but on some global and historical aspects of problems in energy, food, population, etc. This first encounter impressed me very much. Since then, Professor Yamaguchi has kept us impressed all the times.

I am very grateful to dedicate this essay to Professor Yamaguchi on his sixty years birthday.

SOME CONSEQUENCES OF COMPOSITION-DEPENDENCE OF POSSIBLE FINITE-RANGE GRAVITATIONAL FORCE

Yasunori Fujii
Institute of Physics
University of Tokyo, Komaba
Meguro-ku, Tokyo 153
Japan

Abstract

Effects of composition-dependence of the possible finite-range gravitational force are examined, focussing on geophysical measurements and the Eötvös or Galilei experiments. We reasonably expect that new information will soon emerge about coupling to leptons.

A number of theoretical models have been proposed suggesting the presence of an extra force that couples to the matter nearly as weakly as the gravitational force.[1-9] It has a finite force-range given basically by $\lambda \sim (Gm^4c/\hbar^3)^{-1/2}$ which is (G= Newton's constant) of the order of km for a choice $m \sim 1$ GeV/c^2, a typical mass scale of elementary particles. The force would be mediated by a scalar field or a vector field.

A "circumstantial" evidence of a force of this nature has been reported from a geophysical study.[10] It has also been shown that the Eötvös experiment of the original type could exhibit apparent composition dependence being consis tent with the suggested extra force,[6] though the original claim is now somewhat weakend.[11] Different theoretical models predict different composition dependence.

In this paper we discuss consequences of composition dependence in geophysical measurements as well as the Eötvös or Galilei (free-fall) experiments for a variety of models. We point out that, in principle, one has to know the composition of the "weight" of the gravimeter when one compares these two kinds of measurements. In some of the extreme choices of the models, a scalar (vector) field might result, rather unexpectedly, in a repulsive (attractive) force in geophysical measurements, though this possibility seems unlikely to be supported by the Eötvös experiment. We also emphasize that enhanced effects in the Eötvös or Galilei experiments can be expected once the coupling to leptons is included.

We assume that the scalar or vector field couples to a baryon and a lepton with the coupling constants f_B and f_L, respectively. Consider two objects with the masses M_i and M_j, assumed to consist of the elements i and j, respectively. We use the symbols B, L and μ for the baryon number (=averaged mass number), the lepton number (= atomic number), and the mass in units of m_{H1}(=mass of the ${}_1H^1$ atom divided by 1.007825). For a substance made of several elements we use averaged values according to its composition. We ignore chemical binding energies.

The static potential between two objects is then given by

$$V_{ij} = -G_\infty(M_i M_j/r) + V'_{ij} , \qquad (1)$$

where the contribution from the extra force is

$$V'_{ij} = -\varepsilon(f_B^2/r)e^{-r/\lambda} \; N_i H_i N_j H_j , \qquad (2)$$

where N_i and N_j are the numbers of atoms in each object. The constant ε is +1 or −1 according to whether the extra force is due to a scalar or a vector field, with λ the force-range, H_i and H_j being given by

$$H_i = B_i + \xi L_i, \qquad H_j = B_j + \xi L_j, \tag{3}$$

with $\xi = f_L/f_B$.

The potential (1) may also be put into the form

$$V_{ij} = -G_\infty(M_iM_j/r)(1+\alpha_{ij}e^{-r/\lambda}) , \tag{4}$$

where the composition-dependent coefficient α_{ij} is given by

$$\alpha_{ij} = \varepsilon(f_B^2/G_\infty m_{H1}^2)(H_i/\mu_i)(H_j/\mu_j) . \tag{5}$$

Various theoretical models can be parameterized by ε and ξ. A vector field which couples to the baryon number, as proposed by Fischbach et al,[6] or suggested from a Kaluza-Klein theory,[4] corresponds to $\varepsilon = -1$ and $\xi = 0$, while Hayashi and Shirafuji prefer the coupling to the fermion number (=sum of quark number and lepton number), implying $\varepsilon = -1$ and $\xi = \pm 1/3$, based on a point of view that a fundamental force as weakly as gravity hardly distinguishes quarks from leptons.[7] The value $\xi = -1$ may also be chosen correspond ing to the B-L current in Grand Unified Theories.[8] (See also ref. 7 for $\xi = \pm 1$.) A supersymmetric model due to Fayet is still flexible about the precise value of ξ.[9] Glashow advocates a phenomenological approach to determine ξ by trying many different materials.[8]

A scalar theory of Chang et al as derived from spontaneously broken lepton number gives $\varepsilon = +1$, $\xi = m_e/m_q$ in terms of the masses of the electron and the quark.[5] The two-scalar model of spontaneous breaking of scale invariance, on the other hand, allows a more general choice of ξ, because only one of the two scalar fields is directly responsible for generating particle masses.[2]

Let us consider a geophysical measurement using a gravimeter in which a spring hangs a weight that has averaged values B_w, L_w and μ_w. One is thus measuring a finite-range force between this weight and the eartn's crust nearby. We assume that the crust is homogeneous having the averaged density ρ_c and the averaged quantities B_c, L_c and μ_c. An analysis of the data in terms of the potential (4) would give

$$\alpha_{cw} = \varepsilon(f_B^2/G_\infty m_{H1}^2)(H_c/\mu_c)(H_w/\mu_w). \tag{6}$$

One may also carry out the Eötvös or Galilei experi ment on the same kind of crust as assumed above. The coefficient

$$\eta_{ij} = 2\,\frac{(m_G/m_I)_i - (m_G/m_I)_j}{(m_G/m_I)_i + (m_G/m_I)_j}$$

should be given by

$$\eta_{ij}=(3/2)(\lambda/R)(\rho_c/\rho_E)(\varepsilon f_B^2/G_\infty m_{H1}^2)(H_c/\mu_c)\Delta(H/\mu), \tag{7}$$

where $\Delta(H/\mu) = (H_i/\mu_i) - (H_j/\mu_j)$, ρ_E being the density of the earth as a whole, R the earth's radius.

A comparison between (6) and (7) yields

$$\eta_{ij} = (3/2)(\lambda/R)(\rho_c/\rho_E)\alpha_{cw}(H_w/\mu_w)^{-1}\Delta(H/\mu). \tag{8}$$

This clearly shows that one can compare above two types of measurement only depending on the substance of the weight of the gravimeter. How this affects the result, however, depends on the value of ξ. .

In Table I we show examples of H/μ for different values of ξ. For $\xi = 0$, the ratio $H/\mu(=B/\mu)$ varies from element to element only less than 0.1%, whereas it dose up to 3% (excluding H) for $\xi = \pm 1/3$. (I thank Kenji Hayashi for pointing this out to me.) Variability is even larger for larger values of $|\xi|$. Unless $|\xi|$ is extremely small, H/μ is given roughly by $1 + \xi(L/B)$, which may assume both signs if $\xi \sim -2$, because $L/B \sim Z/A$ is distributed around 1/2.

One of the extreme examples is shown for $\xi = -2.2$. Since H_c/μ_c for the crust is then negative, the product with H_w/μ_w is negative for the (nearly pure) tungsten weight (as in the LaCoste-Romberg gravimeter G608 or G20 used in the mine measurements[10]), implying a scalar field for the reported negative α_{cw}. If this is indeed the case, one should observe a positive α_{cw} for an aluminum weight, for example. It seems rather unlikely, however, that this extreme choice is acceptable because it would lead to too large effects in the Eötvös or Galilei experiment, as will be shown shortly.

We now turn to the Eötvös or Galilei experiment. For $\xi = 0$, Eq.(8) combined with the geophysical result $\alpha \sim -10^{-2}$, $\lambda \sim 200$ m gives η always below the level of 0.5×10^{-9}, somewhat inaccessible for the moment. For $\xi \neq 0$, however, η would be much enhanced. To get a maximum effect, one should choose a pair of substances having A/Z as much apart as possible, like H_2O or paraffin ($C_{2n}H_{2n+2}$) on one hand, and W, Au or Pt, on the other hand. The last line of Table I shows an example of $\eta(C_{20}H_{42}, Au)$ as a representative for large n.

The results for $\xi = -2.2$ and -3 seem already too large for $\lambda = 10$ m which is near the shorter end obtained by combining geophysical and laboratory experiments.[10] Other values of ξ seem to offer a good chance to be tested by the experiments already available or planned in the near future.

To conclude we summarize our analysis:

(i) Accepting the result of the geophysical measurements, and if the extra force couples exclusively to baryons, i.e. $\xi = 0$, we find that the expected coefficient η appers still beyond the accuracy 10^{-9} of the the measurements which we can foresee in the immediate future.

(ii) If ξ is not equal to zero, on the other hand, the

experiment would be rather feasible, if one chooses a pair of paraffin and gold, for example. Currently available free-fall measurements (at several institutions including Bureau International Poids & Mesures, Paris, International Lattitude Observatory, Mizusawa and Joint Institute for Laboratory Astrophysics, Boulder) seem to have capability of probing ξ even if λ is rather close to 10 m. It is yet to be seen what accuracy improved versions of the Eötvös experiment of the original type will reach.[12)]
(iii) In the geophysical measurement, using a weight of lighter element might be interesting from the same point of view. Marine experiments has an added advantage that the ratio B/L of water is 1.800 being different from 2.023 of the averaged crust.

I thank Drs. J. Faller, K. Kuroda, L. LaCoste, I. Murata, A. Sakuma, F. Stacey and T. Tsubokawa for having provided valuable information. I also appreiate discussions with Kenji Hayashi and Naoki Onishi.

References

1) Y. Fujii, Ann. Phys. (N.Y.) 69 (1972) 494; Nature Phys. Sci. 234 (1971) 5.
2) Y. Fujii, Phys. Rev. D9 (1974) 874; GRG 13 (1981) 1147.
3) J. Scherk, Phys. Lett. 88B (1979) 265.
4) Y. Fujii, to be published in Prog. Theor. Phys. 75 (1986); UT-Komaba 86-10.
5) D. Chang, R.N. Mohapatra and S. Nussinov, Phys. Rev. Lett. 55 (1985) 2835.
6) E. Fischbach, D. Sudarsky, A. Szafer, C. Talmadge and S.Aronson, Phys. Rev. Lett. 56 (1986) 3.
7) K. Hayashi and T. Shirafuji, preprints Kitasato (86-2,5) and Saitama (TH-86-2,5).
8) S.L. Glashow, Harvard preprint HUTP-86/A012 (1986).
9) P. Fayet, Phys. Lett. 171B (1986) 261.
10) S. Holding, F. Stacey and G. Tuck, Phys. Rev. D33

(1986) 3487: F. Stacey, G. Tuck, G. Moore, S. Holding, B. Goodwin and R. Zhou, University of Queensland preprint, (1986).

11) See Comments in Phys. Rev. Lett. 56 (1986) 2423-2426.

12) P.T. Keyser and J.E. Faller, JILA preprint, (1986): K. Kuroda, private communication.

Table I The [illegible] line shows ξ, while in the following ten lines are shown the values of H/μ for different ξ. In the first column enter substances, where "crust" means a mixture of O(mass ratio 46.6%), Si(27.7), Al(8.1), Fe(5.0), Ca(3.6), Na(2.8), K(2.6), Mg(2.1), Ti(0.4), Ba(0.3), P(0.1), Mn(0.1), F(0.1), S(0.1). We find B_c=21.308, L_c=10.535 and μ_c=21.296. The last line gives the Eötvös-Galilei coefficient η for $C_{20}H_{42}$ and Au. We use α_{cw}= -0.0075, ρ_c/ρ_E=0.47 and λ=10 m. The result should be 20 times larger if λ=200 m.

ξ	2	1	1/3	0	-1/3	-1	-2.2	-3
H	2.976	1.984	1.323	0.9922	0.662	0.000	-1.190	-1.984
C	1.999	1.500	1.167	1.0000	0.834	0.501	-0.099	-0.499
Al	1.964	1.483	1.161	1.0007	0.840	0.519	-0.059	-0.445
Cu	1.914	1.458	1.153	1.0011	0.849	0.545	-0.003	-0.368
Ag	1.872	1.437	1.146	1.0009	0.856	0.565	0.042	-0.306
W	1.805	1.403	1.134	1.0003	0.866	0.598	0.115	-0.207
Au	1.800	1.400	1.134	1.0002	0.867	0.600	0.121	-0.199
H_2O	2.110	1.554	1.184	0.9994	0.814	0.444	-0.222	-0.666
$C_{20}H_{42}$	2.146	1.572	1.190	0.9988	0.808	0.426	-0.263	-0.721
crust	1.990	1.495	1.165	1.0006	0.836	0.506	-0.088	-0.484
$\eta\times10^8$	-0.159	-0.102	-0.041	0.0011	0.057	0.243	2.772	-2.090

Addendum

According to a more precise information from Dr. LaCoste, the gravity responsive element of the LaCoste-Romberg land gravity meter consists of a pivoted beam of tungsten, brass and aluminum. The torques exerted are 23.62(gm-inches) for tungsten, 3.63 for brass and 0.75 for aluminum. The line to be added to Table I corresponding to this weight is:

LCR	1.821	1.411	1.137	1.0004	0.864	0.590	0.098	-0.230

COSMIC TIME GAUGE IN QUANTUM COSMOLOGY AND CHAOTIC INFLATION MODEL

Akio Hosoya
Fermi National Accelerator Laboratory
P. O. Box 500, Batavia, Illinois 60510

and

Department of Physics,
Osaka University
Toyonaka, Japan

ABSTRACT

We propose a cosmic time gauge formalism in quantum cosmology to get an equation for the Schrödinger type. Its application to the chaotic inflaton scenario reveals that the uncertainty in the scale factor grows exponentially as the universe inflates.

I. Introduction

The quantum theory of geometry[1] is still in its infancy and suffers from many conceptual difficulties as well as technical ones. One of the conceptual difficulties comes from general coordinate invariance built in to general relativity. The arbitrariness of the coordinate choice makes the meaning of time obscure.

Previously many people [2] have attempted to identify the "time" in the Wheeler-DeWitt equation.[3] Even in the simplified minisuperspace model[2], the concept of time remains unclear.

The problem of time in quantum cosmology is not only an academic one but a potentially practical one. Many people now believe that there was once an exponentially expanding era in history of universe to solve the horizon, flatness problems etc.[4] At the moment the most probable inflationary universe scenario seems to be the chaotic universe scenario advocated by Linde.[5] There, the initial conditions are essentially given by the consideration of quantum gravity era of universe and the classical motion of the "inflaton" scalar field and the scale factor of the universe are described by the cosmic time parameter. In order to study the quantum era before the classical era of the universe we have to find out the right description of the time development of the wave function of universe.

In this paper dedicated to Professor Yoshio Yamaguchi for his 60th birthday, the author would like to make a proposal which hopefully demystifies the wave equation of universe[6] by reducing quantum cosmology to a down-to-earth Schrödinger equation

$$i\frac{\partial}{\partial t}\psi = \mathcal{H}(t)\psi \ . \tag{1}$$

Here the time is the cosmic time in the Robertson-Walker metric and $\mathcal{H}(t)$ is a time dependent Hamiltonian which we shall define in what follows. An approximate solution is also given for the chaotic universe model.

II. A Model and Hamiltonian Constraint

Let us consider the $\lambda\phi^4$ model in the Robertson-Walker universe having the chaotic inflationary universe scenario[5] in mind. The action is given by

$$\begin{aligned} S = & - \frac{1}{16\pi G}\int\sqrt{-g}d^4xR \\ & + \int\sqrt{-g}d^4x[\frac{1}{2}g^{\mu\nu}\partial_\mu\phi\partial_\nu\phi - \frac{\lambda}{4}\phi^4] \end{aligned} \tag{2}$$

For the metric of closed universe

$$ds^2 = \mathcal{N}^2dt^2 - a^2(t)\left\{\frac{dr^2}{1-r^2} + r^2(d\theta^2 + sin^2\theta d\phi^2)\right\}$$

and a homogeneous scalar field ϕ, the action S in Eq.(2) reduces to

$$S = \int dt\left[\frac{1}{2g^2}\frac{\mathcal{N}}{a}\left\{-\frac{\dot{a}^2a^2}{\mathcal{N}^2} + a^2\right\} + 2\pi^2a^3\mathcal{N}\left\{\frac{1}{2\mathcal{N}^2}\dot{\phi}^2 - \frac{\lambda}{4}\phi^4\right\}\right] \tag{3}$$

Here $\mathcal{N}$ is the lapse function [7] and $g^2 = \frac{2G}{3\pi}$.

By varying the action S with respect to the lapse function $\mathcal{N}$, we obtain the Hamiltonian constraint

$$H = -\frac{g^2\pi_a^2}{2a} - \frac{a}{2g^2} + \frac{1}{2\pi^2a^3}\frac{\pi_\phi^2}{2} + 2\pi^2a^3\cdot\frac{\lambda\phi^4}{4} \approx 0 \ , \tag{4}$$

where π_a and π_ϕ are conjugate momenta to a and ϕ, respectively. In the conventional quantization prescription, this constraint is replaced by an equation for the quantum state

$$H\psi = 0 \tag{5}$$

with $\pi_a \to -\partial/\partial a$ and $\pi_\phi \to -i\partial/\partial\phi$.

The Wheeler-DeWitt equation (5) has some peculiarities. First it does not contain the "time" at all. How can it describe the history of universe? The second, which is intrinsically related to the first, is that it is impossible to construct a positive definite conserved probability, since Eq.(5) is a second order differential equation of the Klein-Gordon type. So we are in trouble with the probabilistic interpretation of the wave function of universe.

III. Time in Quantum Cosmology

One may realize that the passage from the first class constraint(4) to the wave equation is not the only possibility in the quantum theory of constrained systems.[8] The other possibility, which is perhaps more familiar to particle physists, is fixing the gauge, $\chi \approx 0$ in such a way $\{\chi, H\}_{P.B.} \neq 0$.[9] The gauge fixing condition is arbitrary unless $\{\chi, H\}_{P.B.} = 0$. However, the canonical choice is the one which satisfies

$$\{\chi, H\}_{P.B.} = 1 \tag{6}$$

Since in this case there is no Faddeev-Popov complexity. From Eq.(6) it is clear that χ plays a role of time, which is conjugate to the Hamiltonian.

It is perhaps instructive to digress here and recall the old story of the "time operator" in quantum mechanics. Pauli[10] pointed out that the "time operator" T which satisfies the commutation relation with the Hamiltonian, [T,H]=i is actually ill-defined in the Hilbert space. His argument is based on the observation that the lower boundedness of Hamiltonian contradicts with the existence of an energy shift operator $e^{i\epsilon T}$. For example, a singular expression $T = (p^{-1}q + qp^{-1})/2$ would be obtained in the case of the Hamiltonian $H = p^2/2$ for a free point particle which is obviously positive definite. A counter example is $T = p/E$ for an unbounded Hamiltonian $H = p^2/2 + Eq$ which describes a point particle in a uniform electric field, E. Here it has to be pointed out that the Hamiltonian (4) of quantum cosmology is not bounded due to the negative sign in front of the kinetic term of scale factor. This may suggest that quantum cosmology is in a unique position in which the time operator is well-defined and plays a significant role.

Going back to the gauge condition $\chi \approx 0$, let us consider its formal construction so that $\{\chi, H\}_{P.B.} = 1$ holds in a general framework of analytical dynamics. Let a canonical set of dynamical variables be $(p_i, q_i), i = 1, 2, \cdots N$.

A general solution of Hamilton's equations

$$\begin{aligned} \dot{q}_i &= \frac{\partial H}{\partial p_i}, \\ \dot{p}_i &= -\frac{\partial H}{\partial q_i} \\ i &= 1, 2, \cdots N, \end{aligned} \tag{7}$$

contains 2N integration constants corresponding to initial values of p's and q's. One of 2N constants corresponds to the arbitrariness of initial time t_0,

$$\begin{aligned} p_i &= p_i(t - t_0; c_1, c_2, \cdots c_{2N-1}) \\ q_i &= q_i(t - t_0; c_1, c_2, \cdots c_{2N-1}) \quad , \; i = 1, 2, \cdots N. \end{aligned} \tag{8}$$

Eliminating the integrating constants $c_1, c_2 \cdots c_{2N-1}$ we may solve Eq.(8) in terms of $t - t_0$,

$$t - t_0 = T(p_i, q_i) \tag{9}$$

By construction, $T(p, q)$ satisfies the Poisson bracket $\{T, H,\}_{P.B.} = 1$. It may be amusing to reproduce $T's$ in the previous examples by following the procedure (7) – (9). It is now obvious that we can take as a gauge condition.

$$\chi(p, q) = T(p, q) - t \approx 0 \tag{10}$$

It is suggestive to point out that the "time" has been manufactured from dynamical variables through dynamics. This may sound philosophically deep.

Now that we have chosen a gauge, let us consider the dynamics in the restricted phase space under the constraint $H(p, q) = 0$ and the gauge condition $\chi(p, q, t) = 0$. First find a canonical transformation from a set (p_i, q_i), $i = 1, 2, \cdots N$ to a new set $(p*_i, q*_i)$, $i = 1, 2, \cdots N$ and (H, χ), such that

$$\sum_{i=1}^{N} p_i dq_i - H dt = \sum_{i=1}^{N-1} p_i^* dq_i^* + H d\chi - \mathcal{H} dt + d\Phi(q, q^*, \chi, t) \; . \tag{11}$$

Here Φ is a generator of the canonical transformation. It is clear from Eq. (11) that in the restricted phase space the time development is described by a new Hamiltonian

$$\mathcal{H}(p*, q*, t) = \frac{\partial \Phi(q, q*, \chi, t)}{\partial t}\Big|_{q=q(p*,q*,t),H=\chi=0} \tag{12}$$

We are now in a position to quantize our constrained system by employing the new Hamiltonian (12) and setting commutation relations, $[p_i^*, q_j^*] = i\delta_{ij}$. Or in the Schrödinger representation, the time development of the state is dictated by

$$i\partial_t \psi(q^* t) = \mathcal{H}(-i\partial/\partial q^*, t)\psi(q^*, t) \tag{13}$$

There may well remain operator ordering ambiguities for which we do not have any general prescription.

In the next section we are going to apply the quantization method prescribed here to the model given in the previous section.

IV. Quantum Chaotic Universe Model

Let us go back to our model given in Eq.(4). The classical equations for ϕ and a are given by Hamilton's equations with the Hamiltonian (4) and are the standard ones,

$$\begin{aligned} \ddot{\phi} \; + \; 3H\dot{\phi} &= -\frac{\partial V}{\partial \phi} = -\lambda \phi^3 \, , \\ H^2 \; &= \; \frac{\dot{a}^2}{a^2} = \frac{4}{g^2}(\frac{1}{2}\dot{\phi}^2 + V) - \frac{1}{a^2} \, . \end{aligned} \tag{14}$$

(Here we took $\mathcal{N} = 1$.)

In the interesting regime ($M_p \lesssim \phi \lesssim \lambda^{-\frac{1}{4}} M_p$, $\lambda \sim 10^{-12}, M_p$: Planck mass $\sim g^{-1}$), the scalar field is slowly varying, so we can neglect the $\ddot{\phi}$ term in Eq. (14) and the $\dot{\phi}^2$ term in Eq. (15). We can also ignore the spatial curvature term $-1/a^2$ in Eq.(15) in that regime.[5] Then ϕ satisfies a first order linear equation!

$$\dot{\phi} \doteq -\sqrt{\frac{\lambda}{6\pi}} M_p \phi = -\frac{\sqrt{\lambda}}{3g}\phi \; . \tag{15}$$

In the regime under consideration, it is sufficient to consider a simplified Hamiltonian constraint instead of the original one (4),

$$H' = -\frac{g^2\pi_a^2}{2a} - \frac{\sqrt{\lambda}}{3g}\phi\pi_\phi + 2\pi^2 a^3 \frac{\lambda\phi^4}{4} \approx 0\,. \tag{16}$$

From now on we shall confine ourselves to this new constrained system which is supposed to be a good approximation to the original one before the oscillation of the scalar field starts.

For H', Eq. (16) becomes exact. According to the prescription in the previous section, the canonical gauge condition is simply given by

$$\chi = -\frac{3g}{\sqrt{\lambda}} log(\phi/\phi_0) - t \approx 0\,. \tag{17}$$

Namely log ϕ is a "clock"[2].

Perhaps for such a simple system we do not need the heavy artillery of canonical transformation in §3 to get a new Hamiltonian. Eqs. 17 and 18 immediately give

$$\begin{aligned} \mathcal{H}(\pi_a, a, t) &= -\frac{g^2\pi_a^2}{2a} + 2\pi^2 a^3 \frac{\lambda\phi^4}{4}\,, \\ \phi &= \phi_0 e^{-\mu t}\,, \\ \mu &\equiv \frac{\sqrt{\lambda}}{3g} = \sqrt{\frac{\lambda}{6\pi}} M_p^{-1}\,. \end{aligned} \tag{18}$$

Let us choose the symmetrization prescription for the operator ordering, $\pi_a^2/a \rightarrow -a^{-\frac{1}{2}}\,\partial_a a^{-\frac{1}{2}}$ and write $x = \frac{2}{3}a^{\frac{3}{2}}$, $m = g^{-2}$ and

$$k(t) = \frac{9}{4}\pi^2\lambda\phi_0^4 \cdot e^{-4\mu t}\,. \tag{19}$$

We obtain a time dependent Schrödinger equation

$$-i\frac{\partial}{\partial t}\Psi = \left[-\frac{1}{2m}\frac{\partial^2}{\partial x^2} - \frac{k(t)}{2}x^2\right]\Psi \tag{20}$$

Apart from the minus sign on the left hand side, this is a quantum mechanics of an upside-down harmonic oscillator with a time dependent spring constant given by Eq. (20). (We may extend the region of x so that $-\infty < x < \infty$, or set $\psi(0) = 0$. Here we do not discuss the latter possibility).

For $|\mu t| \ll 1$, the potential is just that of the upside-down harmonic oscillator discussed by Guth and Pi[11] in a completely different context. The wave packet initially localized at $x = 0 (a = 0)$ would quickly spread out while the average value of $a = (\frac{3}{2}x)^{2/3}$ would show an exponential growth, corresponding to Linde's inflation.

For $|\mu t| \gtrsim 1$ but within the range $M_p \lesssim \phi \lesssim \lambda^{-\frac{1}{4}} M_p$, the potential flattens out and the growth of $< x^2 > \propto < a^3 >$ becomes proportional to t as one can see in any quantum mechanics text book. Fortunately Eq. (21) is exactly soluble. We just give an expression for $|\psi|^2$ in the case of a gaussian distribution at $t = 0$,

$$|\psi(t)|^2 = \frac{1}{\sqrt{A_1^2 + \alpha^2 A_2^2}} \sqrt{\frac{\alpha}{2\pi}} exp \left[-\frac{\alpha}{A_1^2 + \alpha^2 A_2^2} \frac{4}{9} \left(\frac{a}{g} \right)^3 \right] \tag{21}$$

Here α is an initial spread of the wave function and A_1 and A_2 are solutions of

$$(\frac{d^2}{d\tilde{t}^2} - \omega^2(\tilde{t}))A_i = 0 \quad (i = 1, 2) \ , \tag{22}$$

with

$$\begin{aligned} A_1(0) &= \frac{d}{d\tilde{t}} A_2(0) = 1 \ , \\ \frac{d}{d\tilde{t}} A_1(0) &= A_2(0) = 0 \ , \\ \omega^2(\tilde{t}) &= \lambda \phi_0^4 M_p^{-4} exp \left[-4 \frac{\sqrt{\lambda}}{3\pi} \tilde{t} \right] \ , \\ \tilde{t} = t/g &= \sqrt{\frac{2}{3\pi}} M_p^{-1} t \ , \end{aligned} \tag{23}$$

(24)

(time in Planck's unit).

A's can be expressed by Bessel functions of an imaginary argument. For an almost constant ω (and therefore ϕ), the $A's$ grow exponentially while for a small ω (and therefore large t), $A_1 \approx 1$ and $A_2 \approx \tilde{t}$.

What are implications of our result Eq. (22) for the chaotic universe scenario?

At least one thing is clear; the probability distribution of the scale factor becomes broader while its mean value follows the classical value as time goes on.

This phenomenon is also easily understood from consideration of the uncertainty principle. Roughly speaking, the Hubble expansion rate is conjugate to the scale factor. If we sharply define the scale factor at the Planck time, the scale factor after some time would be extremely uncertain since the rate of the exponential expansion is very much uncertain., Perhaps it is important to point out that in contrast to the scale factor, the distribution of the scalar ϕ becomes more and more sharply peaked around the classical value as time goes on even if we take into account the zero point fluctuation of ϕ.[12] Hence inclusion of inhomogeneous component ϕ will not change a qualitative feature of the distribution of scale factor. It is not clear for the present author whether the broadening distribution of the scale factor alters the qualitative features of the global structure of the universe.[12]

V. Summary

In order to quantize a simplified version of general relativity (minisuperspace model), we have proposed a special gauge choice, $T(p,q) - t \approx 0$. Here the "time" T(p,q) is determined by classical dynamics. Then the quantum mechanics of the universe is formally given by Eq.(1)((13)) with the Hamiltonian (12).

Fortunately enough the problem is exactly solvable in the case of chaotic inflation model if one concentrates on the nonoscillation regime. The probabilistic distribution of the scale factor of universe is explicitly given. It spreads by a huge amount corresponding to the gigantic expansion of universe.

VI. Acknowledgements

The author would like to think Dr. R. Holman for reading the manuscript. He also thanks people in the theory group at Fermilab for their warm hospitality.

REFERENCES

[1] Wheeler, J.A., in "Battle Recontes 1967", p273, DeWitt, C.M. and Wheeler, J.A., eds., Benjamin, New york, (1968).

[2] Blyth, W.F. and Isham, C.J., Phys. Rev. **D11** 768(1975), For reviews, e.g. MacCallum, M.A.H., in "Quantum gravity: an Oxford symposium", Isham,

C.J., Penrose, R., and Sciama, D.W., eds., Claredon Press, Oxford, (1975), Ryan, M.P., "Hamiltonian Cosmology", Springer-Verlag, Heidelberg, (1972).

[3] DeWitt, B.S., Phys. Rev. **160**, 1113 (1967).

[4] Guth, A.H., Phys. Rev.**D23** 347 (1981), Einhorn, M.B., Sato, K.,Nucl. Phys. **180** 385 (1981), Linde, A.D., Phys.Lett. **108B** 389 (1982), Albrecht, A., and Steinhardt, P.J., Phys. Rev. Lett, **48** 1220 (1982).

[5] Linde, A.D., Rep. Prog. Phys. **47** 925 (1984).

[6] Hawking, S. W. and Hartle, J.B., Phys. Rev. **D28** 83(1984), Vilenkin, A., Phys. Rev. **D30** 508 (1984).

[7] Arnowitt, R., Deser, S. and Misner, C.M., in "Gravitation: An Introduction to Current Research", Witten, L. ed., Wiley New York, (1962).

[8] Dirac, P.A.M., "Lectures on Quantum Mechanics", Belfer Graduate School of Science, Yeshiva University, New York, (1964).

[9] Hanson, A., Regge, T. and Teitelboim, C., "Constrained Hamitonian Systems", Accaemia Nazionale dei Lincei, Rome, 1975.

[10] Pauli, W., "Handbuch der Physik", Fluügge ed., Springer Verlag, (1958) vol. 5/1 p60. Gotto, T.,Naka, S. and Yamaguchi, K., Prog. Theor. Phys. **64**, 1, (1981).

[11] Guth, A.H. and Pi, So-Young, Phys. Rev. **D32** 1899 (1985)

[12] Linde, A.D., "Eternally Existing, Self-Reproducing Inflationary Universe", Lebedev Institute Preprint, (1986).

HAWKING RADIATION IN STRING THEORIES

N. Sakai
Department of Physics, Tokyo Institute of Technology
Oh-okayama, Meguro, Tokyo 152, Japan

ABSTRACT

String theories in a uniform gravitational field are studied to examine the Hawking radiation. An upper limit is found for the strength of the possible gravitational field: the corresponding Hawking temperature cannot be larger than the Hagedorn limiting temperature divided by π.

1. Introduction

Since strings are extended objects, string field theories are very soft at short distances compared to point particle field theories[1]. This drastic modification of the short-distance behavior is intimately related to the ultraviolet convergence of string theories. The softness of string theories manifests itself most directly in the extremely large number of vibrational modes each corresponding to one-particle states. Density $\rho(m)$ of one-particle states of free string theories grows exponentially as mass m increases

$$\rho(m) \sim \rho_0 (m\sqrt{\alpha'})^{-a} \, e^{bm\sqrt{\alpha'}} \tag{1.1}$$

where α' is the slope parameter of the string, and constants a, b, and ρ_0 depends on the type of string theories. Consequently string theories possess the limiting temperature (Hagedorn temperature) beyond which thermal equilibrium cannot be achieved[2]

$$T_{Hagedorn} = (b\sqrt{\alpha'})^{-1} \quad . \tag{1.2}$$

One of the most profound results coming out of an interplay between quantum theory and gravity is the Hawking's discovery of thermal radiation from black holes[3]. He showed that a spherical black hole with mass M emits particles as if it were hot bodies with the Hawking temperature

$$T_{Hawking} = (8\pi k_B \ GM)^{-1} \tag{1.3}$$

where k_B is the Boltzmann constant and G is the gravitational constant. The Hawking radiation is determined by the "surface gravity" $(4GM)^{-1}$ near the horizon. For smaller black hole mass M, the surface gravity is stronger because of the smaller horizon size, and hence particle creation is larger (higher Hawking temperature). Thus Hawking radiation can be viewed as a phenomenon of particle creation in a strong gravitational field around the black hole, analogous to the electron-positron pair creation in a strong electromagnetic field. From this viewpoint, a change of vacuum in the strong gravitational field is responcible for the Hawking radiation. In order to clarify the nature of the Hawking radiation in a simpler situation, much efforts are devoted to studies of a uniform gravitational field which is best investigated in the so-called Rindler space[4-10]. Actually the uniform gravitational field can be eliminated by going to a uniformly accelerated frame of reference. However, there is still a nontrivial effect due to the change of vacuum in the Rindler space compared to the usual Minkowski vacuum. The vacuum in the Rindler space contains a thermal spectrum of quanta defined by the Minkowski vacuum. This phenomenon has a close mathematical relation to the more realistic gravitational field around black holes[4-6].

If we consider string theories in a strong gravitational field, we will find that various particles corresponding to the string excitations emerge as the Hawking radiation with the Hawking temperature. String theories are, however, very soft at short distances.

As we increase the strength of the gravitational field, we would expect to obtain a catastrophically large amount of the Hawking radiation near the Hagedorn temperature. This consideration suggests an upper bound for the strength of the background gravitational field in string theories. In this paper we attempt to study string theories viewed from an accelerated observer and in the Rindler space. We shall find that the Hawking temperature be less than the Hagedorn temperature devided by π.

There has been an enormous interest in the search for candidate background metrics (and other classical field configurations) in string theories corresponding to the ground state ("vacuum"). We would like to emphasize that, even if a candidate background metric is given, choice must still be made among various possible vacuum states in string field theories (vacua of second quantized strings). For instance, both the Rindler vacuum and the Minkowski vacuum correspond to the same flat space geometrically, but they are completely different vacuum states for the second quantized string field. This phenomenon is completely analogous to point particle field theories and is inherent to quantum field theories in curved spacetime. It has been repeatedly emphasized that no simple or obvious choice of vacuum state is available in a general curved background metric[4].

2. Uniformly Accelerated Detector Functional

According to the general theory of relativity, a space with a uniform gravitational field is equivalent to a space without any gravitational field viewed from a uniformly accelerated observer. Hence we shall first consider an observer with a uniform proper acceleration in the flat D-dimensional Minkowski space (D=26 for the bosonic string and D=10 for superstring). A convenient parametrization of a part of the flat Minkowski space is given by

$$-ds^2 = -e^{2ay}((dy^0)^2-(dy)^2) + \sum_{i=1}^{D-2}(dx^i)^2 \qquad (2.1)$$

where a is a positive constant and (y^0,x^i,y) are related to ordinary D-dimensional Minkowski coordinates (x^0,x^i,x^{D-1}) by

$$x^0 = \pm\frac{1}{a}\, e^{ay} \sinh ay^0 \quad , \quad x^{D-1} = \pm\frac{1}{a}\, e^{ay} \cosh ay^0 \tag{2.2}$$
$$-\infty < y^0 < \infty \quad , \quad -\infty < y < \infty$$

with the +(-) sign corresponding to right- (left-) Rindler wedge $x^{D-1} > |x^0|$ $(x^{D-1} < -|x^0|)$[4]. The trajectory $y = 1$, $-\infty < y^0 < \infty$ gives the spacetime path of an observer with the uniform proper acceleration a.

Identifying the direction of the uniform acceleration as longitudinal (x^{D-1}), we use the light-cone gauge. The transverse string variables $X^i(\tau,\sigma)$ in open bosonic strings can be decomposed into normal modes

$$X^i(\tau,\sigma) = x^i + 2\alpha' p^i \tau + i\sqrt{2\alpha'} \sum_{n\neq 0} \frac{1}{n}\, \alpha^i_n \cos n\sigma\, e^{-in\tau} \tag{2.3}$$

The first quantization of the string gives the physical particle states labeled by transverse oscillators $I \equiv (i^{(1)}_1 \cdots i^{(1)}_{n_1}; \cdots)$

$$|I> \equiv \prod_{k=1}^{\infty} \frac{1}{\sqrt{n_k!}}\, \alpha^{i^{(k)}_1}_{-k} \cdots \alpha^{i^{(k)}_{n_k}}_{-k} \, |0> \tag{2.4}$$
$$<I|J> = \delta_{I,J}$$

where $\delta_{I,J}$ stands for a symmetrized Kronecker delta symbol. Physical states of closed bosonic strings are labeled by two sets of transverse oscillators (left- and right-movers) and superstrings require fermionic oscillators in addition. We shall denote these labels of physical states collectively by $|I>$. Free string wave functional $\Phi(x)$ is most conveniently expressed in terms of the first quantization Fock space with component wave functions $\phi_I(x)$

$$|\Phi(x)> = \sum_I \phi_I(x)\; |I> \tag{2.5}$$

Second quantization of string amounts to regarding the component wave functions $\phi_I(x)$ as operators creating or annihilating the particle labeled by I.

Following Unruh and DeWitt[8,10], we shall introduce a detector functional M(t) consisting of monopole type detectors $M_I(t)$ coupled to the component fields $\phi_I(x)$ of the string

$$|M(t)> = \sum_I M_I(t) \; |I> \tag{2.6}$$

$$L_{int} = c \sum_I M_I(t) \; \phi_I(x(t)) \tag{2.7}$$

$$= c < M(t) \; | \; \Phi(x(t)) >$$

where c is a small coupling parameter and t is the detector's proper time. Suppose the detector is in its ground state (energy = 0) and the string is in the Minkowski vacuum state $|0_M>$ initially. The amplitude for the system to make a transition to a state with the detector energy E and with the string excited state Ψ is given by

$$ic < E|M_I(0)|0 > \int_{-\infty}^{\infty} dt \; e^{-iEt} < \Psi|\phi_I(x(t))|0_M > \tag{2.8}$$

for sufficiently small c. Assuming that the detector trajectory is stationary and there is no mixing among various component fields,

$$<0_M|\phi_I(x(t))\phi_J(x(t'))|0_M> = <0_M|\phi_I(x(t-t'))\phi_J(x(0))|0_M> \propto \delta_{IJ}$$

We obtain the transition probability P(E) per unit proper time to all possible states with the detector energy E

$$P(E) = c^2 \sum_I < 0|M_I^{+}(0)|E > < E|M_I(0)|0 > \cdot F_I(E) \tag{2.9}$$

$$F_I(E) = \int_{-\infty}^{\infty} dt \; e^{-iEt} < 0_M|\phi_I(x(t))\phi_I(x(0))|0_M > \tag{2.10}$$

We may call $F_I(E)$ as a responce function for the field ϕ_I. In strong gravitational field, all possible particle states are emitted as the Hawking radiation. Hence it is more convenient to use a detector which does not discriminate particle species and polarizations $M_I(t) = M(t)$. Then the transition probability P(E) is given by a "reduced" responce function F(E)

$$P(E) = c^2 \; |< E|M(0)|0 > |^2 \; F(E)$$

$$F(E) = \sum_I F_I(E) \tag{2.11}$$

String theories contain excited states with arbitrarily high spin.

For longitudinal polarizations we need to apply the so-called Fermi-Walker transport matrix in order to account for the rotation of vielbein associated to the accelerated observer relative to the inertial observer. However, physical particles in string theories in the light-cone gauge are all transversely polarized. Hence the responce functions F_I of high spin fields ϕ_I are identical to the scalar field responce function F(E;m) with the same mass m except for the polarization multiplicity. Hence we obtain using the level density $\rho(m)$ at the mass m

$$F(E) = \sum_m \rho(m)\, F(E;m) \tag{2.12}$$

The responce function for the uniformly accelerated detector ($y^0 = t$, $y = 1$, $x = 0$) is given by[9,10]

$$F(E;m) = \int_{-\infty}^{\infty} dt\, e^{-iEt}\, \frac{1}{2\pi}\left(\frac{ma}{4\pi i\ \mathrm{sh}\frac{a}{2}(t-i\varepsilon)}\right)^{\frac{D-2}{2}} K_{\frac{D-2}{2}}\left(\frac{2im}{a}\,\mathrm{sh}\,\frac{at}{2} + \varepsilon\right) \tag{2.13}$$

where K is the modified Bessel function. The asymptotic behavior as $m \to \infty$ can be most easily obtained by deforming the t-integration contour from real axis to a line along Im $t = -2\pi/a$

$$\begin{aligned} F(E;m) &= \int_{-\infty}^{\infty} dt\, e^{-iEt-\frac{\pi}{a}E}\, \frac{1}{2\pi}\left(\frac{ma}{4\pi\mathrm{ch}\,\frac{at}{2}}\right)^{\frac{D-2}{2}} K_{\frac{D-2}{2}}\left(\frac{2m}{a}\,\mathrm{ch}\,\frac{at}{2}\right) \\ &\sim \frac{a}{8\pi}\left(\frac{ma}{4\pi}\right)^{\frac{D-4}{2}} e^{\frac{2m}{a} - \frac{\pi}{a}E} \qquad \text{for } m \to \infty \end{aligned} \tag{2.14}$$

Combining eqs. (1.1), (2.12) and (2.14) we find that the high mass contribution converges only if

$$T_{\mathrm{Hawking}} \equiv \frac{a}{2\pi} < \frac{1}{\pi}\,\frac{1}{b\sqrt{\alpha'}} = \frac{1}{\pi}\,T_{\mathrm{Hagedorn}} \quad . \tag{2.15}$$

As we see in the next section, the Hawking temperature in this case is obtained by replacing the surface gravity $(4GM)^{-1}$ with the acceleration a.

The above conclusion applies for open as well as closed strings.

One should note from eq. (2.14) that the distribution in E deviates significantly from the Boltzmann distribution for large m as pointed out previously[9]. Moreover the "energy" E with respect to the observer's proper time is independent of the mass m contrary to the Minkowski case[10]. These two points are apparently the origin of the fact that the critical Hawking temperature is a factor of π less than the naive expectation (Hagedorn temperature).

In the case of superstring and heterotic string, we can define responce functions for component fermion fields of string in the light-cone gauge. Apart from polarization multiplicities, they are identical to the responce function of a spin 1/2 fermion with the same mass because physical state polarizations are transverse. It is most convenient to take the trace of spinor indices[9] to define a reduced responce function

$$F^{\text{fermion}}(E) = \sum_I \text{tr}\,(\gamma^0 F_I(E)) = \sum_m \rho(m)\,\text{tr}\,(\gamma^0 F^{\text{fermion}}(E;m)) \qquad (2.16)$$

We find that the spin 1/2 responce function shows the same exponential dependence on m as the scalar for $m \to \infty$

$$\text{tr}\,(\gamma^0 F^{\text{fermion}}(E;m)) \sim (\text{tr}\cdot 1)\cdot mF(E;m) \qquad (2.17)$$

Therefore the critical value for the Hawking temperature is the same as eq. (2.15).

3. String Theories in Rindler Space and Vacuum Stress

Thermal property of the responce function of the uniformly accelerated detector is best clarified by quantizing the field in the Rindler space (2.1). The string action in the Rindler space is given in the conformal gauge

$$S = \frac{1}{4\pi\alpha'}\int d\tau d\sigma\,\{e^{2aY}((\partial_\alpha Y^0)^2-(\partial_\alpha Y)^2) - \sum_{i=1}^{D-2}(\partial_\alpha X^i)^2\} \qquad (3.1)$$

Using the light-cone gauge (+(-) refers to the right- (left-) Rindler wedge),

$$Y^{+} \equiv \frac{Y^{0}+Y}{\sqrt{2}} = \frac{1}{\sqrt{2}a} \log[\pm\sqrt{2}a(x^{+}+2\alpha'p^{+}\tau)] \tag{3.2}$$

the first quantization of the string is the same as in the flat Minkowski space except the restriction on values of zero modes $x^{+}\cdot x^{-} < 0$.

In the second quantization, however, we have to choose a vacuum state from various candidate states such as the Minkowski vacuum $|0_M>$ or the Rindler one $|0_R>$. They are defined by using component fields $\phi_I^{(+,x)}$ or $\phi_I^{(+,y)}$ of positive frequency with respect to the Minkowski time x^0 or the Rindler time y^0

$$\phi_I^{(+,x)}|0_M> = 0 \quad , \quad \phi_I^{(+,y)}|0_R> = 0 \tag{3.3}$$

These two vacua are different and the corresponding quanta are related by the Bogoliubov transformation[4-10]. Using the Rindler mode expansion, a manifestly thermal expression can be given for the responce function[9,10]

$$F(E;m) = \frac{1}{e^{2\pi E/a}-1} \frac{2sh\frac{\pi E}{a}}{\pi a} \int \frac{d^{D-2}k}{(2\pi)^{D-2}} K_{i\frac{E}{a}}\left(\frac{\sqrt{m^2+k^2}}{a}\right) \tag{3.4}$$

which exhibits the Hawking temperature

$$T_{Hawking} = a/2\pi \tag{3.5}$$

In order to obtain quantities more directly related to physical observables, there has been an extensive study of vacuum stress in the Rindler vacuum compared to the Minkowski vacuum[4,7-10]. Lagrangian L for a massive scalar field and the corresponding stress tensor $T_{\mu\nu}$ is given by

$$L = \frac{-1}{2}\sqrt{-g}\,[g^{\mu\nu}\partial_\mu\phi\partial_\nu\phi + (m^2+\xi R)\phi^2] \tag{3.6}$$

$$\begin{aligned} T_{\mu\nu} &\equiv \frac{-2}{\sqrt{-g}}\frac{\partial L}{\partial g^{\mu\nu}} \\ &= (1-2\xi)\partial_\mu\phi\partial_\nu\phi - 2\xi\phi_{;\mu\nu}\phi + \xi R_{\mu\nu}\phi^2 \\ &\quad - (\tfrac{1}{2} - 2\xi)g_{\mu\nu}(g^{\rho\sigma}\partial_\rho\phi\partial_\sigma\phi + (m^2+\xi R)\phi^2) \end{aligned} \tag{3.7}$$

The difference of the vacuum stress in the Rindler and the Minkowski vacua can be expressed by certain moments of the responce functions F_D, F_{D-2} and F_{D-4} (D, D-2 and D-4 dimensional ones are needed).

$$\tilde{T}_{\mu\nu} \equiv \langle 0_R|T_{\mu\nu}|0_R\rangle - \langle 0_M|T_{\mu\nu}|0_M\rangle$$

$$\tilde{T}_0{}^0 = \int_0^\infty \frac{dE}{\pi} \Big[(E^2 - (\xi(D-1) - \frac{D-2}{4})(D-2)a^2)\, F_D\,(E;m)$$
$$- (2\xi(2D-5) - (D-3)\,\frac{m^2a^2}{4\pi}\, F_{D-2}\,(E;m)$$
$$- (2\xi - \frac{1}{2})\,\frac{m^4a^2}{8\pi^2}\, F_{D-4}\,(E;m)\Big]$$
$$\sim - (2\xi - \frac{1}{2})\,\frac{4}{\pi}\,(\frac{ma}{4\pi})^{D/2}\, e^{-\frac{2m}{a}} \quad \text{as } m \to \infty \tag{3.8}$$

$$\tilde{T}_i{}^j = \delta_i{}^j \int_0^\infty \frac{dE}{\pi} \Big[-2\pi\, F_{D+2}\,(E;m) - (2\xi - \frac{1}{2})\Big\{\frac{(D-2)^2}{2}\, a^2\, F_D(E;m)$$
$$+ \frac{(D-3)m^2a^2}{2\pi}\, F_{D-2}\,(E;m) + \frac{m^4a^2}{8\pi}\, F_{D-4}\,(E;m)\Big\}\Big]$$
$$\sim - \delta_i{}^j\,(2\xi - \frac{1}{2})\,\frac{4}{\pi}\,(\frac{ma}{4\pi})^{D/2}\, e^{-\frac{2m}{a}}\ , \tag{3.9}$$

whereas the longitudinal component is $O(a/m)$ smaller

$$\tilde{T}_{D-1}{}^{D-1} = \int_0^\infty \frac{dE}{\pi} \Big\{ (-E^2 - \frac{(D-2)(D-2-4\xi)}{4}\, a^2 + m^2)\, F_D\,(E;m)$$
$$- \frac{m^2a^2}{4\pi}\,(D-3-2\xi)\, F_{D-2}\,(E;m) - \frac{m^4a^2}{16\pi^2}\, F_{D-4}\,(E;m)$$
$$+ 2\pi(D-2)\, F_{D+2}\,(E;m)\Big\}$$
$$\sim (2\xi - \frac{1}{2})\,\frac{2a}{m\pi}\,(\frac{ma}{4\pi})^{D/2}\, e^{-\frac{2m}{a}}\ . \tag{3.10}$$

If we sum over all physical particles of bosonic string

$$\tilde{T}^{\text{string}}_{\mu\nu} = \sum_m \rho_{(m)}\, \tilde{T}_{\mu\nu} \tag{3.11}$$

we find that the vacuum stress also shows the same critical behavior as eq. (2.15): the Hawking temperature must be less than the Hagedorn temperature divided by π.

Although we obtained a reasonable result, we should point out two caveats in computing $T_{\mu\nu}$ in string theories: i) There is a field redefinition ambiguity of $g_{\mu\nu}$ (what should we call graviton?). ii) Usual semi-classical treatment of the Hawking radiation as above seems to involve off-shell stress tensor which may be ill-defined in present technology of string theories.

It may well be that more stringent constraints such as modular invariance might be able to decide not only the vacuum metric, but also the vacuum choice of the second quantized string. We hope to study this and other issues more thoroughly.

Acknowledgements

We thank H. Ooguri for valuable suggestions on the boundary condition and stringent constraints of conformal invariance in string theories. Useful discussions with S. Takagi and T. Yoneya are gratefully acknowledged. This work is supported in part by the Grant-in-Aid, Ministry of Education, Science and Culture.

REFERENCES

[1] J. Sherk, Rev. Mod. Phys. 47 (1975) 123; Dual theory, Phys. Reports reprint Volume I. ed. M. Jacob (North-Holland, Amsterdam, 1974).

[2] K. Huang and S. Weinberg, Phys. Rev. Lett. 25 (1970) 895; E. Alvarez, Nucl. Phys. B269 (1986) 596.

[3] S. W. Hawking, Commun. Math. Phys. 43 (1975) 199.

[4] N. D. Birrell and P. C. W. Davies, Quantum fields in curved space (Cambridge Univ. Press, 1982).

[5] S. A. Fulling, Phys. Rev. D7 (1973) 2850; P. C. W. Davies, J. of Phys. A8 (1975) 609.

[6] W. G. Unruh, Phys. Rev. D14 (1976) 870.

[7] P. Candelas and D. Deutsch, Proc. R. Soc. Lond. A354 (1977) 79; ibid. A362 (1978) 251; D. W. Sciama, P. Candelas and D. Deutsch, Adv. Phys. 30 (1981) 327.

[8] B. DeWitt, in General Relativity, eds. S. W. Hawking and W. Israel (Cambridge Univ. Press, 1979).

[9] S. Takagi, Prog. Theor. Phys. 72 (1984) 505; ibid. 74 (1985) 142, 501, 1219.

[10] K. Freese, C. T. Hill, and M. Mueller, Nucl. Phys. B255 (1985) 693; C. T. Hill, Fermilab preprint Pub-85/100-T (1986).

QUANTUM GRAVITY AND SPACETIME STRUCTURE*)

Noboru Nakanishi

Research Institute for Mathematical Sciences
Kyoto University, Kyoto 606
JAPAN

ABSTRACT

From the author's point of view, it is discussed how the fundamental theory should be formulated, and some comments are made on the superstring theory.

1. QUANTUM GRAVITY

General relativity is a brilliantly successful theory of gravity both theoretically and observationally. As far as large-scale phenomena are concerned, general relativity is undoubtedly the correct theory; there is no need for considering any other alternative.

General relativity has two distinguished features. Firstly, the theory is invariant under general coordinate transformations. Secondly, it shows that spacetime structure is not a priori given but determined by matter distribution. Although these two features are mutually closely related, each of them does not necessarily imply the other.

General relativity is, of course, a classical theory, but the fundamental theory should be a quantum theory. Accordingly, the

*) Dedicated to Professor Y. Yamaguchi for celebrating his 60th birthday.

Einstein gravity must be quantized. Since the perturbative approach to it seriously suffers from the non-renormalizable ultraviolet-divergence difficulty, quantum gravity should be formulated in a non-perturbative way. Fortunately, general relativity can be elegantly quantized in the Heisenberg picture. The manifestly covariant canonical formalism of quantum gravity[1)] (MCCFQG) is an outstandingly beautiful quantum theory of gravity formulated in terms of Heisenberg operators. Although this theory has many remarkable properties, I do not review it here but concentrate my attention to the problem related to spacetime structure.

Since general covariance is a local symmetry, one must break it explicitly by introducing "gauge fixing" at quantization. In MCCFQG, the gauge-fixing term is given by

$$\mathcal{L}_{GF} = -\kappa^{-1}\sqrt{-g}\; g^{\mu\nu}\partial_\mu b_\nu, \tag{1}$$

where $g_{\mu\nu}$ and b_ν are the gravitational field and its B field (a covariant vector boson field), respectively, $g = \det g_{\mu\nu}$, and κ denotes the Einstein gravitational constant. Of course, (1) implies the de Donder (or harmonic coordinate) condition

$$\partial_\mu(\sqrt{-g}\; g^{\mu\nu}) = 0. \tag{2}$$

It is the common belief of particle physicists that gauge fixing has no physical meaning because of the gauge independence of the physical S-matrix. I feel, however, that such a standpoint is rather curious from the fundamental point of view. Since quantum theory is more fundamental than classical theory, <u>the quantum Lagrangian density</u>, which necessarily contains a <u>particular</u> gauge-fixing term, <u>must be the most fundamental quantity</u>, as long as the S-matrix-theoretical standpoint is not adopted. In order that this idea be acceptable, there should exist a very <u>distinguished</u> choice of gauge fixing. This is indeed the case as far as quantum gravity is concerned. As mentioned above, the gauge-fixing term (1) is very much distinguished because the remarkably beautiful results of MCCFQG are crucially based on it. I have therefore proposed elsewhere[2)] that the fundamental unification of gravity and particle physics should be made directly at the quantum level on the basis of MCCFQG, in contrast to the traditional approach in which unification is made first at the classical level and then one proceeds to quantization.

"Gauge fixing" is a misleading word in the case of gravity; it is a coordinate condition in the classical general relativity. Observables are explicitly dependent on the choice of a coordinate system in contrast to the gauge choice in the classical gauge theories. Indeed, without knowing what coordinate system one is employing, one can make no predictions from theoretical expressions obtained from general relativity. Furthermore, as Chou pointed out,[3)] matter distribution, which is necessary for setting up the Einstein equation, cannot be properly described without introducing a particular coordinate system. Although general covariance was a powerful guiding principle for constructing general relativity, its physical contents may not be independent of coordinate systems. Indeed, for instance, one cannot define gravitational energy-momentum in a coordinate-independent way. Chou, from classical gravity, and I, from quantum gravity, expect that the de Donder condition is preferential and perhaps physically significant. I have recently conjectured[4)] that the gravitational energy-momentum pseudotensor $t_{\mu\nu}$ is physically sensible under the de Donder condition.

2. SPACETIME STRUCTURE

I discuss the problem of spacetime structure. Although there still remains some uneasiness about the appearance of singularities in the course of gravitational collapse, it may be understood that the classical spacetime structure is correctly described by general relativity. When quantized, however, the metric tensor $g_{\mu\nu}$ becomes an operator, whence the Einstein causality is no longer meaningful. Therefore, one cannot define spacelike hypersurfaces which are needed for the covariant dynamical description of the system. The customary way of avoiding this difficulty is to assume the existence of a background metric. That is, one postulates that $g_{\mu\nu}$ consists of a c-number background metric and a q-number quantum fluctuation. Then one can deal with the latter in the same way as a quantum field in particle physics. I believe, however, that this procedure is too easygoing to be regarded as a method for constructing a fundamental theory. From the very successful point of view of general relativity, it is quite unnatural to adopt the Minkowski metric $\eta_{\mu\nu}$ as the *a priori* background metric. On the other hand, if one supposes that the background metric is determined by the classical Einstein equation, one encounters a very bizarre situation that there exists a classical theory which *precedes* the most fundamental quantum theory. One might suggest a bootstrap-type determination of the background metric, but such an approach is incompatible with the Lagrangian formalism.

In the fundamental formulation of quantum gravity, $g_{\mu\nu}$ should be quantized without introducing any particular c-number metric. This is exactly what is done in MCCFQG. Of course, since canonical quantization is employed there, x^0 plays a very special role, but it is not directly identified with the physical time coordinate. MCCFQG is manifestly invariant under GL(4) at the operator level, and GL(4) is necessarily spontaneously broken up to Lorentz invariance at the level of representation. The relation between x^0 and the time coordinate depends on the way of violating GL(4). Thus the Einstein causality is *not* the first principle but arises as a consequence of spontaneous breakdown of GL(4).

The relevant scale of quantum gravity is, of course, the Planck length $\sqrt{\kappa}(\sim 10^{-33}\text{cm})$. It cannot be overemphasized that it is a scale of extreme smallness beyond any human imagination. Nevertheless, recently, a large number of particle physicists (and cosmologists, as well) treat it like an everyday experience. They make an extraordinarily excessive extrapolation more boldly than describing nuclear structure by Newtonian mechanics. This trend has recently made the superstring theory[5)] very fashionable.

The string model was originally invented as a semi-phenomenological model of hadrons. Forgetting about its bases such as duality and Regge behavior, people now wish to reinterpret this model with some modifications as the most fundamental theory of physics. They make this tremendous scaling-down (by a factor 10^{20}) leaving its basic structure essentially unchanged. The superstring theory therefore has no fundamental principles. Its basic Lagrangian necessarily contains a background metric from the outset. Its formulation is crucially based on perturbation theory (Superstring field theory is still of very primitive stage.). The perturbative finiteness*) is certainly a very interesting result, but, in my opinion, it should not be regarded as the qualification for the fundamental theory.

I believe that the Planck length is nothing but the scale below which the notion of distance loses its meaning because the non-commutativity of $g_{\mu\nu}$ is of order of the Planck length. Accordingly, it is not sensible to consider geometric structure finer than the Planck length. Especially, one should not expect that the most fundamental theory is understandable in terms of classical geometry.**) Quantum gravity is not a theory which is directly accessible by experiments. It should be a theory which is qualitatively different from the theory of particle physics. In order to obtain particle physics from the most fundamental theory, I guess that some kind of "coarse graining" is necessary concerning the degree of freedom of gravitons. Note that the ratio between the reaction time of elementary particles and the macroscopic time, which is regarded as infinity in the definition of the S-matrix, is approximately equal to that between the Planck time and the reaction time of elementary particles. I think that it is totally absurd to discuss the problem of the cosmological

---

*) The perturbative finiteness of the superstring theory can be achieved by replacing a wrong-sign Gaussian integral of energy by a convergent one by hand. One should note that taking a finite part usually endangers the probabilistic interpretability of the theory.

**) By classical geometry, I mean any finite-dimensional geometry formulated in terms of c-numbers.

constant directly at the level of quantum gravity.

3. HIGHER-DIMENSIONAL SPACETIME

Recently, it has become quite fashionable to consider a higher-dimensional spacetime not as a technical device but as a fundamental entity. In the Kaluza-Klein theory, the Kaluza-Klein supergravity, and the superstring theory, one postulates the existence of a D(>4)-dimensional spacetime and then makes D-4 spatial dimensions invisible by an _ad hoc_ procedure called "spontaneous compactification". Those theories, therefore, indicate that the scale of the universe and the Planck length should be dealt with on an equal footing. It seems to me that such an idea is too grotesque to be physically sensible. The fact that our spacetime is of four dimensions is the most well-established thing in physics.*) If there were extra dimensions of the same kind as our spacetime, the universe would probably exhibit their remnant in some way, but in reality there is no evidence for their slightest trace. One must therefore make a very artificial assumption that the compactified dimensions are _uniformly_ orthogonal to our spacetime throughout the universe. Indeed, if the structure of the compactified space varied position to position in the universe, then elementary particles could live only locally, that is, they could not move around the universe, because their fundamental properties are characterized by the _local_ structure of the compactified space. Thus an infinitely fine tuning is necessary to be assumed in order to achieve the desired compactification.

I believe that there are no extra spacetime dimensions of the _same kind_; dimensional extension should be made in a completely different way. Although the superspace formulation of supersymmetry[6] is certainly such a dimensional extension, it is quite incomplete in the sense that one must introduce many constraints from the outset. In this respect,

---

*) If one dares to introduce a spacetime other than the four-dimensional one, it might be more natural to consider a "dimensional number operator" acting on a "dimensional Fock space" rather than choosing any particular number of dimensions.

it is quite worth emphasizing that MCCFQG has a complete sixteen-dimensional superalgebra as its symmetry in spite of the fact that it is a theory in four dimensions. The sixteen-dimensional "supercoordinates" consist of the c-number coordinates x^μ, the gravitational B field b_ρ, the gravitational Faddeev-Popov ghost c^σ and anti-ghost $\bar{c}_\tau$, where the last three are quantum fields. Since this sixteen-dimensional extension of the spacetime symmetry is so natural and so beautiful that I expect it is the only correct way of extending dimensionality.

In the normal situation, physics is progressed by the cooperation of theories and experiments. If theorists must dare to enter the realm non-accessible by experiments, they should avoid introducing any assumptions which are motivated by their desire only. As is suggested by the history of unified field theories, it is quite dangerous to be deeply involved in a theory based on the assumptions which are supported neither experimentally nor theoretically.

REFERENCES

1) Nakanishi, N., Proceedings of the International Symposium on Gauge Theory and Gravitation, ed. by K. Kikkawa, N. Nakanishi, and H. Nariai (Springer-Verlag), 171-183 (1983).
Nakanishi, N., Publications RIMS (Kyoto University), 1095-1137 (1983).
Nakanishi, N. and Ojima, I., Covariant Operator Formalism of Gauge Theories and Quantum Gravity (World Scientific), to be published.
2) Nakanishi, N., Prog. Theor. Phys. Suppl. 86, to be published.
3) Chou, P.-Y., Scientia Sinica (Ser. A) 25, 628 (1982).
4) Nakanishi, N., Prog. Theor. Phys. 75, to be published.
5) Schwarz, J. H., Physics Reports 89,223 (1982).
Green, M. B. and Gross, D. J. (eds.), Unified String Theories, Proceedings of the Workshop on Unified String Theories, Santa Barbara, California (World Scientific, 1986).
6) Wess, J. and Bagger, J., Supersymmetry and Supergravity (Princeton University Press, 1983).

SPACE-TIME DIMENSIONS

Hidezumi Terazawa
Institute for Nuclear Study, University of Tokyo
Midoricho, Tanashi, Tokyo 188
Japan

ABSTRACT

A few possible paradigms which may determine the number of space-time dimensions are presented and discussed.

1. INTRODUCTION

Why is our space-time of 4 dimensions? This question seems to have been addressed many many times in the history of physics (or philosophy). Some people might take this question as metaphysical while others might do as physical. Reviewing the recent remarkable progress in physics and astronomy, it seems that the time has come when we should take this question seriously as semi-physical or "prephysical" and try to provide an answer. In this essay, I shall present a few possible paradigms which may determine the number of space-time dimensions in "prephysics", a new line of physics (or philosophy but not metaphysics) in which some basic assumptions taken as scared ones in ordinary physics are to be reasoned.

2. TWO PARADIGMS

Let us first define the space-time as a place or domain where the space-time metric, $g_{\mu\nu}(x)$, is non-vanishing and finite, i.e.

$$g_{\mu\nu}(x) \neq 0 \quad \text{and} \quad |g_{\mu\nu}(x)| < \infty \quad \text{for} \quad \mu,\nu = 0,1,2,\cdots,N-1, \tag{1}$$

in a mathematical manifold of the space-time coordinates of N dimensions, (x^μ). Here, the space-time metric is defined, as usual in Riemannian geometry, as a function of the space-time coordinates which determines a distance, ds, between two points, (x^μ) and $(x^\mu+dx^\mu)$, in the following relation:

$$ds^2 = g_{\mu\nu}(x)dx^\mu dx^\nu . \tag{2}$$

Then, well-defined is the question to be addressed, "Why is our space-time of 4 dimensions?". There are the following two paradigms in which an answer to the question will be provided:
1) Spontaneous "solidification" of the indefinite dimensional space-time and 2) induced "liquidization" of the definite dimensional one. A very long time ago, the space-time of dimensions higher than 4 was hypothesized by Kaluza[1] and Klein[2] and very much later revived by many others[3] in their supergrand unified theory of all elementary particle forces. A possible compactification of the space-time of N dimensions (N>4) into that of 4 dimensions was also discussed there. However, the reason why the original space-time is a priori of N dimensions has never been presented.

3. "SOLIDIFICATION" OF THE INDEFINITE DIMENSIONAL SPACE-TIME

In the possibility 1), I assert that since there is no reason why the space-time is a priori of a definite number of dimensions, it may be of indefinite dimensions. In other words, the dimension of the space-time, N, may also be a "dynamical" variable. If this is the case, the Einstein theory of general relativity for gravity in 4-dimensional Riemannian geometry may be extended to a theory in which the action integral is given by

$$I_n = \int d^n x\sqrt{-g}(c_n R+L_m) \tag{3}$$

in n-dimensional Riemannian geometry. Here, $g = \det g_{\mu\nu}$, c_n is a "dimensional coefficient" of the $(\text{mass})^{n-2}$ dimension, R is the Riemann curvature scalar and L_m is the Lagrangian density for matter fields. As other ordinary dynamical variables, the number of dimensions of our space-time, N, may therefore be determined either by the "Hamilton principle" of

$$\delta_n I_n = 0 \quad \text{for} \quad n = N \tag{4}$$

or by minimizing the "space-time potential" as

$$-I_n = \min \quad \text{for} \quad n = N. \tag{5}$$

Although the dimensional coefficient, c_n's, are all unknown except for the only one,

$$c_4 = \frac{1}{16\pi G} \tag{6}$$

where G is the Newtonian gravitational constant, the simplest possible and non-trivial example for them is given by

$$c_n = \frac{M^{n-4}}{16\pi G} \tag{7}$$

where M is a constant of the mass dimension.

Suppose the simplest possible case for the topology and other properties of the space-time, i.e. the closed, homogeneous and isotropic n-dimensional space-time (or universe), which is described in the generalized Robertson-Walker metric of

$$ds^2 = -dt^2 + a^2(t)d\Omega_n^2 \tag{8}$$

where a(t) is the "radius" of the universe and Ω_n is the "solid angle" of the n-dimensional space ($d\Omega_2^2 = d\phi^2$, $d\Omega_3^2 = d\theta^2 + \sin^2\theta d\phi^2$, $d\chi_4^2 = d\chi^2 + \sin^2\chi(d\theta^2 + \sin^2\theta d\phi^2), \cdots$). Then, the Riemann curvature scalar is simply given by

$$R = [(n-1)(n-2)-2(n-1)q]H^2 + \frac{(n-1)(n-2)}{a^2} \tag{9}$$

where the Hubble "constant", H, and the deceleration parameter, q, are defined by

$$H = \frac{\dot{a}}{a} \qquad \text{and} \qquad q = -\frac{\ddot{a}a}{\dot{a}^2} \, . \tag{10}$$

Also, the relevant Ricci and Einstein curvature tensors are given by

$$R_{00} = (n-1)qH^2, \quad R_{11} = a^2[(n-2-q)H^2 + \frac{n-2}{a^2}], \quad \cdots \tag{11}$$

and

$$G_{00} = \frac{1}{2}(n-1)(n-2)(H^2 + \frac{1}{a^2}), \quad G_{11} = -\frac{1}{2}a^2\{[(n-2)(n-3)-2(n-2)q]H^2 + \frac{(n-2)(n-3)}{a^2}\}, \cdots, \tag{12}$$

respectively.

For further simplicity, let us suppose a single homogeneous and isotropic scalar field, $\phi(x)$, representing all the matter fields and being described by

$$L_m = -\frac{1}{2} g^{\mu\nu}\partial_\mu\phi\partial_\nu\phi - V(\phi) \tag{13}$$

where $V(\phi)$ is the potential (including the mass term if any). Then, the independent Einstein field equations are for the time-time component

$$c_n(n-1)(n-2)(H^2 + \frac{1}{a^2}) = \frac{1}{2}\dot{\phi}^2 + V(\phi) \tag{14}$$

and for the space-space component

$$c_n\{[(n-2)(n-3)-2(n-2)q]H^2 + \frac{(n-2)(n-3)}{a^2}\} = -\frac{1}{2}\dot{\phi}^2 + V(\phi) \tag{15}$$

while the Euler equation of motion for the matter field is

$$\ddot{\phi} = -(n-1)H\dot{\phi} - V'(\phi) \tag{16}$$

A set of Eqs. (14) and (15) are equivalent to the following set:

$$\dot{\phi}^2 = 2c_n(n-2)[(1+q)H^2 + \frac{1}{a^2}] \tag{17}$$

and

$$V(\phi) = c_n(n-2)[(n-2-q)H^2 + \frac{n-2}{a^2}]. \tag{18}$$

By using Eqs. (9) and (15), the action integral (3) can be transformed into

$$I_n = 2c_n\int d^nx\sqrt{-g}[(n-2-q)H^2 + \frac{n-2}{a^2}]$$

$$= 2c_n\frac{n\pi^{n/2}}{\Gamma(1+\frac{n}{2})}\int dt\ a^{n-1}[(n-2-q)H^2 + \frac{n-2}{a^2}] \quad . \tag{19}$$

Furthermore, by using Eq. (18), it can be transformed into another form of

$$I_n = \frac{2}{n-2}\int d^nx\sqrt{-g}V(\phi)$$

$$= \frac{2}{n-2}\cdot\frac{n\pi^{n/2}}{\Gamma(1+\frac{n}{2})}\int dt\ a^{n-1}V(\phi) \ . \tag{20}$$

Notice that this last relation can be more directly proved by using the relation of

$$c_n(1 - \frac{n}{2})R = \frac{1}{2}\, g^{\mu\nu}\partial_\mu\phi\partial_\nu\phi + \frac{n}{2}\, L_m \tag{21}$$

which can be derived by contracting the general form of Einstein field equations of

$$c_n(R_{\mu\nu} - \frac{1}{2}\, g_{\mu\nu}R) = \frac{1}{2}\, \partial_\mu\phi\partial_\nu\phi + \frac{1}{2}\, g_{\mu\nu}L_m \ . \tag{22}$$

For illustration, assume a simple time-dependence of

$$a = a_0(t/t_0)^{\ell} \tag{23}$$

where a_0 and t_0 are the "present radius" and "present age" of the universe, respectively, and ℓ is a constant power index. Then, the Hubble constant and the deceleration parameter are simply given by

$$H = \frac{\ell}{t} = \frac{H_0}{t/t_0} \quad \text{and} \quad q = -\frac{\ell-1}{\ell} . \tag{24}$$

where H_0 is the "present" Hubble constant. In this simple case, the action integral (19) can be explicitly calculated to be

$$I_n = 2c_n \frac{n\pi^{n/2} a_0^{n-2}}{\Gamma(1+\frac{n}{2})} \left[\frac{r}{1+q} + \frac{(n-2)(1+q)}{(n-2+q)r} \right] \quad \text{for} \quad -1 < q < n-2 \tag{25}$$

where the "present radius-to-time ratio" of the universe, r, is defined by

$$r = a_0/t_0 . \tag{26}$$

For $-1 < q < n-2$, the action integral has the minimum for varing r as

$$(I_n)_r = 4c_n \frac{n\pi^{n/2} a_0^{n-2}}{\Gamma(1+\frac{n}{2})} \sqrt{\frac{n-2}{n-2+q}} \quad \text{at} \quad r = (1+q)\sqrt{\frac{n-2}{n-2+q}} . \tag{27}$$

For further varing q, it has the minimum as

$$(I_n)_{r,q} = 2\sqrt{2}\, c_n \frac{n\pi^{n/2} a_0^{n-2}}{\Gamma(1+\frac{n}{2})} \quad \text{at} \quad q = n-2(-0) \quad \text{and} \quad r = \frac{n-1}{\sqrt{2}} . \tag{28}$$

It is amusing to notice that the "prediction" of q=2 for n=4 is roughly consistent with the observed value of $q = 1.0\pm0.5$ by Sandage[4]. In the simplest example for the dimensional coefficient (7), this minimum value can be reduced to .

$$(I_n)_{r,q} = \left(\frac{\sqrt{2}}{4GM^2}\right)\frac{(\pi^{1/2}Ma_0)^{n-2}}{\Gamma(n/2)} . \qquad (29)$$

This space-time potential, $-(I_n)_{r,q}$, is illustrated in Fig. 1 as a function of the space-time dimension, n, for various values of the parameter, ξ, defined by

$$\xi = \pi^{1/2}Ma_0 . \qquad (30)$$

As is seen in this figure, for $\xi \gtrsim 1$ there exists a stable ground state of the space-time with the dimension larger than two while otherwise there exists none.

What can be deduced from this example is that the dimension of the space-time may also depend on the topology and other properties of the space-time. If this is the case and if the space-time (or universe) were in a

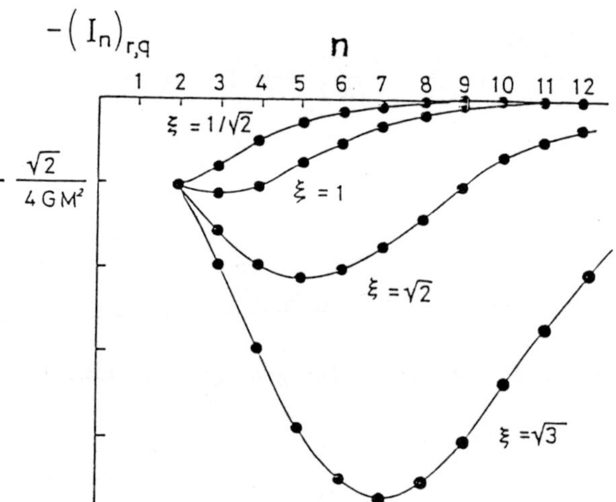

Fig.1

The space-time potential, $-(I_m)_{r,q}$, as a function of the space-time dimension, n, for various values of the parameter, ξ.

quantum state as has been claimed by Hartle and Hawking[5], we might have to worry about not only a disastrous phase transition of our universe from the present topology to a different topology, which has been discussed lately by DeWitt[6], but also an even more disastrous phase transition of our universe from the present 4-dimensional space-time to a new dimensional one. How to explain the stability of our universe against such a phase transition into an infinite number of possible states not only with a different topology but also with a different dimension would become even more difficult and almost impossible.

4. "LIQUIDIZATION" OF THE DEFINITE DIMENSIONAL SPACE-TIME

In the possibility 2), I assert that since the space-time metric is made of some fundamental matters, it may be of the definite dimensions. Many years ago, Phillips[7] proposed to take the gravitational field as a Nambu-Goldstone boson, by following Bjorken[8] and others[9,10] who had taken the electromagnetic field as a Nambu-Goldstone boson, much influenced by the unified spinor model of Nambu and Jona-Lasinio[11] for hadrons. A year later, Sakharov[12] proposed to take gravity as a quantum effect of matters. He even sketched how the Lagrangian of the Riemann curvature scalar in Einstein's general relativity theory of gravity would be reproduced as an effective Lagrangian at low energies. Following Sakharov, many people[13-19] have investigated such theories of "pregeometry", named by Misner, Thorne and Wheeler[20], as are more fundamental than Einstein's general relativity theory of gravity in Riemannian geometry, in which the latter theory can be reproduced as an effective theory at low energies. In fact, we have proposed a simple field theoretical model for pregeometry and first demonstrated that the remarkable idea of Sakharov is really working[16].

If pregeometry is right, all the properties of the space-time may be attributed to those of the matters. In other words, "the matters come first and then the space-time does". In fact, in scalar pregeometry, the simplest possible model of pregeometry, whose

Lagrangian is given by

$$L = F[-\det(\partial_\mu\phi^\alpha\partial_\nu\phi^\alpha)]^{1/2} \tag{31}$$

where F is a constant of the $(\text{mass})^{-4}$ dimension, the space-time metric may be taken as the following vacuum expectation value of the operator-product of fundamental scalar fields, ϕ^α ($\alpha = 0,1,2,\cdots,A-1$):

$$g_{\mu\nu}(x) = F\langle\partial_\mu\phi^\alpha(x)\partial_\nu\phi^\alpha(x)\rangle_0 \qquad \text{for } \mu,\nu = 0,1,2,\cdots,n-1. \tag{32}$$

The dimension of the space-time is determined by the condensation of the fundamental fields as

$$\langle\partial_\mu\phi^\alpha(x)\partial_\nu\phi^\alpha(x)\rangle_0 \neq 0 \qquad \text{for } \mu,\nu = 0,1,2,\cdots,N-1, \tag{33}$$

and, therefore, dependent of the unknown dynamics of the fundamental fields. However, it may be that the more exist the fundamental fields, the more would appear the space-time dimensions. A similar conjecture has already been made in the "color-space correspondence":[16]

$$\begin{matrix} x_0 & x_1 & x_2 & x_3 \\ \updownarrow & \updownarrow & \updownarrow & \updownarrow \\ C_0 & C_1 & C_2 & C_3 \end{matrix} \tag{34}$$

where C_α (α=0,1,2,3) are "chroms", the subquarks[16,21] which are more fundamental constituents of leptons and quarks carrying the color quantum number[22,23] of $SU(4)_C$[24].

5. THE POSSIBLE THIRD PARADIGM

There may exist the third paradigm in which the number of the space-time dimensions may be determined. It is the "field-space symmetry"[25] which combines the internal symmetry of leptons and quarks or subquarks, the fundamental matters, and the external symmetry of the space-time. A long time ago, such a field-space symmetry of SU(4), combining SU(2) of hadron isopsin and O(3) or SU(2) of spin, was first introduced by Wigner[26] and much later extended to SU(6), combining SU(3) of quark flavor and O(3) or SU(2) of spin, by Gürsey and Radicati[27] and by Sakita[28]. It was further extended to supersymmetry by Miyazawa[29], by Volkov and Akulov[30] and by Wess and Zumino[31]. Also, many years ago, the Lorentz symmetry was taken as a local gauge symmetry, similar to the one of Yang and Mills[32], by Utiyama[33] and later by many others[34-40]. About a decade ago, an extension of this gauge symmetry of SL(2,C) to a field-space symmetry of SL(6,C) including SU(3) of quark flavor was made by Salam[41] and others[3,42-44] to invent a supergrand unified gauge theory of all elementary particle forces. Such a maximum extension of field- space symmetry is either $GL(32N_g,C)$ or $GL(12+2N_g,C)$ for leptons and quarks or for subquarks where N_g is the total number of generations (or families) of leptons and quarks[25]. It certainly contains not only the standard gauge symmetry of SU(3)×SU(2)×U(1) where SU(3) is the color gauge symmetry of quarks in quantum chromodynamics[45,46] and SU(2)×U(1) is the electroweak gauge symmetry of leptons and quarks in the unified gauge theory of Glashow[47], Salam[48] and Weinberg[49] but also the horizontal gauge symmetry of $SU(N_g)$[25]. On the other hand, a supersymmetric extension of the field-space gauge symmetry, an extended supergravity[50,51], has also been made up to SU(8)[52] and even larger, which certainly contains not only SU(3)×SU(2)×U(1) but also the grand unified gauge symmetry of SU(5)[53] and some horizontal gauge symmetry.

Given such a field-space gauge symmetry, the dimension of the space-time may be determined if an additional constraint such as asymptotic freedom[54,55] or "temporary freedom and asymptotic catastrophe"[56], permanent confinement or "temporary confinement and asymptotic liberation"[57], and so on. The simplest and most instructive example for demonstrating how it works is to remember that the quantum electrodynamics in the 2-dimensional space-time is asymptotically free while that in the higher dimensional one is not.

This third paradigm is, however, weaker than the first two since such an additional constraint is rather artificial or mundane without any direct relation with the origin of the space-time and matter.

In conclusion, it is perfectly conceivable that the real reason why the number of space-time dimensions is determined may lie elsewhere out of the paradigms discussed in this article. I, however, believe that there exists the reason why. Please find it.

Note added:

The related subject has recently been discussed by Sakharov[58] from a different viewpoint. A preliminary version of the present work has been presented in Terazawa, H. An essay presented in honor of the 60th birthday of E. S. Fradkin, to be published in "Quantum Field Theories and Quantum Statistics" (ed Batalin, I. A. & Vilkovisky, G. A.) (Adam Hilger, London, 1986).

REFERENCES

1. Kaluza, T., Sitzker. Press. Akad. Wiss. K1, 966 (1921).
2. Klein, O., Z. Physik 37, 895 (1926).
3. Cho, Y.M. & Freund, P.G.O., Phys. Rev. D12, 1711 (1975).
4. Sandage, A., Astrophys. J. 178, 1 & 25 (1972).
5. Hartle, J.B. & Hawking, S.W., Phys. Rev. D28, 2960 (1983).
6. DeWitt, B.S., in Proc. 3rd Seminar on Quantum Gravity, Moscow, Oct. 23-25, 1984 (eds Markov, M.A. & West, P.C.) (World Scientific, Singapore, 1985) p.103.
7. Phillips, P.R., Phys. Rev. 146, 966 (1966).
8. Bjorken, J.D., Ann. Phys. (N.Y.) 24, 174 (1963).
9. Bialynicki-Birula, I., Phys. Rev. 130, 465 (1963).
10. Guralnik, G.S., Phys. Rev. 136, B1404 (1964).
11. Nambu, Y. & Jona-Lasinio, G., Phys. Rev. 122, 345 (1961).
12. Sakharov, A.D., Dokl. Akad. Nauk SSSR 177, 70 (1967) [Sov. Phys. JETP 12, 1040 (1968)].
13. Ohanian, H.C., Phys. Rev. 184, 1305 (1969).
14. Dürr, H.P., Gen. Rel. Grav. 4, 29 (1973).

15. Adler, S.L., Phys. Rev. 14, 379 (1976); Phys. Rev. Lett. 44, 1567 (1980); Phys. Lett. 95B, 241 (1980); Adler, S.L., Lieberman, J., Ng, Y. & Tsao, H.S., Phys. Rev. D14, 359 (1976).

16. Terazawa, H., Chikashige, Y., Akama, K. & Matsuki, T., Phys. Rev. D15, 1181 (1977); Terazawa, H., Phys. Rev. D16, 2373 (1977); D22, 184 (1980); D22, 1037 (1980); D22, 2921 (1980); in Proc. 3rd Marcel Grossmann Meeting on General Relativity (ed Hu Ning) (Science Press & North Holland, Beijing, 1983) p.239; in Proc. Topical Symposium on High Energy Physics (ed Eguchi, T. & Yamaguchi, Y.) (World Scientific, Singapore, 1983) p.173; Phys. Lett. 133B, 57 (1983); in Quantum Gravity (ed Markov, M.A. & West, P.C.) (Plenum, London & New York, 1984) p.47; in Proc. XXII Int. Conf. on High Energy Physics (ed Meyer, A. & Wieczorek, E.) (Akademie der Wissenschaften der DDR, Zeuthen, 1984) Vol.I, p.63; in Proc. 3rd Seminar on Quantum Gravity, Moscow, Oct. 23-25, 1984 (ed Markov, M.A. & West, P.C.) (World Scientific, Singapore, 1985) p.236; in Proc. Mexican School on Particles and Fields (ed Jusio, J.L.) (American Institute of Physics, New York, 1986); Akama, K., Chikashige, Y., Matsuki, T. & Terazawa, H., Prog. Theor. Phys. 60, 868 (1978); Akama, K., Prog. Theor. Phys. 60, 1900 (1978); Phys. Rev. D24, 3073 (1981); Terazawa, H. & Akama, K., Phys. Lett. 96B, 276 (1980); 97B, 81 (1980); Akama, K. & Terazawa, H., Gen. Rel. Grav. 15, 201 (1983).

17. Atkatz, D., Phys. Rev. D17, 1972 (1978).

18. Hasslacher, B. & Mottola, E., Phys. Lett. 95B, 237 (1980).

19. Zee, A., Phys. Rev. D23, 858 (1981).

20. Misner, C.W., Thorne, K.S. & Wheeler, J.A., in Gravitation (Freeman, San Francisco, 1973) p.1203.

21. Terazawa, H., Chikashige, Y. & Akama, K., Phys. Rev. D15, 480 (1977).

22. Han, M.-Y. & Nambu, Y., Phys. Rev. 139, B1006 (1965).

23. Miyamoto, Y., Prog. Theor. Phys. Extra Number, 185 (1965).

24. Pati, J.C. & Salam, A., Phys. Rev. D10, 275 (1974).

25. Terazawa, H., Prog. Theor. Phys. 58, 1276 (1977).
26. Wigner, E., Phys. Rev. 51, 105 (1937).
27. Gürsey, F. & Radicati, L.A., Phys. Rev. Lett. 13, 173 (1964).
28. Sakita, B., Phys. Rev. 136, B1756 (1964).
29. Miyazawa, H., Prog. Theor. Phys. 36, 1266 (1966).
30. Volkov, D.V. & Akulov, V.P., Phys. Lett. 46B, 109 (1973).
31. Wess, J. & Zumino, B., Nucl. Phys. B70, 39 (1974).
32. Yang, C.N. & Mills, R.L., Phys. Rev. 96, 191 (1954).
33. Utiyama, R., Phys. Rev. 101, 1597 (1956).
34. Kibble, T.W.B., J. Math. Phys. 2, 212 (1961).
35. Sciama, D.W., in Recent Developments in General Relativity (Pergamon, New York, 1962).
36. Schwinger, J., Phys. Rev. 130, 1253 (1963).
37. Carmeli, M., Nucl. Phys. 38B, 621 (1972); Phys. Rev. D14, 1727 (1976); D14, 2518 (1976).
38. Yang, C.N., Phys. Rev. Lett. 33, 445 (1974).
39. Cho, Y.M., Phys. Rev. D14, 252 (1976); D14, 3335 (1976).
40. Carmeli, M. & Malin, S., Ann. Phys. (N.Y.) 103, 208 (1977).
41. Salam, A., in Quantum Gravity (ed Isham, C.J., Penrose, R. & Sciama, D.W.) (Clarendon Press, Oxford, 1975) p.500; Ann. N.Y. Acad. Sci. 294, 12 (1977).
42. Isham, C., Salam, A. & Strathdee, J., Phys. Rev. D8, 2600 (1973).
43. Cho, Y.M., Phys. Rev. D14, 3341 (1976).
44. Carmeli, M. & Kaye, M., Nuovo Cimento 34B, 225 (1976).
45. Nambu, Y., in Preludés in Theoretical Physics (ed de Shalit, A., Feshbach, H. & Van Hove, L.) (North-Holland, Amsterdam, 1966) p.133.
46. Fritzsch, H. & Gell-Mann, M., in Proc. XVI Int. Conf. on High Energy Physics (ed Robert, A.) (NAL, Batavia, 1973) Vol.2, p.135.
47. Glashow, S.L., Nucl. Phys. 22, 579 (1961).
48. Salam, A., in Elementary Particle Physics (ed Svartholm, N.) (Almqvist & Wiksell, Stockholm, 1968) p.367.

49. Weinberg, S., Phys. Rev. Lett. 19, 1264 (1967).

50. Gell-Mann, M., A talk presented at the Washington Meeting of the American Physical Society, Washington, April, 1977.

51. Salam, A., in Proc. XIX Int. Conf. on High Energy Physics (ed Homma, S., Kawaguchi, M. & Miyazawa, H.) (Phys. Soc. Japan, Tokyo, 1979) p.617.

52. Curtwright, T.C. & Freund, P.G.O., in Supergravity (ed van Nieuwenhuizen, P. & Freedman, D.Z.) (North-Holland, Amsterdam, 1979) p.197.

53. Georgi, H. & Glashow, S.L., Phys. Rev. Lett. 32, 438 (1974).

54. Gross, D.J. & Wilczek, F., Phys. Rev. Lett. 30, 1343 (1973).

55. Politzer, H.D., Phys. Rev. Lett. 30, 1346 (1973).

56. Terazawa, H., Chikashige, Y. & Akama, K., Prog. Theor. Phys. 61, 1515 (1979).

57. Terazawa, H., Akama, K. & Chikashige, Y., Prog. Theor. Phys. 60, 1521 (1978).

58. Sakharov, A.D., Zh. Eksp. Teor. Fiz. 87, 375 (1984) [Sov. Phys. JETP 60, 214 (1984)].

VI. OTHERS

THE INVERSE PROBLEM IN THE COUPLING CONSTANT

Khosrow CHADAN
Laboratoire de Physique Théorique et Hautes Energies
Laboratoire associé au CNRS
Université de Paris XI, 91405 Orsay, France

In this work, done in collaboration with Reido KOBAYASI from Science University of Tokyo, we shall treat the inverse problem in the coupling constant in the general case where the potential may be singular at the origin. We shall give the complete solution without any arbitrary constant. For preliminaries, notations and references, we refer the reader to[1-2].

We consider the equation

$$\phi''(g;x) = gV(x)\phi(g;x), \qquad x \in [0,R], \tag{1}$$

with Dirichlet boundary conditions at both ends

$$\phi(g;0) = 0, \quad \phi'(g;0) = 1 \quad ; \quad \phi(g;R) = 0. \tag{2}$$

The conditions at $x = 0$ determine the solution in a unique way for all g, and the one at $x = R$ gives then the spectrum in the coupling constant.

The potential is supposed to be real, purely attractive : $V(x) \leq 0$, continuous, except perhaps at the origin where $xV(x)$ must be integrable, and globally $L^{1/2}$. All these can be summarized in

$$\int_0^R x|V(x)|dx < \infty \quad , \tag{3}$$

$$I = \int_0^R [-V(x)]^{1/2} dx < \infty \quad . \tag{4}$$

As in[2], by making the Liouville transformation :

$$x \to Z = Z(x) = \int_0^x [-V(t)]^{1/2} dt \quad , \tag{5}$$

$$\phi(x) \to \psi(Z) = \lambda[-V(x)]^{1/4} \phi(x) \quad , \tag{6}$$

where λ is an arbitrary (positive) constant, we find, in $Z \in [0, I]$

$$\ddot{\psi}(Z) + g\psi(Z) = \tilde{V}(Z)\psi(Z) \quad , \qquad \dot{} = \frac{d}{dZ} \quad , \tag{7}$$

$$\tilde{V}(Z) = \ddot{F}(Z)/F(Z) , \qquad F(Z) = [-V(x)]^{1/4} \quad . \tag{8}$$

It is to equation (7) that we applied in[2] the Gel'fand-

Levitan method, assuming $V(x)$, and therefore $\tilde{V}(Z)$, none zero and regular at the origin. In the general case, $\tilde{V}(Z)$ has a Z^{-2} singularity at the origin $Z = 0$, as can be seen by considering the example where $V(x) \sim x^{\alpha}$ for $x \to 0$, α either positive, or negative and greater than -2, and we find that $\tilde{V}(Z) \sim \Lambda Z^{-2}$, with $\Lambda > -1/4$. We shall therefore admit, as our assumption, that

$$\tilde{V}(Z) = \frac{\Lambda}{Z^2} + \tilde{V}_1(Z) \ , \qquad \Lambda > -\frac{1}{4} \ , \tag{9}$$

$$\int_0^I Z|\tilde{V}_1(Z)|dZ < \infty \ , \tag{10}$$

and we define ℓ to be the largest of the roots of $\ell^2 + \ell = \Lambda$, that is

$$\ell = (-1 + \sqrt{1+4\Lambda})/2 > -\frac{1}{2} \ . \tag{11}$$

It is well-known that (9), (10) and (11) are the conditions under which the Gel'fand-Levitan formalism applies in general for the continuum as well as the discrete case[3]. Note also that under our assumptions on the potential V, the function ψ satisfies also Dirichlet boundary conditions $(g = G^2)$

$$\psi(G;0) = 0 \ , \qquad \psi(G;I) = 0 \ . \tag{12}$$

The first condition determines ψ up to a multiplicative constant, and the second the spectrum. We shall denote by $g_n = G_n^2$, $G_n > 0$, the eigenvalues, and normalize ψ (quite independently of (6) and λ) in such a way as to become identical, when $\tilde{V}_1$ is zero, to the "free" solution, ψ_o, of the equation

$$\ddot{\psi}_o + g\psi_o = \frac{\Lambda}{Z^2}\,\psi_o \; ; \qquad \psi_o(0) = 0 \; , \tag{13}$$

for which we take[4,5]

$$\psi_o(G;Z) = (2/G)^{\ell+\frac{1}{2}}\,\Gamma(\ell+\tfrac{3}{2})\,Z^{1/2}\,J_{\ell+\frac{1}{2}}(GZ) \underset{Z\to 0}{\simeq} Z^{\ell+1} . \tag{14}$$

where $g = G^2$, and J_ν is the Bessel function.

At $Z = I$, we shall impose to ψ_o, like ψ, the Dirichlet condition, so that the "free" spectrum $g_{no} = G_{no}^2$ is given by :

$$\psi_o(G_{no},I) = 0 \to J_{\ell+\frac{1}{2}}(G_{no}I) = 0 \; . \tag{15}$$

The G_{no} are therefore given by the (positive) zeros of Bessel functions divided by I.

With the precise normalization of ψ and ψ_o given above, that is to say[4]

$$\lim_{Z\to 0} Z^{-\ell-1}\ \psi(Z) = \lim_{Z\to 0} Z^{-\ell-1}\ \psi_o(Z) = 1 ,$$

we know that both functions, for each Z, are entire functions of g of order 1/2 and type Z,[7] and, when $g\to\infty$, the main terms of the asymptotic behaviours of ψ and ψ_o are the same[5]

$$\psi = \psi_o + \frac{e^{|ImG|Z}}{|G|^{\ell+1}}\, o(1) \quad , \tag{16a}$$

$$\psi_o = \frac{2^{\ell+1}}{\pi^{1/2}}\,\Gamma(\ell+\tfrac{3}{2})[G^{-\ell-1}\ \sin(GZ-\tfrac{\ell\pi}{2}) + \frac{e^{|ImG|Z}}{|G|^{\ell+2}}\, 0(1)]. \tag{16b}$$

The eigenvalues $g_n = G_n^2$ being real and positive, and, when $n\to\infty$, tending to the "free" eigenvalues $g_{no} = G_{no}^2$, where $G_{no}I$ are zeros of Bessel functions, we get then, for n large, from (16a and b) and[5]

$$g_n = g_{no} + n\ o_n(1), \qquad g_{no} = \frac{n^2\pi^2}{I^2} + An + B + \ldots \tag{17}$$

In the same way, for the normalizing constants $\tilde{C}'_n$ and $\tilde{C}'_{no}$ of eigenfunctions ψ_n and ψ_{no}, using (14), (16a and b) and[6], we find, for large n, the asymptotic behaviour

$$(\tilde{C}'_n)^{-1} = \int_0^I \psi_n^2\, dZ = (\tilde{C}'_{no})^{-1} + \frac{o_n(1)}{n^{2\ell+2}} , \tag{18a}$$

$$(\tilde{C}'_{no})^{-1} = \frac{2^{2\ell+1}\,\Gamma^2(\ell+\frac{3}{2})}{G_{no}^{2\ell+1}} \int_0^I Z\, J^2_{\ell+\frac{1}{2}}(G_{no}Z)\,dZ =$$

$$= \frac{2^{2\ell}\,\Gamma^2(\ell+\frac{3}{2})\,I^2}{G_{no}^{2\ell+1}}\,[J'_{\ell+\frac{1}{2}}(G_{no}I)]^2 \underset{n\to\infty}{=}$$

$$\frac{2^{2\ell+1}\,\Gamma^2(\ell+\frac{3}{2})\,I^{2\ell+3}}{\pi(n\pi)^{2\ell+2}} + \frac{0(1)}{n^{2\ell+3}}\,. \tag{18b}$$

As a first step, we shall solve now the inverse problem related to (7), with the spectral data $\{g_n;\tilde{C}'_n\}$. However, the spectral data at our disposal are those of the initial problem, (1), that is to say $\{g_n;C'_n\}$, where[2]

$$(C'_n)^{-1} = -\int_0^R V(x)\,\phi_n^2(X)\,dx. \tag{19}$$

For finding $\tilde{C}'_n$, we shall use first (6). ϕ and ψ being uniquely defined, and proportional to each other, we have

$$\lambda = \lim_{x\to 0} \frac{\psi(Z(x))}{[-V(x)]^{1/4}\,\phi(x)} = \lim_{x\to 0} \frac{[Z(x)]^{\ell+1}}{[-V(x)]^{1/4}\,x}\,. \tag{20}$$

The parameter λ is therefore independent of g, a very important fact. Now, combining (6) and (18a and b), we find :

$$C'_n = \lambda^2 \tilde{C}'_n \underset{n\to\infty}{=} \lambda^2 \frac{\pi^{2\ell+3} \; n^{2\ell+2}}{2^{2\ell+1} \; \Gamma^2(\ell+\frac{3}{2}) I^{2\ell+3}} + \ldots \qquad (21)$$

The strategy is now similar to that of[2]. From the asymptotic behaviour of g_n for large n, (17), we determine the length of the fundamental interval in Z, I. From the last part of (21), we can find first the value of ℓ from the exponent of n, and then λ^2. The first part of (21) gives us then $\tilde{C}'_n$. We have now all the ingredients for writing the integral equation of Gel'fand-Levitan attached to (7) [3]

$$\tilde{K}(Z,Z') + \tilde{H}(Z,Z') + \int_0^Z \tilde{K}(Z,Z'')\tilde{H}(Z'',Z')dZ'' = 0, \qquad (22)$$

$$\tilde{H}(Z,Z') = \int [d\rho(g)-d\rho_o(g)] \; \psi_o(G;Z)\psi_o(G;Z') \;, \qquad (23)$$

$$d\rho(g) = \sum_n (\tilde{C}'_n) \; \delta(g-g_n) \;, \quad d\rho_o(g) = \sum_n (\tilde{C}'_{no}) \; \delta(g-g_{no}) . \qquad (24)$$

We have then the potential and ψ by

$$\tilde{V}_1(Z) = -2(d/dZ)\tilde{K}(Z,Z) \;, \qquad (25)$$

$$\psi = \psi_o(G;Z) + \int_0^Z \tilde{K}(Z,Z')\psi_o(G;Z')dZ' . \qquad (26)$$

For going from $\tilde{V}_1(Z)$ and ψ to V(x), we use now (6), (1), and the first part of (2), which gives us, by making g = 0,[8]

$$\omega(Z) = \psi(0;Z) = \lambda\,[-V(x)]^{1/4}\,\phi(0;x) = \lambda\,[-V(x)]^{1/4}\,x. \tag{27}$$

On the other hand, we have, from (5) and (27)

$$dZ = [-V(x)]^{1/2}\,dx = \omega^2(Z)\,dx/(x^2\lambda^2)\ . \tag{28}$$

In this relation, $\omega(Z)$ is known, and is obtained by making $G = 0$ in (26)

$$\omega(Z) = \psi(0;Z) = Z^{\ell+1} + \int_0^Z \tilde{K}(Z,Z')\,Z'^{\,\ell+1}\,dZ'\ . \tag{29}$$

Everything is therefore known in (28), and we obtain, by integrating it, and taking into account the fact that, for $Z = I$, $x = R$

$$x = \frac{R}{1+R\lambda^2 \int_Z^I \frac{dZ'}{\omega^2(Z')}}\ . \tag{30}$$

This formula is quite general and does not make any reference to the value of the potential at the origin. We can now proceed along similar lines as in[2]. We invert the above relation to obtain Z as function of x, and then the potential by $V(x) = -\,(dZ/dx)^2$.

It can be easily seen that we recover the results of[2] by assuming that V is finite at the origin, and has value $-V_o$ there. This gives $Z \sim V_o^{1/2} x$, $\ell = 0$, and $\lambda = V_o^{1/4}$, which, when used in (30), gives back that V is indeed finite at the origin, and has the value $-V_o$. Finally, we have established (30) under the technical assumptions (9) and (10). We shall see in detail, in a paper under preparation, that (30) is quite general, and is valid under (3) and (4).

REFERENCES

1) K. CHADAN, C.R.Ac.Sc.Paris, 299, série II, 1984, p. 271-274.

2) K. CHADAN and H. GROSSE, C.R.Ac.Sc.Paris, 299, série II, p. 1305-1308.

3) B.M. LEVITAN and M.G. GASYMOV, Russian Mathematical Surveys (English translation), 19, 1964, pp. 1-63.

4) R.G. NEWTON, Scattering Theory of Waves and Particles, Springer-Verlag, New York, 1982, chap. 12 and formulae (12.131, 132, 134 and 137).

5) M. ABRAMOVITZ and I. STEGUN, Handbook of Mathematical Functions, Dover, New York, 1965, chaps. 9 and 10. We shall follow the definitions of this book throughout.

6) Ref. 5, formulae 11.4.5, 9.2.11, and 9.5.12.

7) More precisely, an even entire function in the variable $G = g^{1/2}$, of order 1 and type Z.

8) If we make $g = 0$ in (6), and use (7), we find $\ddot{\psi}/\psi = \ddot{F}/F$. An obvious solution is therefore $F = [-V]^{1/4}$ itself. The other, independent solution, which vanishes at the origin $Z = 0$, is proportional to $F(Z) \int^Z dZ'/F^2(Z') = F(Z)x = [-V(x)]^{1/4}\, x$.

STRUCTURE OF THE HILBERT SPACE OF THE MACROSCOPIC SYSTEM AND THE THEORY OF MEASUREMENT

Reijiro Fukuda

Department of Physics, Faculty of Science and Technology,
Keio University, Yokohama, 223 Japan

Recently there are hot discussions on the problem of measurement in quantum theory. It is a surprise that such a firmly established theory receives a lot of debates at the fundamental level.[1)]

We address ourselves here to the following question: If the system of the object plus macroscopic apparatus is described by the usual rule of the quantum mechanics, then does the interaction between the object (with microscopic degrees of freedom) and the apparatus (with macroscopic degrees of freedom) bring the pure state of the object into the mixed state? This is equivalent to the reduction of the wave packet and is of course the central problem of the measuring theory.

We discuss here the same subject from a novel point of view and give a possible answer in the affirmative. The essential ingredient is the structure of the Hilbert space of the macroscopic system. Let us describe our macroscopic system by the second quantized bosonic field operator $\phi(x)$ and its conjugate operator $\pi(x)=\frac{\hbar}{i}\partial/\partial\phi(x)$. The macroscopic nature is represented by the limit $V=\infty$ with N/V=fixed where V is the volume and N the particle number of the system.

With any macroscopic system associated are the extensive and intensive variables. In the limit $V=\infty$ there do not exist in the Hilbert space the operators corresponding to the extensive variables since its matrix elements are infinite. We divide the intensive variables into two classes. The class I consists of operators which involve averaging of the local variables over the macroscopic (compared with the atomic scale) region. Examples are $\int d^3x\,\phi(x)/V$, $\int d^3x\,\pi(x)/V$ etc.. The class II includes the rest of the intensive operators, $\phi(x)$, for instance. Now we state the theorem:

Theorem For a macroscopic system , all the intensive variables of the class I lose fluctuations.

Mathematically this is nothing but the central limit theorem and is proved by the stationary phase method. Let the Schrodinger wave functional of the system at time t be given as $\langle\phi|\Psi\rangle_{t_0} \equiv \Psi_{t_0}[\phi]$. Then at later time it is given in terms of the functional path integral formula by

$$\Psi_t[\phi] = \int K_{t,t_0}(\phi,\phi_0)\,\Psi[\phi_0][d\phi_0] \tag{1}$$

The evolution kernal K is rewritten as a sum of the terms each of which is a contribution to K under the condition that the arbitrarily chosen operator $A(\phi)$ of class I has a given value $a(t')$ for $t_0 < t' < t$. Thus we use an identical transformation as , with L the Lagrangian of the system,

$$K = \iint [da][d\phi']\exp\left\{i\int_{t_0}^{t} L\,dt'\right\}\prod_{t_0<t'<t}\delta\big(A(\phi'(t')) - a(t')\big) \tag{2}$$

After ϕ' integration it has the form

$$K = \int [da]\,\exp\{iV\Gamma(a)\}\times C(a). \tag{3}$$

The factor V in the phase is a consequence of the fact that A belongs to class I. As $V\to\infty$, $a(t)$ is determined by the c-number equation of motion

$$\frac{\partial\Gamma(a)}{\partial a(t)} = 0. \tag{4}$$

It is in general a non-local equation in space and time. We call each solution as trajectory. Now we observe the following important points.

1) There does not exist any finite operator that connects different trajectories since the corresponding wave functionals have infinite phase differences. The overlap of these states vanishes as some power of $1/\sqrt{V}$.

2) Each value of a(t) specifies one Hilbert space and it is class II operators that operate on this Hilbert space.

3) The operator which is canonically conjugate to the class I operator is an extensive operator. (For example $\int d^3x\,\pi(x)$ is conjugate to

$\int d^3x\phi(x)/V$.) Since any extensive operator ceases to exist in the limit, all the quantum degrees of freedom of class I are lost. The Hamiltonian does not exist as aquantum operator and the time development is not given by a unitary transformation. This resolves the unitarity problem in the measuring theory.[2)]

The next task is of course the discussion of the measuring process. The measurement is performed by reading off the values of the macroscopic variables such as the position of the needl , the grain density or the current density etc.. They all correspond to the operators involving the average over the region of macroscopic size, i.e. they belong to class I. They have no fluctuations and this fact brings the measuring process in quantum theory to the same level of the Newtonian mechanics.

Consider the detector system measuring the operator Λ of the object and assume that the eigen-state of the operator Λ be $|\lambda_n\rangle$ with the eigen-value λ_n. Let the state of the total system before interaction between object and detector be

$$|\Psi\rangle = \Big(\sum_n C_n|\lambda_n\rangle\Big)|\psi\rangle \tag{5}$$

where $|\psi\rangle$ refers to the detector. In order to observe Λ we first choose the apparatus whose interaction with the object is given by the interaction Hamiltonian $H^I[\Lambda,\phi,\pi]$ or the interaction Lagrangian $L^I[\Lambda,\phi,\dot{\phi}]$. In Eq.(1) the kernal K is given by

$$K = \int[d\phi']\exp\{i\int_{t_0}^{t} L\,dt'\} \tag{6}$$

where L is the Lagrangian of the apparatus system. Here we replace L by $L+L^I+L^O$, where L^O is the Lagrangian of the object. We make the approximation that the coefficients c_n's are given functions of t. Then the wave functional at time t is given by

$$\langle\phi|\Psi\rangle_t = \sum_n c_n(t)\int[d\phi']\exp\Big\{i\int_{t_0}^{t}dt'\big(L(\phi)+L^I[\lambda_n,\phi,\dot{\phi}]\big)\Big\}\langle\phi_0|\Psi\rangle_{t_0}[d\phi_0]$$

A macroscopic variable of the apparatus chosen in the measuring process is is written as A. It is actually a c-number a(t) but in this case depends on n since for different n we get different equation of motion for a(t),

$$\frac{\partial \Gamma(\lambda_n, a)}{\partial a(t)} = 0. \tag{7}$$

We write the solution to this equation as $a_n(t)$. There is a unique co-rrespondence between n and the solution $a_n(t)$ hence the state of the detector. This is one of the requriements for the detector. If we write the state after the interaction as

$$|\Psi\rangle = \sum c_n |\lambda_n\rangle |\psi_n\rangle \equiv \sum c_n |n\rangle,$$

then there is an infinite phase difference $\exp i\{V(\Gamma(\lambda_n, a_n) - \Gamma(\lambda_{n'}, a_{n'}))\}$ between two states $|n\rangle$ and $|n'\rangle$. There does not exist any finite operator that connects n and n'. This is another requirement for the detector.

We have seen that the elucidation of the structure of the Hilbert space of the macroscopic system leads to a possible and natural solution of the problem of measurement. Basic observations are the c-number character of the class I operators and its consequence concernig the structure of the Hilbert space. Our mechanism discussed here is quite differet from the theory introducing many Hilbert spaces.[3)]

Finally we present the prediction of our theory. Our c-number equation (4) can be tested for the macroscopic system. Let us take two macroscopic bodies with particle co-ordinate x_i ($i=1\sim N$) or y_j ($j=1\sim N'$) and assume the potential $V(x_i - y_j)$. We introduce the center of mass and relative co-ordinates of each body as $x_i = X + x_i'$, $y_j = Y + y_j'$. Define $R_\pm = X \pm Y$ then $R_\pm$ belong to the class I. The c-number equations of motion for $R_\pm$ are given in the limit $N, N' \to \infty$ as

$$\ddot{R}_+ = 0,$$

$$\mu \ddot{R}_- + NN' \nabla V(R_-) + \frac{NN'}{6}(\langle \bar{x}'^2 \rangle + \langle \bar{y}'^2 \rangle) \nabla \nabla^2 V(R_-) = 0,$$

where $\mu^{-1} = (Nm)^{-1} + (N'm)^{-1}$ with m the particle mass. We have introduced $\bar{x}'^2 = \frac{1}{N}\sum_{i=1}^{N} x_i' \cdot x_i'$ and $\langle \cdots \rangle$ is the quantum average in the relative co-ordinates of each body. We have assumed that $|R_-| \gg |x'|, |y'|$. The above equations are deterministic yet the quantum effects are contained in $\langle \bar{x}'^2 \rangle$ or $\langle \bar{y}'^2 \rangle$. They may be used as equations of motion of the system of two large molecules or nuclei.

REFERENCES

1) See the following book for example:
Foundation of Quantum Mechanics, Proc. of International School of Physics, E.Fermi - course IL, B.d'Espagnat ed. (Academic Press, New York and London, 1971).

2) Furry, W.H., Boulder Lectures in Theoretical Physics, vol.8A (1965), (University of Colorado Press, 1966).
Wigner, E.P., Amer. Jour. Phys. 31, 6 (1963).
Fine, A., Phys. Rev. D2, 2783 (1970).

3) Everett, III, H., Rev. Mod. Phys. 29, 454 (1957).
Machida, S. and Namiki, M., Prog. Theor. Phys. 63, 1457, 1833 (1980).

A SCREWY DIVE AND RISE FOR A MAGNETIC MONOPOLE

Tadashi Watanabe
Department of Physics, Kobe University, Kobe 657, JAPAN

ABSTRACT

We discuss the behaviour of a point charge moving around a magnetic monopole. In addition to the total angular momentum, one more constant vector of the motion is found out, which resembles the Runge-Lenz (R-L) vector in the hydrogen atom. By calculating the Poisson brackets, classical commutation relations are obtained among the extended R-L and the angular momentum vectors. Then, a dynamical symmetry group of a non-compact nature is shown to emerge through those 6 generators.

§1. Trajectories of a point charge around a monopole

It is well-known that the non-relativistic equation of motion around a monopole with a magnetic charge g is solved for a point particle with a electric charge e and a mass m. No bound state appears. The total angular momentum $\vec{J}$ of the system consists of two parts, the orbital one $\vec{L}$ of the particle and the one $\vec{\Lambda}$ of the electromagnetic field:

$$\vec{J} = \vec{L} + \vec{\Lambda}, \qquad \vec{L} = \vec{r} \times \vec{p} \qquad \text{and} \qquad \vec{\Lambda} = -\eta \cdot \vec{r}/r, \tag{1}$$

where $\eta = ge/c$ and η takes a value $\frac{1}{2}\hbar$ for the Dirac's unit charge.[1)] By virtue of the conservation of the angular momentum $\vec{J}$, an electric charge can only move on a conical surface, at the apex of which the monopole is situated (Fig.1). The semi-apex angle θ is given by

$$\theta = arccos(-\eta/J), \quad \text{for } \eta < 0. \tag{2}$$

For η>0, the angle is taken to be π-θ. The constant vector $\vec{J}$ lies along the axis inside (outside) the cone for η <0 (>0).

The trajectory is easier to follow on the development surface than on the conical one.[2)] It is a straight line on the former so that a geodesic on the latter. The point particle comes once to the perihelion (peripole) P. Then, winding clockwise around the angular momentum vector $\vec{J}$, it passes once more again the 1st, 2nd,...and $2n$th aphelion (apole) A_1, A_2,... and A_{2n}, respectively. The index $2n$ of the last aphelion satisfies the following condition:

$$\tfrac{1}{2}\, cosec\theta - 1 \leqq 2n < \tfrac{1}{2}\, cosec\theta + 1, \qquad \text{for } \eta < 0. \tag{3}$$

In a word, a point charge makes a screwy dive and rise for a magnetic monopole. It, however, cannot reach the monopole except the case that its impact parameter happens to vanish. Then, a phenomenon called Rosenbaum's paradox takes place.[3)]

§2. A generalization of the Runge-Lenz vector

In the hydrogen atom, or more generally in a system under the inverse-square law of force, the angular momentum $\vec{L}$ is a constant of the motion, since the hamiltonian possesses rotational invariance. The symmetry is enough to cause the orbit to lie in some plane through a focus of the conic, but it is not enough to protect the orbit from any apsidal motion. Besides the geometrical one, there is one more symmetry characterized by a constant vector of the motion, under the name of Runge-Lenz (R-L) vector.[4)] It lies along the major axis, pointing from a focus to the perihelion. The conservation law of the R-L vector prohibits the perihelion from precessing. Then, the particle revolves around its parent millennially like a sleeping top. This owes to the inverse-square force so that the symmetry is of a dynamical one.

In the case of a monopole-electric charge system, we find out also two constant vectors: the total angular momentum $\vec{J}$ and a Runge-Lenz like vector $\vec{M}$. The former assigns some conical surface for the charge to move on as stated before, whereas the latter protects aphelia from precessions. More specifically, every aphelion lies on a plane Σ which is determined by the perihelial vector $\vec{r}_P$ and the angular momentum $\vec{J}$. The plane is characterized by a normal vector $\vec{M}$ to Σ. It may be said that the additional conservation law appears owing to the inverse-square force as well.

The Runge-Lenz vector in question is expressed in the following form having the dimension of length:

$$\vec{M} = \frac{1}{\eta r p}\left[\cos\phi \cdot J\vec{J} \times \vec{r} + \sin\phi\, \vec{J} \times (\vec{J} \times \vec{r})\right], \tag{4}$$

$$\left.\begin{aligned} &\text{where } \phi = (\chi - \tfrac{1}{2}\pi)/\sin\theta, \qquad \theta = \arccos(-\eta/J), \\ &\text{and } \chi = \operatorname{arccot}(\vec{r}\cdot\vec{p} \,/\, |\vec{r}\times\vec{p}|). \end{aligned}\right\} \tag{5}$$

The radius vector $\vec{r}(t)$ of the point charge has the origin at the apex of the conical surface (see Fig.1). On revolving by the azimuthal angle ϕ above, it reaches the perihelion P to become $\vec{r}_P$. The area of the parallelogram made with the first aphelial $\vec{r}_A$ and the perihelial vectors $\vec{r}_P$ is $N(\theta)$ times the square of magnitude of the R-L vector $\vec{M}$, with

$$N[\theta(J)] = \cos^2\theta \sin(2\theta)/\cos(\pi \sin\theta). \tag{6}$$

The R-L vector in Eq.(4) is seen to be in fact a constant of the motion and to point the same direction as $\vec{r}_A \times \vec{r}_P$. Three vectors of $\vec{J}\times\vec{r}$, $\vec{J}$ and $\vec{J}\times(\vec{J}\times\vec{r})$ makes the right-handed system.

§3. A dynamical symmetry

The hydrogen atom has an $O(3)\times O(3)$ symmetry, where 6 generators consist of the angular momentum L_i and the R-L vector M_i. Pauli has derived energy spectra of it purely by algebraic methods for the dynamical symmetry group.[5)]

We shall study the classical commutation relations among J_i and M_i in the monopole-electric charge system by calculating the Poisson brackets. The hamiltonian of it is given by

$$H = \frac{1}{2m}(\vec{\Pi} - \frac{e}{c}\vec{A})^2, \tag{7}$$

where the canonical momentum $\vec{\Pi} = \vec{p} + \frac{e}{c}\vec{A}$. The vector potential $\vec{A}$ involves the notorious Dirac's string[1] as implied in the expression,

$$\vec{H} = g\vec{r}/r^3 = \vec{\nabla}\times\vec{A} + \vec{G}(string). \tag{8}$$

However, we can calculate the Poisson brackets without referring to the explicit expression of $\vec{A}$. Furthermore, we need not be worried about the string in case of non-vanishing impact parameter.

Commutation relations are obtained as follows:

$$\{J_i, J_j\}_{PB} = e_{ijk}J_k, \tag{9}$$

$$\{J_i, M_j\}_{PB} = e_{ijk}M_k \tag{10}$$

and $$\{M_i, M_j\}_{PB} = C_M e_{ijk}J_k, \tag{11}$$

where $$C_M(E,J) = -\frac{1}{2mE}\cdot\frac{J^2+L^2}{\eta^2} < 0, \tag{12}$$

with E the total energy $E=p^2/2m$ and $L^2=|\vec{r}\times\vec{p}|^2=J^2-\eta^2$.

§4. Concluding remarks

The commutation relation (11) is much more complicated than the corresponding one in the hydrogen atom. The coefficient C_M is negative definite, which suggests that the symmetry associated with our system is of a non-compact group. This ought to relate to the fact the system has no bound state. Though much complicated, it is interesting to investigate the representation of the dynamical symmetry group prescribed through Eqs.(9)∿(11).

As another problem left to be solved, what is the meaning of the number of aphelia from the quantum mechanical point of view?

In table 1, we summarize each set of two constant vectors of the motion in the hydrogen atom and in the monopole-electric charge system.

Table 1; Comparison between two systems

	Hydrogen atom	g — e system
Hamiltonian	$H = \frac{1}{2m}\vec{p}^2 - \frac{e^2}{r}$	$H = \frac{1}{2m}(\vec{\Pi} - \frac{e}{c}\vec{A})^2$
Energy	$E(a) = -\frac{e^2}{2a}$	$E(p) = \frac{p^2}{2m}$
Angular momentum	$\vec{L} = \vec{r}\times\vec{p}$	$\vec{J} = \vec{r}\times\vec{p} - \frac{eg}{c}\frac{\vec{r}}{r}$
	$L^2(a,\varepsilon) = me^2a(1-\varepsilon^2)$	$J^2(\theta) = (-\frac{eg}{c})^2 sec^2\theta$
Runge-Lenz vector	$\vec{M} = \frac{\vec{p}\times\vec{L}}{m} - e^2\frac{\vec{r}}{r}$	$\vec{M} = \frac{1}{\eta rp}[cos\phi\ J\ \vec{J}\times\vec{r} + sin\phi\ \vec{J}\times(\vec{J}\times\vec{r})]$
	$M^2(\varepsilon) = e^4\varepsilon^2$	$M^2(\theta,\sigma) = \sigma^2/N(\theta)$
	$\vec{L}\cdot\vec{M} = 0$	$\vec{J}\cdot\vec{M} = 0$
Commutation relations	$\{M_i, M_j\}_{PB} = C_H\varepsilon_{ijk}L_k$	$\{M_i, M_j\}_{PB} = C_M\varepsilon_{ijk}J_k$
	$C_H(E) = -\frac{2E}{m}$	$C_M(E, J) = -\frac{1}{2mE}\frac{J^2+L^2}{\eta^2}$

	a: semimajor axis	$\sigma = r_P r_A sin(2\theta)$: the area of the fundamental parallelogram
	ε: eccentricity	

Now, since 1982, we, a research group in Kobe University, have been seeking for magnetic monopoles by using superconducting coils connected to SQUID's.[6] Just like an electric charge described here, we have made a screwy dive for a monopole but have not yet reached the treasure.

REFERENCES

1) Dirac, P.A.M., Proc. Roy. Soc. (London) A133, 60 (1931).
2) A detailed account is given in: Nakamura, K., *The Bulletin of Arts and Science, Meiji University* (in Japanese) #181, p.1 (Meiji University, Tokyo, 1985).
3) Rosenbaum, D., Phys. Rev. 147, 891 (1966).
4) Runge, C., *Vektoranalysis* 1, p.70 (S. Hirzel, Leipzig, 1919); Lenz, W., Z.Physik 24, 197 (1924).
5) Pauli, W., Z.Physik 36, 336 (1926).
6) Ebisu, T. and Watanabe, T., J.Phys.G: Nucl.Phys. 11, 883 (1985).

Fig.1; Screwy orbit of an electric charge around a magnetic monopole

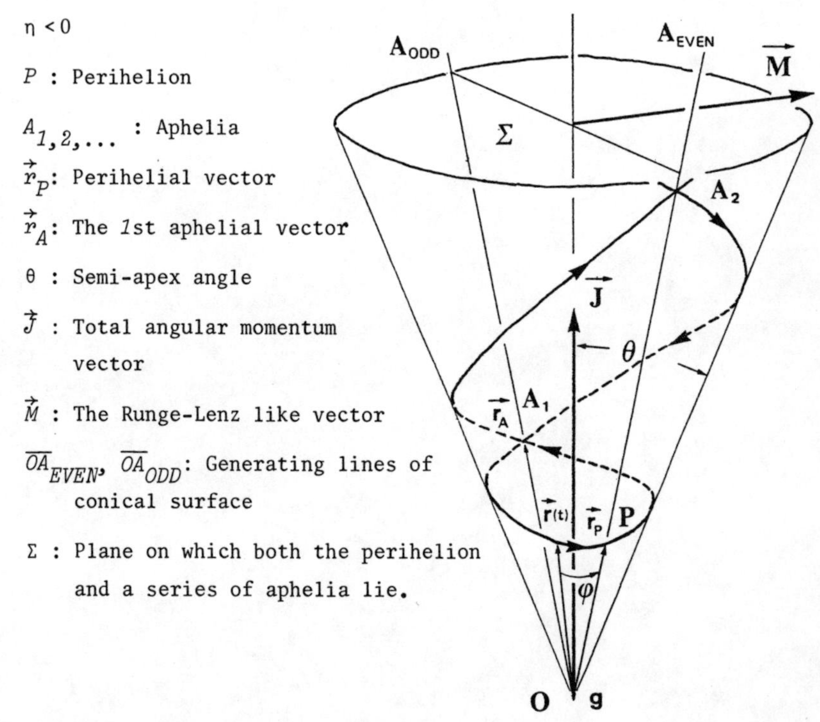

REFERENCES

1) Dirac, P.A.M., Proc. Roy. Soc. (London) A133, 60 (1931).
2) A detailed account is given in: Nakamura, K., The Bulletin of Arts and Science, Meiji University (in Japanese) 183, p.1 (Meiji University, Tokyo, 1985).
3) Rosenbaum, D., Phys. Rev. 147, 891 (1966).
4) Runge, C., Vektoranalysis I, p.70 (S. Hirzel, Leipzig, 1919); Lenz, W., Z. Physik 24, 197 (1924).
5) Pauli, W., Z. Physik 36, 336 (1926).
6) Ohishi, T. and Watanabe, T., J. Phys. Soc. Jpn. 11, 865 (1956).

Fig.1) Screwy orbit of an electric charge around a magnetic monopole.

(gMe < 0)

P : Perihelion

A_n : Aphelion

$\vec{r}_P$: Perihelion vector

$\vec{r}_1$: The 1st aphelion vector

Θ : Semi-apex angle

$\vec{J}$: Total angular momentum vector

$\vec{K}$: The Runge-Lenz like vector

$\overline{OA}$, $\overline{OA_1}$, ... : Generating lines of conical surface

Σ : Plane on which both the perihelion and a series of aphelia lie.

AUTHOR INDEX